Aviation 2018

Transportation Research Record: Journal of the Transportation Research Board Volume 2672, Issue 23

ISBN: 978-1-5108-8242-3

Printed from e-media with permission by:

Curran Associates, Inc.
57 Morehouse Lane
Red Hook, NY 12571

Some format issues inherent in the e-media version may also appear in this print version.

Published by Sage Publications USA

ISBN (Print) 978-1-5108-8242-3
ISBN 2018 Printed Set (All Issues) 978-1-5108-7735-1

Printed by Curran Associates, Inc. (2019)

Additional copies of this publication are available from:

Curran Associates, Inc.
57 Morehouse Lane
Red Hook, NY 12571 USA
Phone: 845-758-0400
Fax: 845-758-2633
Email: curran@proceedings.com
Web: www.proceedings.com

Volume 2672 Number 23 December 2018

TRR
JOURNAL OF THE TRANSPORTATION RESEARCH BOARD

Contents

Aviation

Articles

2018 FRED BURGGRAF AWARD: Aviation

Article

Transportation Officials, Washington, D.C.
K. Jane Williams, *Acting Administrator, Federal Transit Administration, U.S. Department of Transportation*
Paul Zukunft *(Admiral, U.S. Coast Guard), Commandant, U.S. Coast Guard, Washington, D.C.*

---

* Membership as of February 2019.

TRB

TRANSPORTATION RESEARCH BOARD

CORE SPONSORS

Transportation Departments of the 50 States and the District of Columbia

Federal Government

U.S. Department of Transportation
- Federal Aviation Administration (FAA)
- Federal Highway Administration
- Federal Motor Carrier Safety Administration (FMCSA)
- Federal Railroad Administration (FRA)
- Federal Transit Administration
- National Highway Traffic Safety Administration (NHTSA)
- Office of the Assistant Secretary of Transportation for Research and Technology (OST-R)

U.S. Air Force Civil Engineer Center (AFCEC)
U.S. Coast Guard (USCG)
U.S. Department of Energy
U.S. Department of the Interior (DOI)
U.S. Environmental Protection Agency (EPA)

Nongovernmental Organizations

American Public Transportation Association (APTA)
Association of American Railroads (AAR)
California Air Resources Board

The National Academies of
SCIENCES · ENGINEERING · MEDICINE

The **National Academy of Sciences** was established in 1863 by an Act of Congress, signed by President Lincoln, as a private, nongovernmental institution to advise the nation on issues related to science and technology. Members are elected by their peers for outstanding contributions to research. Dr. Marcia McNutt is president.

The **National Academy of Engineering** was established in 1964 under the charter of the National Academy of Sciences to bring the practices of engineering to advising the nation. Members are elected by their peers for extraordinary contributions to engineering. Dr. C. D. Mote, Jr., is president.

The **National Academy of Medicine** (formerly the Institute of Medicine) was established in 1970 under the charter of the National Academy of Sciences to advise the nation on medical and health issues. Members are elected by their peers for distinguished contributions to medicine and health. Dr. Victor J. Dzau is president.

The three Academies work together as the National Academies of Sciences, Engineering, and Medicine to provide independent, objective analysis and advice to the nation and conduct other activities to solve complex problems and inform public policy decisions. The Academies also encourage education and research, recognize outstanding contributions to knowledge, and increase public understanding in matters of science, engineering, and medicine.

Learn more about the National Academies of Sciences, Engineering, and Medicine at **www.national-academies.org.**

The **Transportation Research Board** is one of seven major programs of the National Academies of Sciences, Engineering, and Medicine. The mission of the Transportation Research Board is to increase the benefits that transportation contributes to society by providing leadership in transportation innovation and progress through research and information exchange, conducted within a setting that is objective, interdisciplinary, and multimodal. The Board's varied activities annually engage about 7,000 engineers, scientists, and other transportation researchers and practitioners from the public and private sectors and academia, all of whom contribute their expertise in the public interest. The program is supported by state transportation departments, federal agencies including the component administrations of the U.S. Department of Transportation, and other organizations and individuals interested in the development of transportation.

Learn more about the Transportation Research Board at **www.TRB.org**.

Article

TRR
JOURNAL OF THE TRANSPORTATION RESEARCH BOARD

Transportation Research Record
2018, Vol. 2672(23) 1–7

DOI: 10.1177/0361198118758336
journals.sagepub.com/home/trr
SAGE

A Tale of Two Cities: The Impact of Airline Mergers and Consolidation at London and New York

Nigel Dennis[1] and David Pitfield[1]

Abstract

This paper considers the changes to airline networks, service patterns, and competition that have taken place as a result of recent airline mergers on both sides of the North Atlantic as well as through transatlantic alliances. Capacity, frequency and the competitive position are studied at London and New York with the use of schedule data within different markets in which measures of market concentration are evaluated. International Civil Aviation Organization data is employed to examine load factors on international routes, and UK Civil Aviation Authority data to consider the distribution of traffic between airports in London. It is shown that the effectiveness of the hubs has increased, with enhanced efficiency for surviving airlines, through fewer competitors, an enlarged network and greater control of capacity. Potential concerns are identified however, regarding passenger choice, pricing, and service options that suggest the industry is moving toward an oligopoly. Smaller cities are also seen to be the losers from consolidation with slot divestments favoring increased service in the dense markets, with many regional links being axed altogether. The paper supplements the literature on airline consolidation, with a particular focus on the two biggest markets in the world—London and New York—which demonstrate some similar but also some different issues. Both airline network impacts and choice, and service for local consumers are considered.

> It was the best of times, it was the worst of times, it was the age of wisdom, it was the age of foolishness . . .
>
> Charles Dickens, *A Tale of Two Cities*, 1859

The airline industry has seen major consolidation over the last decade, particularly in Europe and North America. In Europe, mergers have taken place between British Airways, Iberia, bmi British Midland, and Aer Lingus to form International Airlines Group (IAG) (*1*), while Lufthansa has acquired Swiss, Austrian, and more recently Brussels Airlines. In the United States (U.S.), Delta and Northwest set off the current round of consolidation in 2008 followed by United–Continental, Southwest–Air Tran, American–US Airways and most recently Alaska–Virgin America.

On the North Atlantic, the market has become concentrated around four joint ventures in which there were once many competing rival airlines. These mirror the three global alliance groups, plus Delta with Virgin Atlantic, which provides another alternative for traffic to and from the UK (*2*).

It was therefore considered interesting to assess the impact of these changes on air service provision at London and New York. These are still the two biggest air travel markets in the world and are served by a wide range of carriers through several different airports. London has six airports designated by the International Air Transport Association (IATA): Heathrow, Gatwick, Stansted, Luton, London City, and Southend. New York has five: John F Kennedy (JFK), Newark, La Guardia, Westchester County, and Newburgh-Stewart. Unlike dominated hubs such as Frankfurt, Amsterdam, Atlanta, or Dallas Fort Worth where there has been little change as a result of the industry restructuring, London and New York offer the opportunity to review the impact on a broader and more diverse market. Another development at New York was the slot swap between Delta and US Airways in 2012 (*3*), which gave Delta an increased presence at JFK and La Guardia (in exchange for US Airways strengthening its position at Washington DC Reagan).

The period from 2005 to 2015 was chosen for study; 2005 provides a good representation of the position after the industry had recovered from the terrorist attacks of 9/11 in 2001, but before the economic crisis and the "credit crunch" hit in

[1]Department of Planning and Transport, University of Westminster, London, UK

Corresponding Author:
Address correspondence to Nigel Dennis: dennisn@westminster.ac.uk

2007. With the exception of Air France/KLM, the major merger activity also took place after this point. The year 2015 was selected because complete data was not yet available for 2016 from most sources at the time the research was undertaken. It also post-dated most of the recent industry consolidation.

Several different sources were used to analyze the developments. Innovata schedule data, which provides capacity and frequency information by route and airline, were collected for the second week in July 2005 and July 2015 at the London and New York airports. This is considered representative of the peak season schedule without holiday period adjustments. International Civil Aviation Organization (ICAO) data were obtained for the years 2005 and 2015 to study changes in the international traffic and capacity utilization by route and airline. UK Civil Aviation Authority (CAA) traffic data were used to study the London airport market share by route.

There are always risks with the representativeness of comparative static analyses, but it is felt that the findings vindicate the methods used as they appear intuitively correct and do not suggest the existence of different trends in intervening years, or the need to investigate alternative principal causal factors besides mergers and consolidation.

Market Concentration

The first stage of the analysis aimed to investigate the impact of airline mergers and transatlantic alliances on market concentration in London and New York. The Innovata data for July 2005 and July 2015 showed operations and seats per week by airline and alliance by route. This was turned into percentages to enable indices of market concentration to be calculated on both bases. Using the Herfindahl–Hirschman Index (HHI) to investigate the share by airline involved the aggregation of the percentages for each route for the airline examined. It should be noted that this involved the identification of airlines operating as franchisees for another, such as GB Airways and British Mediterranean for British Airways, or subsidiaries such as Envoy Air owned by American Airlines.

It is a standard and well-established means of measuring the acceptability of mergers and alliances and of their impact on market share and competition (*4, 5*). The calculation would show that an industry or sector totally dominated by one firm would score 10,000. The index is given by

$$\text{HHI} = S_1^2 + S_2^2 + S_3^2 + S_4^2 + \cdots + S_n^2$$

where S_i^2 is the market share of individual firms i (airlines here) and has been used by the UK Office of Fair Trading, the U.S. Department of Justice (DOJ) and the U.S. Federal Trade Commission. Values of HHI <1,000 are said to represent an unconcentrated and very competitive market whereas market values >1,800 represent concentration and a relative absence of competition. If mergers and alliances were to raise the HHI above 1,800 then the merger would be challenged, however when the statistic is already above 1,800, any proposed merger should not result in an increase of more than 100. Alternative guides suggest these absolute numbers can change, for example, the U.S. DOJ uses 2,500 as an indication of a concentrated market.

Table 1. HHI Results, Operations/Week, London and New York, 2005, 2015

	Airport	2005	% of operations	2015	% of operations
London					
	LCY	1,699	6.6	2,758	7.0
	LGW	1,979	22.6	2,449	27.1
	LHR	2,030	47.2	2,738	43.4
	LTN	4,595	7.0	2,957	8.7
	STN	3,944	16.6	5,875	13.8
New York					
	EWR	4,922	36.2	5,413	32.7
	JFK	1,704	29.0	1,987	36.7
	LGA	2,062	34.8	2,745	30.6

Source: Analysis of Innovata schedules data for second week of July.

Table 2. HHI Results, Seats/Week, London and New York, 2005, 2015

	Airport	2005	% of seats	2015	% of seats
London					
	LCY	1,548	2.4	2,774	3.2
	LGW	1,584	20.1	2,180	26.5
	LHR	1,857	55.1	2,289	48.4
	LTN	4,380	6.5	2,833	8.3
	STN	4,472	15.9	6,374	13.6
New York					
	EWR	4,283	32.9	4,813	29.7
	JFK	1,381	40.0	1,386	46.8
	LGA	1,633	27.1	2,258	23.5

Source: Analysis of Innovata schedules data for second week of July.

Table 1 shows the HHI results on the basis of frequency, measured as operations per week, for the London area airports and the New York area airports for 2005 and 2015. Table 2 shows the results for the same airports on the basis of capacity, measured as seats per week. The airports for London are London City (LCY), London Gatwick (LGW), London Heathrow (LHR), London Luton (LTN) and London Stansted (STN). For New York, they are New York Newark (EWR), New York JFK and New York La Guardia (LGA).

The comparison of these airports for each year is interesting showing that in 2005 measured in frequency, LCY and JFK were similar with relatively low values, whereas LTN and EWR were similar with higher values. Looking at capacity, EWR was similar to both LTN and STN, and JFK had the lowest concentration measured by both capacity and

Table 3. HHI Results, Operations/Week, Airline Groups, London and New York, 2005, 2015

	Airport	2005	% of operations	2015	% of operations
London					
	LCY	1,804	6.6	2,758	7.0
	LGW	1,979	22.6	2,503	27.1
	LHR	2,037	47.2	2,919	43.4
	LTN	4,595	7.0	2,957	8.7
	STN	3,944	16.6	5,875	13.8
New York					
	EWR	4,922	36.2	5,434	32.7
	JFK	1,705	29.0	2,025	36.7
	LGA	2,062	34.8	3,125	30.6

Source: Analysis of Innovata schedules data for second week of July.

Table 4. HHI Results, Seats/Week, Airline Groups, London and New York, 2005, 2015

	Airport	2005	% of seats	2015	% of seats
London					
	LCY	1,683	2.4	2,774	3.2
	LGW	1,584	20.1	2,226	26.5
	LHR	1,861	55.1	2,421	48.4
	LTN	4,380	6.5	2,833	8.3
	STN	4,472	15.9	6,374	13.6
New York					
	EWR	4,283	32.9	4,837	29.7
	JFK	1,384	40.0	1,425	46.8
	LGA	1,633	27.1	2,604	23.5

Source: Analysis of Innovata schedules data for second week of July.

frequency. In 2015, JFK remains the lowest in frequency and capacity terms, and STN and EWR still have similar high scores with STN the highest.

However, the instructive comparison is between the years. This shows that all airports had more concentrated traffic in 2015 with the exception of LTN. This was less concentrated as Ryanair and Wizz Air had greater frequency and capacity than in 2005 when easyJet dominated. The figures for JFK show that this concentration is least pronounced for the HHI measure of capacity.

Consequently, it can be seen that the effectiveness of hubs has increased as the traffic has concentrated in fewer airlines and competition has declined. At congested airports, such as LHR, LGW, and LGA, airlines may well reduce links to smaller markets, as well as changing aircraft gauge to maintain key international slots; in turn, this may affect services from other airports in the system, but there are still implications for consumer welfare (*6*).

Tables 3 and 4 show the HHI calculations when, in both 2005 and 2015, Air France and KLM are grouped together. In 2015, the other major groupings are of American Airlines and US Airways who merged in late 2013 but operated as separate subsidiaries until late 2015, and the airlines in the IAG: British Airways, Iberia and Vueling. Aer Lingus did not join IAG until after July 2015. In addition, there is the Lufthansa group that includes German Wings, Austrian and Swiss. Obviously, not all airports are affected and so their calculations are the same as shown in tables 1 and 2. It was expected that these calculations would indicate increased levels of concentration by comparison to tables 1 and 2 and that was largely the case.

In every case in which the groupings have been taken into account, except for London City, concentration is seen to be proportionately greater in 2015 than in 2005 by comparison to the airline analysis between the years. It is conclusive that the HHI has shown a greater inclination toward oligopoly as time has moved on. For London City in 2005, the operation of KLM and CityJet, then owned by Air France, is not mirrored in 2015 as CityJet was sold in 2014, so the increased concentration over the decade, although marked, is not as large as the airline analysis shown in tables 1 and 2 in which groupings are not taken into account. It is noteworthy that for New York JFK, the HHI scores are mostly below 1,800.

It may also be interesting to examine alliance traffic across the North Atlantic where anti-trust immunity has had profound impacts (*7*). Consequently, for the traffic from London, destinations were restricted to Canada and the U.S. In the other direction, destinations in Europe were selected along with Moscow. These results are shown in tables 5 and 6. In London airports, in both years, oneworld is the chief player with over 60% of the share. However, in New York, in 2005, SkyTeam has over 50%, whereas Star Alliance is the largest alliance in 2015. New York is less concentrated, and the HHI calculations indicate that. The switch of Continental from SkyTeam to Star has led to three more evenly matched alliance groups in New

Table 5. HHI Results, Operations/Week, Transatlantic Alliances, London and New York, 2005, 2015

		2005	2015
London			
	All airports	4,561	4,665
New York			
	All airports	4,159	3,484

Source: Analysis of Innovata schedules data for second week of July.

Table 6. HHI Results, Seats/Week, Transatlantic Alliances, London and New York, 2005, 2015

		2005	2015
London			
	All airports	4,728	5,055
New York			
	All airports	4,035	3,470

Source: Analysis of Innovata schedules data for second week of July.

Table 7. Change in Operations at London Heathrow on Routes Duplicated between British Airways (BA) and bmi (BD), 2005, 2015

	2005			2015		
	Frequencies/week	Seats/week	Average aircraft size (seats)	Frequencies/week	Seats/week	Average aircraft size (seats)
BA	545	85,300	157	541	84,344	156
BD	343	48,697	142	0	0	na
other IAG	42	7,560	180	56	10,004	179
IAG total	930	14,557	152	597	94,348	158
Others	224	39,404	176	251	38,969	155
Total	1154	180,961	157	848	133,317	157

Note: BA = British Airways; BD = bmi British Midland; IAG = International Airlines Group.
Routes from London: Aberdeen, Amsterdam, Brussels, Edinburgh, Glasgow, Madrid, Manchester, Milan Linate, Mumbai, Nice, Paris CDG.
Source: Analysis of Innovata schedules data for second week of July.

York in 2015, which has reduced concentration in New York, whereas it has increased in London over the same period. Of course, a wider or different focus on transatlantic markets beyond London would yield somewhat different conclusions and some work by the authors, not reported here, shows this in terms of alliance domination. This could be the subject of further research.

Traffic and Network Developments

To consider in more detail the impact of the major changes in industry structure and slot allocation, several case studies have been conducted for London and New York. Previous research has identified differences in airline merger outcomes between Europe and the US (*8*).

Impact on Individual Routes at London Heathrow of British Airways Merger with bmi British Midland

British Midland had built up a short-haul network at London Heathrow over many years, adding the "shuttle" routes to Belfast, Edinburgh, and Glasgow in 1982–1984 to a few regional services it had long operated such as Birmingham, East Midlands, Leeds/Bradford, and Teesside. After that, expansion followed to the major European destinations such as Paris, Brussels, Amsterdam, and Dublin. Further European cities with a business travel focus were subsequently added including Frankfurt and Zurich. After Lufthansa acquired SAS's interest in bmi however, direct competition with the Lufthansa group was dropped in favor of more distant cities dominated by other alliances such as Milan and Madrid. There were also a few leisure routes at relatively low frequencies to Alicante, Naples, Nice, Palma, and Venice, plus a lone long-haul service to Mumbai.

Thus in 2005 bmi served 21 destinations from Heathrow with a total of 535 weekly frequencies. Of these 21 routes, British Airways competed from Heathrow on 11 of them (545 frequencies) and other carriers contested 7 (314 frequencies). On eight routes, bmi was the sole operator. By 2015, British Airways was serving 17 of these places from Heathrow with a total of 697 frequencies.

Table 7 considers only those routes that were duplicated by British Airways and bmi in 2005. Here the changes over the period 2005–2015 are more pronounced.

The key points are that British Airways has largely removed former bmi frequencies and capacity altogether on routes where the airlines competed—actually offering fewer flights and seats than British Airways alone did in 2005. The beneficiaries in terms of market share have been easyJet in particular and other Low Cost Carriers based at alternative London airports. British Airways has taken over the former bmi services from Heathrow to Belfast City, Dublin, Hanover, Leeds Bradford, Palma, and Venice. Inverness has been also subsequently reinstated. Former bmi slots have been used by BA for more profitable opportunities elsewhere such as the long-haul market.

Other carriers on these routes (e.g., Air France, Alitalia, Virgin Atlantic) have increased frequency slightly but reduced aircraft size, also giving fewer seats in total. The main factor behind the increase in the other carriers' frequency is the 63 weekly slots that British Airways had to give up to a new entrant willing to serve Aberdeen and Edinburgh from Heathrow. These were used by Virgin Atlantic's Little Red in 2015, and now by flybe. This has still failed to generate more competition at Heathrow overall however. Virgin Atlantic has indeed been one of the main losers from bmi's demise as it received around a quarter of its transatlantic traffic from a code-share arrangement (*9*).

Table 8 analyzes the load factors for the calendar years 2005 and 2015 using ICAO data. These are only available for international routes so only seven of the 11 routes in Table 7 are covered. It shows British Airways' load factor on these routes has improved dramatically, but the other carriers have done even better. In 2005 load factors were on the low side, suggesting bmi was generating excess capacity and competition—bmi itself only managed a 62% load factor. In 2015 load factors are much more in line with global averages

Table 8. Change in Load Factor at London on International Routes Duplicated between British Airways (BA) and bmi (BD), 2005, 2015

	2005	2015
	% Load factor	% Load factor
BA	68.1	76.0
BD	62.2	na
Other IAG	77.1	68.7
IAG	67.3	74.9
Others	70.3	83.4
Total	68.0	77.5
BA systemwide	76.3	81.5
IATA scheduled	74.8	80.4

Note: BA = British Airways; BD = bmi British Midland; IAG = International Airlines Group; IATA = International Air Transport Association.
Routes from London: Amsterdam, Brussels, Madrid, Milan Linate, Mumbai, Nice, Paris CDG.
Source: ICAO (BA figures include LGW and LHR, but not LCY).

Table 9. Change in Market Share of London Airports on Routes Duplicated between British Airways (BA) and bmi (BD), 2005, 2015

	2005		2015	
	Scheduled passengers	Market share (%)	Scheduled passengers	Market share (%)
London City	583,416	2.9	1,603,178	7.8
Gatwick	3,291,794	16.3	4,425,408	21.6
Heathrow	12,678,597	62.9	10,703,996	52.3
Luton	2,220,175	11.0	1,528,828	7.5
Southend	0	0.0	214,566	1.0
Stansted	1,368,812	6.8	1,973,703	9.7
Total	20,142,794	100.0	20,449,679	100.0

Note: Routes from London: Aberdeen, Amsterdam, Brussels, Edinburgh, Glasgow, Madrid, Manchester, Milan Linate, Mumbai, Nice, Paris CDG.
Source: UK Civil Aviation Authority.

(short haul is generally lower than long haul for most airlines). The "Other IAG" category is essentially Iberia, which seems to be running excess capacity in 2015; the ICAO figures do not allow London Heathrow and London Gatwick to be separated out, it is likely that Heathrow alone would have shown a stronger performance in 2015.

Table 9 examines the split of traffic between London airports in 2005 and 2015 on the routes formerly served by both British Airways and bmi from Heathrow. The total traffic on these routes has hardly changed over 10 years but Heathrow (and Luton) have lost ground with all the others gaining market share. This shows that changing airline strategies have reduced the role of Heathrow on short-haul routes. We are perhaps seeing a segmentation of business passengers to London City, connecting passengers to Heathrow and leisure passengers to the other airports.

The implication of the various data above is that British Airways has prioritized yields over volume—and probably also connecting passengers over local traffic—after bmi was absorbed, on the combined network. Many previous Heathrow customers (presumably the lower yield point-to-point ones) are now having to use other London airports. Other research has shown that nearby airport pairs are imperfect substitutes of each other (*10*), and frequency choices of network airlines are not affected by competition from low-cost airlines operating in nearby secondary airports (*11*).

Impact on Individual Routes at New York Newark of the Merger between Continental and United

The routes from Newark that were served by both Continental and United in 2005 have been compared with United's service in 2015 (Table 10). There are only five such routes, which are to other United hubs (Chicago O'Hare, Denver, Los Angeles, San Francisco, and Washington Dulles).

In this case, the merger appears to have produced a positive outcome as total capacity has increased, total frequency has stayed the same, and average aircraft size has got larger. Other airlines flying these routes have also increased seat capacity and aircraft size but with a slight decline in frequency. This is one of the "silver linings" that can emerge from mergers in the longer term (*12*).

This is because these are the hub–hub routes for the enlarged United and hence critical to the new network. The other United routes from Newark in 2015 were only served by Continental originally and are now served by United instead, so there is not much change on those. This conforms to other research suggesting connectivity has increased following the United–Continental merger (*13*).

Impact on Individual Routes at New York La Guardia of the Slot Swap between US Airways and Delta

At New York La Guardia, a rather different situation can be seen that is more analogous to the London one. US Airways and Delta arranged a slot swap whereby Delta gained US Airways slots at La Guardia while US Airways acquired Delta slots at Washington National together with a cash payment (*14*). The transaction was viewed negatively by many stakeholders (*15*).

As in London, Delta was obliged to cede some slots as part of the package to introduce new competition.

This has had a negative impact on travelers who were previously served by US Airways at La Guardia but are no longer part of the American–US Airways network. Table 11 shows that 10 routes have been dropped altogether. These typically each operated about three times per day to regional points in the northeast U.S. using small aircraft (e.g., Albany,

Table 10. Change in Operations at New York Newark on Routes Duplicated between Continental (CO) and United (UA), 2005, 2015

	2005			2015		
	Frequencies/week	Seats/week	Average aircraft size (seats)	Frequencies/week	Seats/week	Average aircraft size (seats)
CO	209	25,001	120	0	0	Na
UA	158	17,590	111	366	52,739	144
CO+UA	367	42,591	116	366	52,739	144
Others	124	12,890	104	109	13,451	123
Total	491	55,481	113	475	66,190	139

Note: CO = Continental; UA = United.
Routes from New York: Chicago O'Hare, Denver, Los Angeles, San Francisco, Washington Dulles.
Source: Analysis of Innovata schedules data for second week of July.

Table 11. Change in Operations at New York La Guardia on Routes No Longer Served by US Airways after the Slot Swap with Delta

	2005			2015		
	Frequencies/week	Seats/week	Average aircraft size (seats)	Frequencies/week	Seats/week	Average aircraft size (seats)
US routes lost	176	6,628	38	0	0	na
US transferred to DL	269	11,705	44	163	9,840	60
US and DL/NW transferred to DL	253	12,651	50	117	7,544	64

Note: AA = American Airlines; DL = Delta; NW = Northwest; US = US Airways.
Source: Analysis of Innovata schedules data for second week of July.

Ithaca, Providence). On the eight routes that were served by US Airways alone in 2005 but are now served only by Delta, frequencies and capacity have both been slashed, although aircraft size has become larger. This suggests the consumer has seen a reduced level of service with the potential for higher fares and load factors for the airline. These are slightly larger cities than the first group, such as Buffalo, Louisville, Rochester, and Burlington VT. Some may be "rust belt" locations with declining demand, but others are clearly not. A further five destinations were served by both US Airways and Delta in 2005 (or Northwest with which Delta merged in 2008) but reduced to only Delta service by 2015. These have seen even more drastic cutbacks as the former US Airways capacity has been lost altogether and frequencies have more than halved. These are also significant centers: Charleston SC, Greenville/Spartanburg, Indianapolis, Manchester NH, and Portland ME. Small- and medium-sized cities have been clearly identified as the big losers (*16*).

It is fair to say that there are other places which were served by both US Airways and Delta in 2005 and are still served by US Airways (or American) and Delta in 2015 so the competitive position on those markets has not changed greatly. Delta has launched new service to large cities that neither it nor US Airways served from La Guardia before. New carriers have taken advantage of the slot divestments (16 pairs) to launch service on the busier routes such as WestJet to Toronto (somewhat ironically code-shared with Delta). Southwest and Virgin America also gained some slots following the American–US Airways merger (*17*).

Conclusions

Airline market concentration has greatly increased over the time period 2005–2015 at all airports in London and New York (except Luton where the growth of Wizz and Ryanair has diluted the previous dominance of easyJet) as shown by the HHI analysis. This is in line with expectations. The two main hub airports, Heathrow and Newark, have become heavily dominated by British Airways and United respectively. Gatwick is now the most important airport in easyJet's system, while Stansted has become a massive fortress for Ryanair. JFK is still a relatively dispersed market among airlines, whereas La Guardia has become dominated by Delta. When the airline groupings are considered, this accentuates the concentration at all airports except London City where Air France/KLM has become a weaker player. The oneworld alliance is now overwhelmingly dominant at Heathrow and with the loss of bmi, other alliances are focusing on their alternative hubs in Europe. There is an overall tendency toward greater monopoly power with its consequences for pricing and consumer choice.

An analysis of operations, load factors and market shares shows that the change in Newark has been positive in terms of increases in frequency and seat capacity on the duplicated routes between United and Continental as these are between the hubs of the merged airlines. The hub network has also been maintained intact.

At Heathrow and La Guardia however there is more evidence of benefits to the airlines in reduced capacity and higher load factors on duplicated routes, which may have negative implications for consumers. It is notable that small cities have been the big losers from these changes, seeing their slots raided to provide the divestments required by regulators. In contrast, more services are available on the dense routes and in the case of Heathrow to long-haul destinations. It is debatable as to whether the slot divestments have been an entirely positive experience for consumers. Arguably, this has created more competition and frequency in which there was already plenty—once other airlines and airports are taken into account—but reduced capacity or lost service altogether on thinner monopoly markets, which may have heavily depended on these links. It also may have reduced the ability to connect from these smaller cities to the wider domestic and international networks, something that is rarely considered in competition assessments, which focus on competition between carriers on direct routes. Without the divestment requirements, the dominant airlines may have sustained more of these services to the smaller cities. In London, point-to-point passengers seeking competitive fares are increasingly being forced to alternative airports with resultant inconvenience in overall journey times, while British Airways and Heathrow concentrate on connecting and high-yield traffic. These are significant findings.

In the transatlantic market from London, concentration has increased partly through the BA–American grouping, but also the tie up between Delta and Virgin Atlantic, which has significantly reduced the number of independent competitors. At New York however this is less marked as the split between the alliances has become more even, at least as far as direct flights are concerned. An area that merits further research is whether the decline in competition from rival airlines offering indirect routings on a city pair where other carriers have direct service, has led to higher fares on the direct flights, or elimination of discounts for connecting services in which they are now competing against a direct flight of the same merged airline or alliance group.

An awareness of these findings can inform airline and airport reactions to the market, in particular, whether regional links should be reinstated, and also whether greater or different regulatory intervention is needed. Returning to where this paper started with the quotation from Charles Dickens, mergers and consolidation have led to the best of times for many of the airlines involved in the London and New York markets, but the worst of times for some of the passengers. Regulatory approval and amelioration measures, which may have appeared wise at the time, may yet come to be seen as foolish with the evolution of the markets and the benefit of hindsight.

Acknowledgments

The assistance of Alistair Rivers, Client Services Manager, FlightGlobal is much appreciated in sourcing the Innovata schedules and capacity data.

References

1. Kingsley-Jones, M. On the Lookout. *Airline Business*, Vol. 32, No. 5, 2016, pp. 16–20.
2. Russell, E., G. Dunn, and D. Perry. Virgin-Delta Joint Venture Has Legs. *Airline Business*, Vol. 29, No. 1, 2013, p. 8.
3. McDonald, J. B. Delta and US Air Convince DOT to Approve LaGuardia-Reagan Slot Swap, and DOJ Continues to Investigate at National. *Jones Day*, Feb. 6, 2012. http://www.lexology.com/library/detail.aspx?g=3cb6c072-c2f0-4565-93e2-5da432c1020d.
4. Stiglitz, J. E. *Economics*. W. W. Norton, New York, 1993.
5. Pitfield, D. E. The Impact on Traffic, Market Shares and Concentration of Airline Alliances on Selected European: US Routes. *Journal of Air Transport Management*, Vol. 13, No. 4, 2007, pp. 192–202.
6. Brueckner, J. K., and E. Pels. European Airline Mergers, Alliance Consolidation and Consumer Welfare. *Journal of Air Transport Management*, Vol. 11, No. 1, 2005, pp. 27–41.
7. Gillespie, W., and O. M. Richard. *Antitrust Immunity and International Airline Alliances*. Economic Analysis Group Discussion Paper, EAG 11-1. US Department of Justice, 2011.
8. Schosser, M., and A. Wittmer. Cost and Revenue Synergies in Airline Mergers: Examining Geographical Differences. *Journal of Air Transport Management*, Vol. 47, 2015, pp. 142–153.
9. Gulliver. In Which British Airways Buys a Money-Losing Airline to the Detriment of Its Own Passengers. *The Economist*, Apr. 1, 2012. https://www.economist.com/blogs/gulliver/2012/04/ba-and-bmi.
10. Lijesen, M., and C. Behrens. The Spatial Scope of Airline Competition. *Transportation Research E*, Vol. 99, 2017, pp. 1–13.
11. Fageda, X. What Hurts the Dominant Airline at Hub Airports? *Transportation Research E*, Vol. 70, 2014, pp. 177–189.
12. Steven, A. B., A. Alamdar Yazdi, and M. Dresner. Mergers and Service Quality in the Airline Industry: A Silver Lining for Air Travellers? *Transportation Research E*, Vol. 89, 2016, pp. 1–13.
13. Jean, G. A., and G. Lohmann. Revisiting the Airline Business Model Spectrum: The Influence of Post-Global Financial Crisis and Airline Mergers in the US (2011–2013). *Research in Transportation Business and Management*, Vol. 21, 2016, pp. 76–83.
14. Ranson, L. Slot Swap Deal Clears First Hurdle. *Airline Business*, Vol. 27, No. 11, 2011, p. 16.
15. Centre for Aviation. A Brake on Competition: The Ongoing Conundrum of Airport Slots. Sept. 8, 2011. https://centreforaviation.com/insights/analysis/a-brake-on-competition-the-ongoing-conundrum-of-airport-slots-58335. Accessed Nov. 6, 2017.
16. Cranky Flier. Winners and Losers in Delta's Big Expansion at New York's La Guardia Airport. Dec. 19, 2011. http://crankyflier.com/2011/12/19/winners-and-losers-in-deltas-big-expansion-at-new-yorks-la-guardia-airport/. Accessed Nov. 6, 2017.
17. Jacobs, K., and D. Bartz. Southwest, Virgin Win Waiver to Purchase La Guardia Slots, Reuters Business News. Dec. 6, 2013. http://www.reuters.com/article/us-southwest-virgin-idUSBRE9B501Q20131206. Accessed Jul. 26, 2017.

The Standing Committee on Aviation Economics and Forecasting (AV040) peer-reviewed this paper (18-01058).

Article

TRR
JOURNAL OF THE TRANSPORTATION RESEARCH BOARD

Transportation Research Record
2018, Vol. 2672(23) 8–14

DOI: 10.1177/0361198118758983
journals.sagepub.com/home/trr
SAGE

Exploiting a Natural Hub: Turning a Stopover into a Destination

Leah Goldsberry[1] and Adam R. Scavette[1]

Abstract

In the mature aviation system of today, it is difficult to establish new hubs that focus solely on transfer traffic. This paper identifies a new type of hub—a natural tourism hub—one at which an airline and the surrounding metropolitan area can simultaneously benefit from a transportation hub and accompanying tourist destination, respectively. The study aims to identify existing airports for these stopover locations that are located on highly trafficked international flight routes. Using Iceland as an example, this country's success in optimizing its stopover location to promote tourism and gain airline passenger demand is examined. The analysis is carried out by implementing a k-means clustering algorithm on total distance added for stopover locations, as well as flight leg symmetry to identify existing airports that are geographically located in an optimal stopover path for international routes across the Atlantic and Pacific oceans, and between Europe and East Asia. Airports in the clusters that minimize total added distance are then observed, and the clusters are ordered based on how symmetric the two flight legs of a stopover journey at an airport in that cluster tend to be. In addition, three airports near the top of this list are analyzed as potential stopover locations. In using this algorithm, not only is it possible to forecast which hubs may become major tourist destinations, but also to identify how airlines can shape people's perception of their location as a tourist destination.

As demand for service in the transportation industry has grown considerably, there has been a shift in focus toward economies of scale. Airline carriers began research on optimizing airport locations to use their transportation network as an advantage over competitors (*1*). Hub-and-spoke (HS) networks emerged from this research, presenting the concept of hub locations that offer efficient passenger routing. Instead of a fully connected network in which point-to-point routes are offered for all origin destination paths, the HS model directs flights through major hub locations that branch off to the different destinations (*2*). Research has shown that utilization of such a model helps reduce service costs, increase passenger frequencies, and reduce uncertainty in demand (*1*). As such, airlines are largely concerned with optimizing such hub locations so as to fully exploit these opportunities for economies of scale.

There are various aspects to consider when determining a hub location. In the continental United States, characteristics such as demographics, climate, and economics all affect the probability of a city hosting a hub airport. One of the most important factors, however, is the city's geographic location (*3*) and the presence of a lot of people with relatively high incomes (*4*). In already competitive airline markets, many healthy hubs are already established; globally and domestically the networks are well covered with options to get between an origin and destination through multiple potential hubs. With every increase in hub location, travelers have more choices among airline carriers and routings they can take to connect to their destination, creating a more competitive environment. New entrants need to offer a better or cheaper product, or a different kind of one to succeed (*5*). An approach is proposed by which new entrants can compete with entrenched airline carriers by borrowing the hubbing concept but using it not to maximize passenger loads but instead to offer a new destination for stopover travel. In this paper, an alternate perspective to hub location identification is offered such that it is possible to forecast markets that can ultimately be successful.

The starting point is the perspective of identifying natural hubs that can be used as a stopover location. A natural hub is defined as a location near the paths of highly trafficked international flight routes. A stop at such a hub would not drastically affect the flight route between origin and final destination. The criteria for a hub such as this are similar to those of a traditional hub in that economic, demographics, and climatic factors affect the hub's success. However, it is slightly different with regard to geographic criteria in that this kind of hub's focus is connecting nations. Not only that, it is promoting its location as an intermediary destination for those traveling internationally. Previously, destination airports for existing tourist locations were attractive for

[1]University of Pennsylvania, Philadelphia, PA

Corresponding Author:
Address correspondence to Adam R. Scavette: scavette@seas.upenn.edu

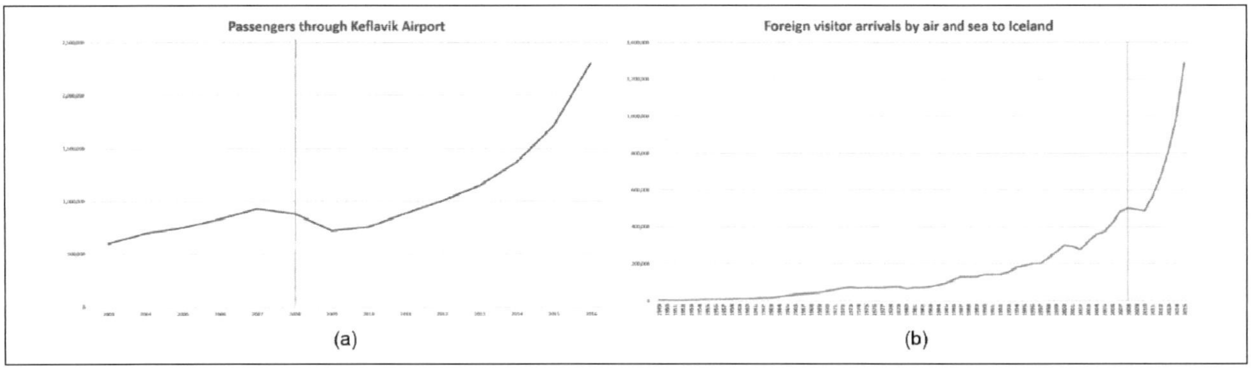

Figure 1. Increase in passengers through KEF from 2003 to 2015 (*a*) and increase in foreign visitors by both air and sea arrival to Iceland from 1950 to 2015 (*b*), both denoting when the financial crisis occurred.

hubbing given the many customers generated (*3*). In this approach, the focus is shifted from the final destination to the connecting destination as a way of promoting tourism. By pulling customers through the hub using a stopover policy, it is possible to change the perception of the location and it begins to be seen as a tourist destination.

This paper will first discuss how Iceland successfully used its location to promote a stopover policy that generated millions of new passengers. It would have been very difficult for airline professionals to accurately forecast the success of both Keflavik International Airport (KEF) and Iceland's tourism industry using traditional time-series models based on the airport's history of passengers or enplanements. A model is offered that identifies airports in an optimal geographic location for hubbing and it will show how the development of these hubs can be forecasted. Finally, three airports are identified particularly and their similarities with Iceland and potential for a successful stopover policy are discussed.

Iceland as a Natural Hub

When the financial crisis hit in 2008, Iceland was affected significantly. Between 2000 and 2008, the finance sector in Iceland had grown rapidly. Due to the large influx of foreign funds, at one point the three largest banks in Iceland held nearly 10 times more than Iceland's economy (as measured by Gross Domestic Product). Thus, when the financial crash occurred, there was a major banking crisis, the housing market collapsed, and there was increased unemployment. The Icelandic Krona collapsed by 60% by the end of 2008 (*6*). To help improve Iceland's failing economy, the Icelandic Tourist Board pushed marketing initiatives to promote Iceland as a tourist destination (*7*). From this came the stopover policy that Iceland's major airline, Icelandair, promoted at the main airport, KEF. This policy allows for up to seven nights of free layover for transatlantic flights. Icelandair connects 27 cities in Europe with 12 cities in North America (*8*).

Since the promotion of this stopover policy, Iceland's tourist economy has boomed. Tourism has increased from 18.8% in 2012 to 31% in 2015, such that it is now the country's top industry (*7*). It contributes nearly 5% to the country's gross domestic product and, as of 2014, 12% of the labor force of Iceland is employed in the tourist industry. Additionally, as seen in Figure 1*b*, air arrivals to KEF have tripled since 2008 (*9, 10*). In 2016 more than 5 million passengers traveled through KEF, in comparison with fewer than 1 million in 2008 (Figure 1*a*).

There are a few key features of Iceland that allow this model to work so well. The most notable is Iceland's geographic location. KEF is 2,600 miles from New York City and 1,200 miles from London, putting it about midway between the two. In addition, stopping at Iceland only adds 322 extra miles, a low number relative to a transatlantic flight (*10*). KEF is also well located within the largest city in Iceland, where almost two-thirds of Iceland's total population live. Thus, the city is well suited to accommodate tourists with regard to staff, amenities, and activities. Furthermore, the local airlines place strong emphasis on growing the surrounding area for tourism. Icelandair Group runs Icelandair Hotels and tour operator Iceland Travel (*7*). WOW air, which was founded in 2011 and also implements a stopover policy, provides online flight, hotel, and car booking services to its customers, a unique offering that has helped grow the airline (*11*). Through exploiting its geographic location, Iceland has effectively changed travelers' perceptions of the nation from somewhere they probably would not have considered visiting to one of it being a major destination (*8*).

Model for Identifying Optimal Natural Hubs

Data Source

This section of the paper focuses on identifying geographically optimal airports for stopovers. Locations of major world airports (airport codes, latitude values, longitude values) were obtained from the OpenFlights Database (*12*). The initial goal was to identify hubs that could serve as feeders to

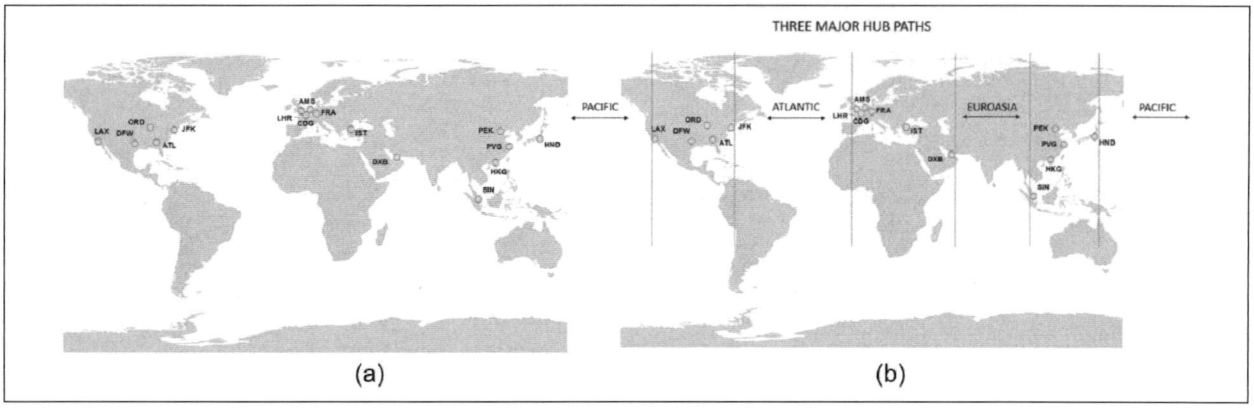

Figure 2. Map of world hubs (*a*) and the flight paths identified (*b*).

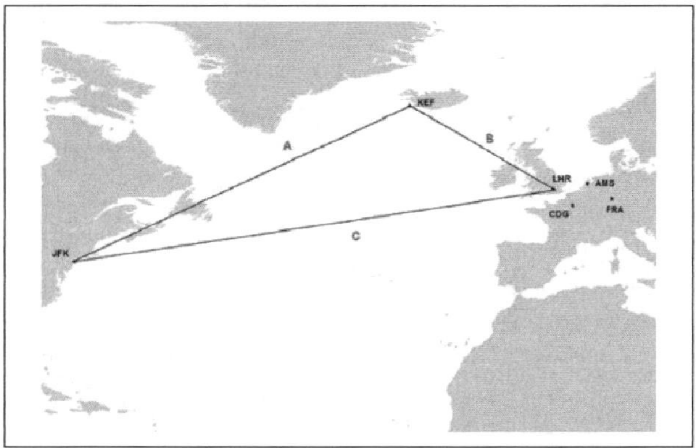

Figure 3. KEF stopover diagram.

the potential geographically optimal stopover airports. The top 16 hubs in the world by total passenger traffic in 2016 were identified (*13*). These hub airports, plotted in Figure 2*a*, are: Singapore (SIN), New York City (JFK), Amsterdam (AMS), Shanghai (PVG), Frankfurt (FRA), Istanbul (IST), Dallas–Fort Worth (DFW), Paris (CDG), Hong Kong (HKG), Los Angeles (LAX), London Heathrow (LHR), Tokyo (HND), Chicago (ORD), Dubai (DXB), Beijing (PEK), and Atlanta (ATL).

From these 16 hubs, three major "flight paths" were identified, as seen in Figure 2*b*: Atlantic, EuroAsia, and Pacific. The Atlantic path comprises all flights with origins in North America (LAX/DFW/ORD/ATL/JFK) and destinations in Europe and the Middle East (LHR/AMS/CDG/FRA/IST/DXB) for a total of 30 origin region to destination region (O–D) pairs. The EuroAsia path comprises all flights with origins in Europe and the Middle East (LHR/AMS/CDG/FRA/IST/DXB) and destinations in Asia (SIN/HKG/PVG/PEK/HND) for a total of 30 O–D pairs. Finally, the Pacific path comprises all flights with origins in Asia (SIN/HKG/PVG/PEK/HND) and destinations in North America (LAX/DFW/ORD/ATL/JFK) for a total of 25 O–D pairs. Throughout the analysis "flight paths" are identified as these three sets of O–D pairs.

Variable Generation

Observations of interest in this study are potential stopover airports in a given flight path. In creating the dataset all airports west of the easternmost hub and east of the westernmost hub on a given flight path were removed. For example, in the Atlantic path all airports west of JFK and all airports east of LHR were removed. The cutoff points for all three flight paths are expressed in Figure 2*b*.

For each of the potential stopover airports in the three flight paths two different distance variables for every O–D pair were generated: added distance and leg symmetry. For each potential stopover and O–D pair in a given path, "added distance" was calculated as the difference between making a stopover at an airport (A and B) and flying directly between the O–D pair (C); added distance is thus $A + B - C$.

For each airport in a given path, "leg symmetry" was calculated as $|A - B|$, the absolute value of the difference between the distances of the legs of a stopover. Thus, the leg symmetry for the stopover, KEF, depicted in Figure 3 for the O–D pair JFK–LHR, was calculated. JFK to KEF (A) is 2,593 miles and KEF to LHR (B) is 1,180 miles, so the symmetry is 1,413 miles.

However, there are 29 other O–D pairs on the Atlantic flight path, so the added distance and leg symmetry for making a stopover at KEF for those pairs was calculated as well (e.g., ATL to FRA, LAX to AMS, etc.). Then, the total added distance and total leg symmetry for the 30 O–D pairs were calculated and the results totaled. Thereafter, for every potential stopover in the dataset the same process was used—the O–D pairs depend on the flight path that the potential stopover airport is located in. Finally, every observation has two variables: total added distance and total leg symmetry.

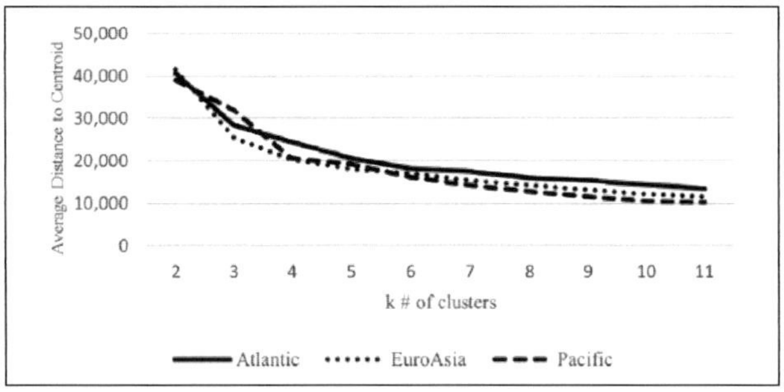

Figure 4. Average distance to centroid by *k* number of clusters.

$$\text{Total Added Distance} = \sum_{\#\text{O-D PAIRS IN PATH}} A + B - C$$

$$\text{Total Leg Symmetry} = \sum_{\#\text{O-D PAIRS IN PATH}} |A - B|$$

Throughout this study, distances between two airports were calculated using the geodesic distance between a pair of points on the surface of the Earth, with an accurate ellipsoidal model of the Earth (*14*). It is understood that true flight paths do not follow the direct measures as calculated by this method. However, it is felt that this calculation serves as a close approximation.

Model and Results

The procedure used to identify geographically optimal hubs is k-means clustering. *k*-means clustering attempts to group observations together based on feature similarity according to *k* number of clusters, where *k* is a number that is chosen a priori (*15*). The clustering was based on two variables throughout the analysis: total added distance and total leg symmetry. Three separate k-means procedures for each of the flight paths identified were run, in which the individual observations were potential stopover airports.

To find the k parameter, the elbow point method was used (*16*). This measures the average distance between all observations and the cluster centroid to which they are assigned based on the k number of clusters. The elbow point is the point at which the average distance levels off, or in simpler terms, the point at which adding more clusters offers little more explained distinction between the clusters. Figure 4 shows the elbow point plots for the three different paths. Ten was chosen as the value for *k* across the analysis as it seems to level off at that point for all three paths.

In the identification procedure, symmetry and added distance are generally minimized. An ideal stopover should not be too far out of the way of a typical path between hubs, as calculated by the "added distance" variable. Furthermore, it is generally preferable to passengers when the leg lengths of the two legs on a stopover are close to equal, because a break in the middle of a trip is generally better than a stopover very close to either the origin or destination. This conclusion was made after observing the success of KEF in promoting stopovers as the airport is approximately halfway across many transatlantic flights.

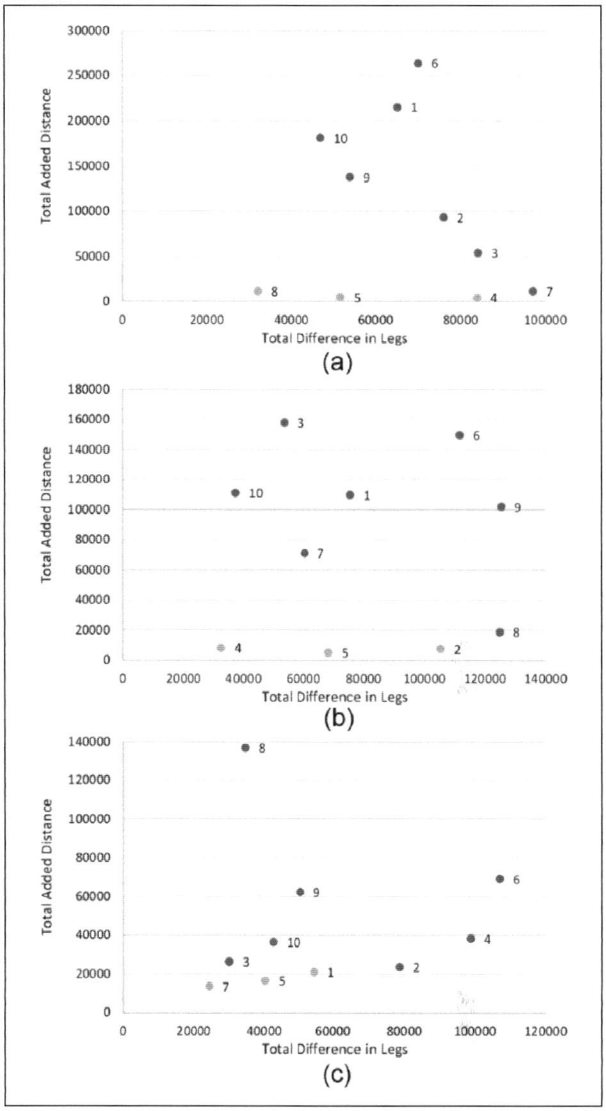

Figure 5. K-means centroid plots for the Atlantic (*a*), EuroAsia (*b*), and Pacific (*c*) flight paths. Selected clusters for analysis are highlighted in yellow.

After running k-means procedures on the three different flight paths, the cluster centroids were plotted to gain a sense of the distribution of the clusters related to the two variables of interest. Figure 5 plots these centroids for the Atlantic (*a*), EuroAsia (*b*), and Pacific (*c*) flight paths, respectively. For the analysis, it is more important to minimize total added distance than total leg symmetry. Therefore, for each of the flight

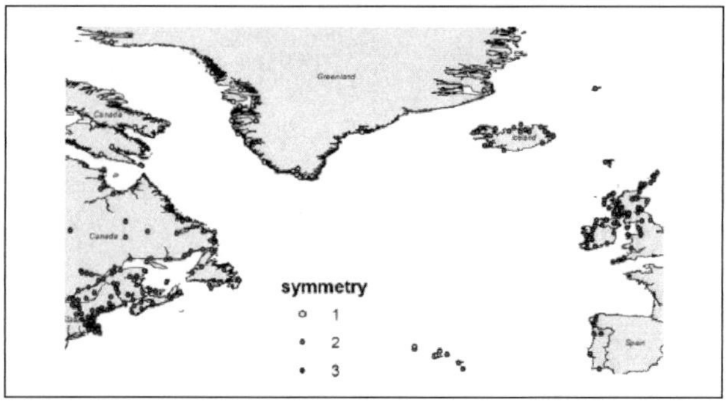

Figure 6. Geographically optimal clusters for Atlantic flight paths.

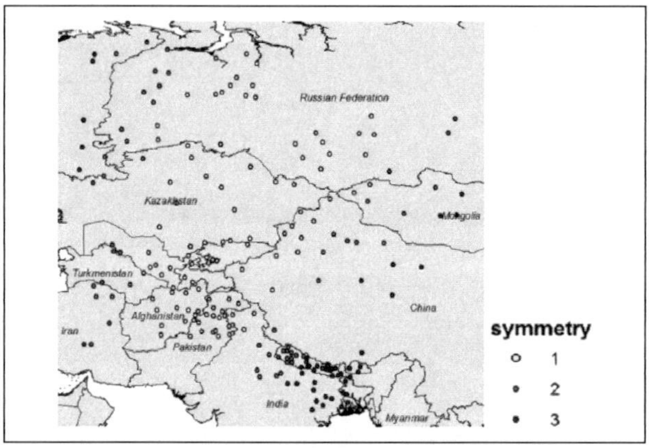

Figure 7. Geographically optimal clusters for EuroAsia flight paths.

paths the three clusters that minimize total added distance are reported, and they are ordered according to their level of leg symmetry, 1 being the most symmetric and 3 the least symmetric. In Figure 5*a*, it can be seen that clusters 8, 5, and 4 minimize total added distance and at the same time have progressively lower levels of total leg symmetry. Therefore, those three clusters were chosen as the "geographically optimal" clusters for stopover airports and ordered according to symmetry, so that 8, 5, and 4 become clusters 1, 2, and 3 for the Atlantic path. Following a similar procedure, clusters 4, 5, and 2 were chosen to become clusters 1, 2, and 3 for the EuroAsia flight path, as seen in Figure 5*b*. Finally, clusters 7, 5, and 1 were chosen to become clusters 1, 2, and 3 for the Pacific flight path, as seen in Figure 5*c*.

The optimal airports with cluster identifiers (based on symmetry) have been mapped for the Atlantic (Figure 6), EuroAsia (Figure 7), and Pacific (Figure 8) flight paths respectively. In Figure 6, it can be seen that Cluster 1 is centered on Greenland and Northeast Canada and Cluster 2 contains airports in Iceland, Newfoundland, and Nova Scotia. With regard to the Atlantic flight path, Cluster 1 contains 34 airports, Cluster 2 has 74, and Cluster 3 has 142. In Figure 7,

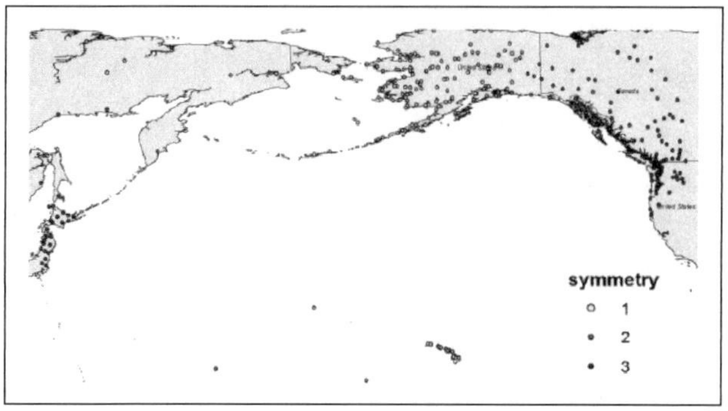

Figure 8. Geographically optimal clusters for Pacific flight paths.

it can be seen that Cluster 1 is centered on Kazakhstan, Pakistan, Afghanistan, and Russia and Cluster 2 is more sparsely concentrated in Iran, India, and Western China. For the EuroAsia flight path, Cluster 1 contains 103 airports, Cluster 2 has 55, and Cluster 3 has 53. In Figure 8, it can be seen that Cluster 1 is centered primarily on Alaska and Hawaii, whereas Cluster 2 is mainly in Yukon Territory, Canada. With regard to the Pacific flight path, Cluster 1 contains 178 airports, Cluster 2 has 64, and Cluster 3 has 112.

The quantitative analysis in this section has identified airports that can be reached over many hub to hub journeys across major flight paths without adding much distance to the trip. Furthermore, these clusters are separated by the level of symmetry that the two legs (before and after the stopover) exhibit. Although this analysis is by no means complete, it offers a picture of which airports can be accessed with ease by many passengers flying between these hubs as stopovers as an alternative to one long-haul flight between the hubs.

Model Limitations

This study on geographically optimal airports for stopovers considers only distance-based measures for optimality. It does not consider the existing infrastructure, population features, or the desirability of the airport locations for potential stopover tourists. Some of the most symmetric optimal airports that have been identified in this analysis would be highly unfeasible for stopover journeys for a variety of reasons. For the Atlantic flight path, most of the optimal airports in the most symmetric cluster are located in Greenland, which does not contain the infrastructure necessary for a large tourism industry at the moment. Furthermore, with regard to the EuroAsia flight path, many of the most optimal and symmetric airports lie in politically unstable countries such as Pakistan, Iran, and Afghanistan, which would be unattractive for many potential tourists. Extensions of the study should consider clustering potential stopover airports based on demographic, economic, and political variables as well as distance variables.

Natural Hub Profiles

Transatlantic

With regard to the Atlantic flight path, the airport highlighted, Saint John's International Airport (YYT) in Saint John's, Canada, falls into Cluster 2 in the analysis. With a metropolitan population of over 200,000, the city and surrounding area offer the infrastructure and a coastline environment that could support an increased level of tourism. The airport is well positioned to serve transatlantic flights as it currently serves Dublin, Ireland in 4 h and 30 min and London in 5 h (both nonstop flights operated by WestJet). Furthermore, it serves several major North American cities such as Toronto, Montreal, Orlando, and Tampa (17). YYT is positioned in a very similar level of symmetry as KEF, which as has already been noted has gained success as a stopover airport. YYT is roughly 5 h to Europe and 3 h to many east coast airports and KEF is 5 h to North America and roughly 3 h away from the many European airports that it serves.

Based on figures from 2016, the province of Newfoundland and Labrador (the Saint John's metro area contains nearly 50% of the province's population) has experienced a modest growth in tourism at a rate of 4.1% from air passengers (*17*). Furthermore, accommodation occupancy remained steady at 50.2% for the year with an average hotel stay costing roughly $139, which is nearly average for the entire country of Canada (*18*). As of March 2017, the unemployment rate in Saint John's stands at 8.9%, over two percentage points higher than Canada's unemployment rate of 6.7%. The local economy in Saint John's is highly sensitive to the performance of the oil industry, which has been somewhat volatile in the past few years (*19*). The local economy might benefit from developing the airport and city as a stopover destination for transatlantic flights. This could tighten the labor market at the same time diversifying it away from a volatile industry.

EuroAsia

The airport considered for the EuroAsia flight path is Tribhuvan International Airport (KTM) in Kathmandu, Nepal. This airport is part of Cluster 3 for symmetry. Although some airports in Nepal fall into Cluster 2, KTM is the only international airport in Nepal, and the only airport with a runway large enough for such airplanes. Additionally, KTM is very closely located to Cluster 2. Currently, KTM serves about 3 million passengers a year, with international flights to South Asia and the Middle East. Its only current connection with Europe is through IST. Thus, if KTM can incentivize more travelers, they have the potential to create more connecting flight paths.

Kathmandu is the capital city of Nepal and, as such, it is the largest metropolitan area in the country. It is a beautiful area in the Himalayan region, hosting palaces, mansions, and gardens. The location has a subtropical climate, allowing for year-round warm weather. In addition, tourism is an important industry for Nepal, contributing largely to income levels in the city. Thus, both the city and the airport have aligned incentives to increase tourism to the area, which may be done through a successful stopover policy.

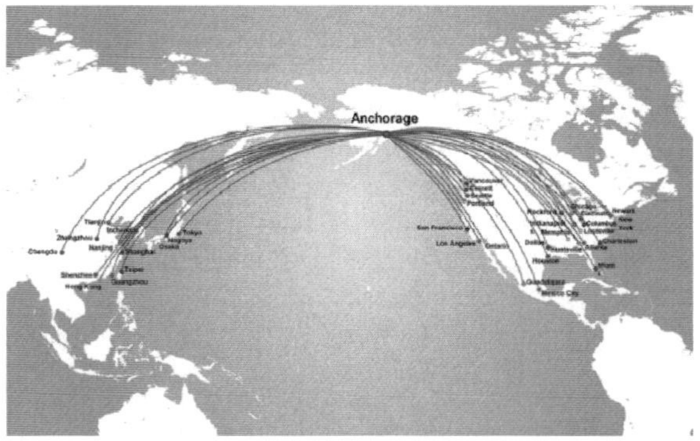

Figure 9. Passenger and cargo routes served by ANC (passenger routes only to North American regions).

Transpacific

With regard to the Pacific flight path, the airport chosen for consideration is Ted Stevens Anchorage International Airport (ANC) in Anchorage, Alaska. This airport falls into Cluster 1 of the analysis, meaning it is conveniently situated between major airports in North America and Asia. As a preliminary observation, the features of ANC were compared with those of KEF. ANC currently has three runways, one more than KEF. This suggests that ANC can cope with a similar capacity to KEF. Additionally, ANC currently has about 2.5 million passengers passing through annually. Before heavily promoting its stopover policy in 2009, KEF dealt with a similar number of passengers, around 2 million annually. Now, in 2016, KEF moves more than 6 million passengers through its airport annually (*10*). This shows how ANC has a large margin of growth given it can implement a successful stopover policy. In addition, just as with Reykjavik, Anchorage has the largest population in Alaska, allowing for the infrastructure and manpower to handle increased tourism.

When the current routes ANC serves were considered, there were only flights for passengers to locations in North America. However, when looking at its cargo routes, ANC clearly uses its geographic location for a competitive edge in cargo transport (*20*). The map in Figure 9 showing the cargo locations ANC serves looks similar to what it might be expected an optimal hub location would look like. Thus, it gives the impression that ANC would be very successful in implementing this policy. In looking at why ANC does not currently offer international travel, it can be seen that the increased airplane range for nonstop flights and the opening of the airspace after 1989 dramatically reduced the demand at ANC for international passenger traffic. Thus, even though ANC does not currently offer international flights, the airport and the surrounding city are very much interested in creating passenger demand. Given the creative spin and competitive edge that a stopover policy offers, implementing this at ANC could be an excellent way of increasing demand for the airport and, as a result, increasing tourism and boosting the economy in the surrounding area.

Conclusion

Airport locations that are geographically optimal for hubbing for international flights were identified using k-means clustering based on total added distance and total leg symmetry. The analysis may be used to successfully forecast the future growth of airports along high-activity international flight paths. These airports were grouped based on paths located in the Atlantic, Pacific, and EuroAsia flight routes. Given the many airports identified, however, future work could be to utilize features other than geography, such as economy, weather, and population, to further filter out the airports most suitable for implementing successful stopover policies. Given the novelty of this idea, it will be interesting to see how popular it becomes and which airports can successfully implement the policy.

Acknowledgments

The authors thank their professor, Dr Megan Ryerson, for all her help with and support for this research. In addition, they thank their teaching assistant, Max Li, for his support in writing this paper.

References

1. Campbell, J. F., and M. E. O'Kelly. Twenty-Five Years of Hub Location Research. *Transportation Science*, Vol. 46, No. 2, 2012, pp. 153–169.
2. Brueckner, J. K., and Y. Zhang. A Model of Scheduling in Airline Networks: How a Hub-and-Spoke System Affects Flight Frequency, Fares and Welfare. *Journal of Transport Economics and Policy*, Vol. 35, No.2, 2001, pp. 195–222.
3. Huston, J. H., and R. V. Butler. The Location of Airline Hubs. *Southern Economic Journal*, Vol. 57, No. 4, 1991, pp. 975–981.
4. Hansen, M. Airline Competition in a Hub-Dominated Environment: An Application of Noncooperative Game Theory. *Transportation Research Part B*, Vol. 24B, No. 1, 1990, pp. 27–43.
5. Dewine, M. *Airline Hubs: Fair Competition or Predatory Pricing, Hearing Before the Committee on the Judiciary, U.S. Senate*. Diane Publishing Company, 1998. Airline hubs, fair competition or predatory pricing?: hearing before the Subcommittee on Antitrust, Business Rights, and Competition of the Committee on the Judiciary, United States Senate, One Hundred Fifth Congress, second session ... April 1, 1998, Volume 4. Vol. 105, No. 659 of *S. hrg*, United States Congress.
6. Lewis, M. Wall Street on the Tundra. *Vanity Fair*. April 2009. https://www.vanityfair.com/culture/2009/04/iceland200904. Accessed April 1, 2017.
7. Scheivachman, A. Iceland and the Trials of 21st-Century Tourism. Skift, 2016. https://skift.com/iceland-tourism/. Accessed April 1, 2017.
8. Lund, K. A., K. Loftsdttir, and M. Leonard. More than a Stopover: Analysing the Postcolonial Image of Iceland as a Gateway Destination. *Tourist Studies*, Vol. 17, No. 2, 2017 pp. 144–163. http://journals.sagepub.com/doi/10.1177/1468797616659951. Accessed April 1, 2017.
9. Fouche, G., and R. Sigurdardottir. Iceland's Tourism Boom Fuels Recovery, Lays Bare Locals' Plight. Reuters, June 13, 2016. Accessed April 1, 2017. https://www.reuters.com/article/us-insight-iceland-economy/icelands-tourism-boom-fuels-recovery-lays-bare-locals-plight-idUSKCN0YZ1H5.
10. Icelandic Tourist Board. Tourism in Iceland in Figures. May 2016. https://www.ferdamalastofa.is/en/recearch-and-statistics/tourism-in-iceland-in-figures. Accessed April 1, 2017.
11. Weller, C. How a Tiny Budget Airline Single-Handedly Turned Iceland into America's Favorite Vacation Spot. Business Insider, October 26, 2016. http://www.businessinsider.com/wow-air-ceo-made-iceland-irresistible-2016-10. Accessed April 1, 2017.
12. Open Flights Database. Airport, Airline and Route Data. https://openflights.org/data.html. Accessed March 2017.
13. Zhang, B. Ranked: The 16 Busiest Airports in the World. Business Insider, April 2016. http://www.businessinsider.com/16-busiest-airports-in-the-world-2016-4. Accessed April 1, 2017.
14. Vincenty, T. Direct and Inverse Solutions of Geodesics on the Ellipsoid with Application of Nested Equations. *Empire Survey Review*, Vol. 23, No. 176, p. 88.
15. Kanungo, T., D. M. Mount, N. S. Netanyahu, C. D. Piatko, R. Silverman, and A. Y. Wu. An Efficient K-Means Clustering Algorithm: Analysis and Implementation. *IEEE Transactions on Pattern Analysis and Machine Intelligence*, Vol. 24, No. 7, 2002, pp. 881–892.
16. Ketchen, D. J. Jr., and C. L. Shook. The Application of Cluster Analysis in Strategic Management Research: An Analysis and Critique. *Strategic Management Journal*, Vol. 17, No. 6, 1996, pp. 441–458.
17. Newfoundland and Labrador Department of Tourism, Culture, Industry and Innovation. http://www.tcii.gov.nl.ca/. Accessed April 1, 2017.
18. Tourism Industry Association of Canada. Monthly Average Daily Rate of Hotels in Canada from 2015 to 2016 (in Canadian Dollars). Statista: The Statistics Portal. https://www.statista.com/statistics/422489/monthly-average-daily-rate-of-hotels-canada/. Accessed April 1, 2017.
19. Department of Community Services. *State of the Economy: St. John's Metro*. Economic Development, Culture and Partnerships, City of St. John's, 2017.
20. Wysocky, K. Anchorage Int'l Airport Serves as Pit Stop for Global Cargo Carriers. Airport Improvement, January 2015. http://www.airportimprovement.com/article/anchorage-intl-airport-serves-pit-stop-global-cargo-carriers. Accessed April 1, 2017.

The Standing Committee on Aviation Economics and Forecasting (AV040) peer-reviewed this paper (18-04270).

TRR
JOURNAL OF THE TRANSPORTATION RESEARCH BOARD

Article

Transportation Research Record
2018, Vol. 2672(23) 15–19

DOI: 10.1177/0361198118758682
journals.sagepub.com/home/trr
SAGE

Investigating the Pricing Impacts of the American Airlines and US Airways Merger

Yunzhou Zhang[1] and Linda Nozick[1]

Abstract

In 2013, American Airlines merged with US Airways creating the largest airline in the world. This paper analyzes this merger's effect on domestic airline ticket pricing using the Department of Transportation's 10% ticket sample. This analysis confirmed insights that have been identified in prior analyses including in which there is heightened competition, prices are lower. It also demonstrates again that mergers can be correlated with rising prices. What is perhaps novel is that although legacy carriers generally increased prices post-merger, low-cost carriers did not appear to have that same freedom. Rather, they lowered prices, and those fare reductions were largest in markets for which American Airlines and US Airways played a larger role pre-merger.

The Airline Deregulation Act of 1978 largely removed U.S. governmental controls over the airline industry for fares and market entry. Since then, more than 50 mergers have taken place in the United States including several mega-mergers, like the merger between Delta and Northwest or United Airlines with Continental. These mergers have dramatically reshaped the airline industry. Out of 12 major airlines operating at the beginning of the last decade, only four of them are still in operation. These four airlines control about 70% of the domestic passenger commercial flight market share based on an analysis of the 10% ticket sample from the U.S. Department of Transportation (*1*). To preserve competition in the passenger airline market so as to protect consumers, the Department of Justice has the power to block mergers under antitrust regulation.

The recent mega-merger between American Airlines and US Airways in 2013 created the largest airline company in the world, measured by revenue, fleet size, and passenger-kilometers-flown; therefore, understanding what changes in fares occurred post-merger and the factors that are particularly correlated with those changes is particularly important. This paper gives an analysis of fare impacts using the 10% ticket survey on legacy and low-cost carriers stemming from the merger, as well as investigates how those changes are influenced when the market is subject to slot constraints. We adopt the idea of potential competition, as described by Kwoka and Shumilkina (*2*), and test whether the fare impacts they identified in the 1987 merger between US Airways and Piedmont are also evident in this recent merger.

There have been a number of studies that have investigated airfare impacts from a variety of causes. Luo used data from 1983 and empirically investigated contestability in the airline industry and concluded that they were imperfectly contestable (*3*). They also showed that both actual and potential competitors reduce fares. Several studies showed, using different data over different time frames, that when low-cost carriers enter a route, prices fall (*4–7*). Morrison and Winston revealed that when a low-cost carrier exited a market, prices rose (*8*). Several studies again showed that when a low-cost carrier provides service on a route, prices fall on that route (*9–11*). Studies (*9–11*) also showed that prices decline on routes that have a low-cost carrier providing service at either the origin or the destination of the route (*12*) showed that prices fall when a new carrier is anticipated on a route, and fall again when that entry materializes, using fare prices over the evolution of Southwest Airlines. Kwoka et al. examined the impact of low-cost carriers on price, and how those impacts are influenced by the magnitude of service they provide (*13*).

Our focus is an empirical analysis of the pricing impacts of the US Airways and American Airlines merger. Hence analyses in the literature that focus on quantifying the impact of mergers on fare price are particularly relevant. For

[1]School of Civil and Environmental Engineering, Cornell University, Hollister Hall Ithaca, NY

Corresponding Author:
Address correspondence to Linda Nozick: lkn3@cornell.edu

example, Kwoka and Shumilkina examined the impacts of the US Airways and Piedmont merger in 1987 (*2*). They found that the magnitude of fare increases was affected by whether either or both carriers provided service on the route prior to the merger, and secondarily, if either or both provided service to either the origin or destination airports on the route. Luo focused on understanding the impact on fares of the Northwest and Delta Airlines merger (*3*). They concluded that if both carriers provided service on a route prior to the merger, the fare impacts were smaller.

The next section describes the data, variables and models used for this analysis. The third section describes the empirical results found. The final section gives conclusions and opportunities for future research.

Data, Variables Considered, and Models

We make use of the Department of Transportation's Airline Origin and Destination Survey (*1*). This survey is a 10% sample, undertaken on a quarterly basis, of passenger tickets. We focused on tickets that met the following four criteria. First, we only used tickets that were non-stop. There are important differences in service when comparing non-stop to single and multi-stop routes, which would have been important to address, had they been included. Second, we focused on tickets that cost more than \$10 and less than \$2,000. These bounds are the same as were used in Kwoka and Shumilkina (*2*). Other authors have suggested other airfare bounds. For example, Luo used a lower bound of \$25 and no upper bound (*3*). Third, we only included fare records from four legacy carriers and seven low-cost airlines. The four legacy carriers are US Airways, American Airlines, Delta, and United Airlines. The seven low-cost carriers considered are Jet Blue, Frontier, Southwest, Spirit, Allegiant, Sun Country, and Virgin Airlines. Finally, we only considered tickets for which the ticketing carrier and the operating carrier are the same.

We made use of 9 fields in the DB1B database: market, year, quarter, ticketing carrier, origin airport, destination airport, passengers, market fare, and market distance. With the date and the market information, we identified whether the origin or destination airports were slot constrained. We also computed the Herfindahl–Hirschman Index (HHI) for each market in the appropriate quarter. HHI is a widely accepted measurement of market concentration.

We also made use of six dummy variables to describe whether US Airways or American Airlines or both was providing service on that particular route at that time, as well as whether one, the other, or both were potential competitors. This yielded the following six variables: two-zero, one-one, zero-two, one-zero, zero-one and zero-zero. For example, two-zero indicates that both American Airlines and US Airways were providing service on the route; hence neither was a potential competitor on that route (because they were both incumbents). In contrast, zero-two indicates that neither was providing service on the route, however both were providing service on other routes that used either the origin or destination airports, or both, for the route. This is the same as described by Kwoka and Shumilkina (*2*).

The model, using ordinary least squares, ignoring the concepts of incumbent and potential competition pre- and post-merger, is given in Equation 1 below.

$$
\begin{aligned}
\ln\left(\text{fare}_{ijt}\right) = {} & \beta_0 + \beta_1 \ln\left(\text{Dist}_{ij}\right) + \beta_2 \ln(\text{HHI}_{ijt}) + \beta_3 \text{Slots}_{ij} \\
& + \beta_4 \text{Post} + \beta_5 \text{OLeg} + \beta_6 \text{LCC} + \beta_7 Q_2 + \beta_8 Q_3 \\
& + \beta_9 Q_4 + \beta_{10} \text{OLeg} * \text{Post} + \beta_{11} \text{LCC} * \text{Post}
\end{aligned} \tag{1}
$$

where

fare_{ijt} = one-way fare from origin airport *i* to destination airport *j* in time period *t*

Dist_{ij} = distance between origin airport *i* and destination airport *j*,

HHI_{ijt} = Herfindahl–Hirschman Index in time period *t* on the direct route from airport *i* to airport *j*

Slots_{ij} = dummy variable that equals 1 if airport *i* or airport *j* or both are slot constrained (where JFK, Newark, LaGuardia, and Reagan National are defined as slot constrained)

Post = dummy variable that equals 1 if time period *t* is after the merger, and 0 otherwise

OLeg = dummy variable that equals 1 if the carrier is either United Airlines or Delta Airlines, and 0 otherwise

LCC = dummy variable that equals 1 if the carrier is one of the seven low-cost carriers and 0 otherwise

Q_2, Q_3, and Q_4 = dummy variables that indicate if the record is for quarters 2, 3 or 4, respectively.

The baseline model given in Equation 1 is extended to include the concept of potential competition as given in Equation 2 below.

$$
\begin{aligned}
\ln\left(\text{fare}_{ijt}\right) = {} & \beta_0 + \beta_1 \ln\left(\text{Dist}_{ij}\right) + \beta_2 \ln(\text{HHI}_{ijt}) + \\
& \beta_3 \text{Slots}_{ij} + \beta_4 \text{Post} + \beta_5 \text{OLeg} + \\
& \beta_6 \text{LCC} + \beta_7 Q_2 + \beta_8 Q_3 + \beta_9 Q_4 + \\
& \beta_{10} \text{OLeg} * \text{Post} + \beta_{11} \text{LCC} * \text{Post} + \\
& \beta_{12} \text{Pot}_{ijt} + \beta_{13} \text{INC2POT0}_{ijt} + \\
& \beta_{14} \text{INC1POT1}_{ijt} + \beta_{15} \text{INC0POT2}_{ijt} + \\
& \beta_{16} \text{INC2POT0}_{ijt} * \text{POST} * \text{OLeg} + \\
& \beta_{17} \text{INC1POT1}_{ijt} * \text{POST} * \text{OLeg} + \\
& \beta_{18} \text{INC0POT2}_{ijt} * \text{POST} * \text{OLeg} + \\
& \beta_{19} \text{INC2POT0}_{ijt} * \text{POST} * \text{LCC} + \\
& \beta_{20} \text{INC1POT1}_{ijt} * \text{POST} * \text{LCC} + \\
& \beta_{21} \text{INC0POT2}_{ijt} * \text{POST} * \text{LCC}
\end{aligned} \tag{2}
$$

where

Pot_{ijt} = number of carriers that provide service to airport *i* or airport *j* or both, in time period *t* but do not provide service from airport *i* to airport *j*

$\text{INC}m\text{POT}n_{ijt}$ = dummy variable that takes on a value of 1 if there are *m* incumbents and *n* potential entrants in the

Table 1. Estimated Coefficients

Variable	AY 11-15 base model (AY11-16 base model)	AY 11-15 extended model (AY11-16 extended model)
Intercept	2.24*** (2.12***)	1.81*** (1.7***)
$\ln(Dist_{ij})$	0.446*** (0.468***)	0.472*** (0.488***)
$\ln(HHI_{ijt})$	0.436*** (0.537***)	0.39*** (0.505***)
$Slots_{ij}$	0.07*** (0.073**)	0.079*** (0.080***)
Post	0.096*** (0.02)	0.126*** (0.038**)
OLeg	−0.039** (−0.039**)	−0.0444*** (−0.042**)
LCC	−0.521*** (−0.533***)	−0.532*** (−0.542***)
OLeg * Post	−0.023 (0.050**)	0.662*** (0.669***)
LCC * Post	−0.289*** (−0.272***)	−0.5*** (−0.598***)
Q_2	−0.022** (−0.049***)	−0.020* (−0.046***)
Q_3	−0.072*** (−0.063***)	−0.090*** (−0.058***)
Q_4	−0.058*** (−0.0263**)	−0.076*** (−0.019*)
Pot_{ijt}		−0.0429*** (−0.040***)
$INC2POT0_{ijt}$		0.565*** (0.561***)
$INC1POT1_{ijt}$		0.596*** (0.595***)
$INC0POT2_{ijt}$		0.618*** (0.613***)
$INC2POT0_{ijt}$ * POST * OLeg		−0.694*** (−0.649***)
$INC1POT1_{ijt}$ * POST * OLeg		−0.668*** (−0.649***)
$INC0POT2_{ijt}$ * POST * OLeg		−0.652*** (−0.635***)
$INC2POT0_{ijt}$ * POST * LCC		0.0339 (0.190***)
$INC1POT1_{ijt}$ * POST * LCC		0.298*** (0.422***)
$INC0POT2_{ijt}$ * POST * LCC		0.321*** (0.454***)
Adjusted R^2	0.279 (0.273)	0.31 (0.317)

Note: (*, **, ***significant at .01, .001, and .0001, respectively).

market from origin airport i to destination airport j in time period t, and 0 otherwise.

The full set of $INCmPOTn_{ijt}$ variables has six different categories (2), but we only use four of them in these models. There are effectively no markets that fit the description of $INC1POT0_{ijt}$ and $INC0POT1_{ijt}$, hence they are discarded.

Notice that in the extended model, we attempted to understand how the presence of US Airways or American Airlines, or both, affected prices post-merger when they were either an actual or a potential competitor prior to the merger, and how these impacts might have varied if the competing carrier was either a legacy carrier (United Airlines or Delta) or a low-cost carrier.

Results

Both models were fit using 2011 and 2015 as well as 2011 and 2016 data. Both of these periods include pre- and post-merger fare data. Both periods were explicitly considered separately in an attempt to understand whether the impacts of the merger were more persistent or transitive in nature. The coefficients for all four models are given in Table 1. Notice that many of the coefficients are very similar across both sets of models, implying that the general trends persisted over the 2-year period.

Consistent with other studies, we found that higher HHI, which implies higher industry concentration, is associated with higher fares. At a national scale and very approximately, the coefficient on HHI implies that there is approximately a 40–50% increase in fares when the number of carriers declines from five equal-size carriers to a single carrier. We also found, again similar to other studies, that prices are generally about 7% higher at slot-constrained airports, and that prices are generally higher for legacy carriers in contrast to low-cost carriers.

The coefficient for the variable post was positive in all four models indicating that prices rose after the merger. In only one model was the significance for that coefficient 0.15. In contrast, in the other three models, the significance level was better than 0.001. In comparison with 2011, the merged carrier's fares were about 12% higher in 2015, but that difference was only about 4% in 2016 (in comparison with 2011). Not unexpectedly, based on the extended model, fares decreased with an increasing number of potential entrants (about 4% per additional potential entrant).

Table 2, using the extended model, gives the percent change in fares (for 2015 and 2016 in comparison with 2011) for the merged carrier, the two legacy carriers, and low-cost carriers. For the other legacy carriers and the low-cost carriers, the estimates are different based on the role American Airlines and US Airways played in the market prior to the merger.

Table 2. Changes in Fares Relative to 2011

Carrier group	2015	2016
Merged carrier: American Airlines	13.43%	3.87%
United/Delta w/o AA/USAir in Mkt	119.90%	102.79%
United/Delta with (2,0)	9.86%	5.97%
United/Delta with (1,1)	12.75%	5.97%
United/Delta with (0,2)	14.57%	7.47%
LCC w/o AA/USAir	−31.20%	−42.88%
LCC with (2,0)	−28.83%	−30.93%
LCC with (1,1)	−7.32%	−12.89%
LCC with (0,2)	−5.16%	−10.06%

Note: (incumbent, potential entrant). LCC = low cost carrier.

On routes in which US Airways and American Airlines were not an incumbent or a potential entrant Delta and United Airlines raised prices and low-cost carriers reduced prices on average (about 120% and −30%, in 2015 in comparison with 2011, for example).

The increase in fares by United Airlines and Delta in these markets, on the surface, is quite interesting. About 1% of the markets fit into this category, that is, in which United Airlines or Delta, or both, operated but neither American Airlines nor US Airways were an incumbent or potential entrant. In 2011, the average airline fares in these markets were about 65% higher in comparison with all markets. In 2015, they were about 125% higher and in 2016, they were 138% higher. Overwhelmingly these markets connect two small airports. Further, in more than 90% of these markets there is only one carrier.

Markets in which American Airlines or US Airways, or both, were an incumbent or a potential entrant, low-cost carriers reduced prices post-merger and those reductions were highest when both American Airlines and US Airways were incumbents, and lowest when they were both potential entrants. More specifically, when both were incumbents, prices fell by about 29% in 2015 (in comparison with 2011), and the comparison between 2016 prices and 2011 was essentially the same. When both American Airlines and US Airways were potential entrants, the decline in 2015 (in comparison with 2011) was only about 5%. This decline did continue, however, into 2016 with a total average decline of about 10% from 2011 to 2016.

In contrast, for other legacy carriers (Delta and United Airlines), on routes where US Airways or American Airlines, or both, played a role pre-merger, prices rose in 2015 but those increases did not fully persist into 2016, with the increases from 2011 to 2016 being about half of those estimated from 2011 to 2015. For example, when both US Airways and American Airlines were incumbents, fares of the other legacy carriers rose by about 10% in 2015 (in comparison with 2011). When both carriers were potential entrants, fares rose by about 15% (in comparison with 2011). These same increases were about 6% and 7.5% in 2016 in comparison with 2011. The presence of US Airways and American Airlines as incumbents was more effective at suppressing price increases by these legacy carriers than when they were both potential entrants.

As an example of the dichotomy between legacy- and low-cost carriers' price-setting behavior, in 2011 the average (one-way) fare from Albuquerque, NM (ABQ) to San Francisco, CA (SFO) was $159 on Southwest and it declined to $154 in 2015. For that same market, the average fare rose on United Airlines from $170 in 2011 to $199 in 2015. As another example, in 2011 the fare from Nantucket, MA (ACK) to Washington, DC (DCA) on US Airways was $198 and it jumped to $243 in 2015. In contrast, on Jet Blue, the fare declined from $233 in 2011 to $104 in 2015.

Conclusions

This paper describes an analysis that makes use of the 10% ticket sample to understand the impact of the mega-merger between US Airways and American Airlines on ticket prices. In particular, and of particular importance in mergers, this analysis focused on the role of competition pre- and post-merger on these impacts. This analysis shows that the merged airline was able to raise its fares. In addition, where a legacy carrier was in the market and US Airways or American Airlines, or both, did not play a role (as either an incumbent or a potential entrant) prices also rose. Markets in which US Airways or American Airlines, or both, did play a role, fares for these legacy carriers still rose but by less than when these carriers were present. For low-cost carriers in the equivalent situations, prices declined. Markets in which American Airlines or US Airways, or both, did play a role (incumbent or potential entrant), prices declined for low-cost carriers, and those declines were larger in markets when American Airlines or US Airways, or both, were actual incumbents instead of potential competitors. It is worth noting that these findings are consistent with Luo, who focused on the merger of Delta and Northwest Airlines (*3*). Luo also found that, post-merger, the fares for legacy carriers were less affected than those associated with low-cost carriers. Further, they also found that the competitive structure of the market had an impact on the change in fares post-merger.

It would be useful to extend this research in several different directions. First, these models ignored the role that hubs and focus cities play in airline networks. There is extensive empirical evidence to suggest that hubs do affect fare prices (see for example, *15*), hence integrating this into the statistical models would be very useful. Second, these models focused on direct service only. Extending the models to include one-hop and multi-hop tickets is very important. Third, there is some evidence that when competing carriers provide service at nearby airports that also suppresses prices (*14,15*). Our focus for potential competition is on carriers that service one or both airports in a market, but that do not serve the market itself. Extending that definition to include nearby airports is likely to be valuable. Finally, there is evidence to suggest that the connection between price and the level of concentration in a market is more complex than suggested by the models in this paper. Evans et al. found that market concentration and prices are interdependent (*16*). To address this feedback between price and market concentration, Cao et al., for example, included log arithmetic and geometric averages of the metropolitan statistical area's population for airport cities, total passengers on a route, and so forth. (*17*).

References

1. U.S. Department of Transporation, BTS. https://www.transtats.bts.gov/. Accessed June 28, 2017.
2. Kwoka, J., and E. Shumilkina. The Price Effect of Eliminating Potential Competition: Evidence from an Airline Merger. *The Journal of Industrial Economics*, Vol. 58, No. 4, 2010, pp. 767–792.
3. Luo, D. The Price Effects of the Delta/Northwest Airline Merger. *Review of Industrial Organization*, Vol. 44, No. 1, 2014, pp. 27–48.
4. Strassmann, D. L. Potential Competition in the Deregulated Airlines. *The Review of Economics and Statistics*, Vol. 72, 1990, pp. 696–702.
5. Whinston, M. D., and S. C. Collins. Entry and Competitive Structure in Deregulated Airline Markets: An Event Study Analysis of People Express. *RAND Journal of Economics*, Vol. 23, No. 4, 1992, pp. 445–462.
6. Bennett, R. D., and J. M. Craun. *The Airline Deregulation Evolution Continues: The Southwest Effect*. US Department of Transportation, Washington, D.C., 1993.
7. Windle, R. J., and M. E. Dresner. The Short and Long Run Effects of Entry on US Domestic Air Routes. *Transportation Journal*, Vol. 35, No. 2, 1995, pp. 14–25.
8. Morrison, S., and C. Winston. *The Evolution of the Airline Industry*. The Brookings Institution, Washington, D.C., 1995.
9. Dresner, M., J.-S. C. Lin, and R. Windle. The Impact of Low-Cost Carriers on Airport and Route Competition. *Journal of Transport Economics and Policy*, Vol. 30, No. 3, 1996, pp. 309–328.
10. Richards, K. The Effect of Southwest Airlines on U.S. Airline Markets. *Research in Transportation Economics*, Vol. 4, 1996, pp. 33–47.
11. Morrison, S. A. Actual and Adjacent and Potential Competition: Estimating the Full Effect of Southwest Airlines. *Journal of Transport Economics and Policy*, Vol. 35, No. 2, 2001, pp. 240–255.
12. Goolsbee, A., and C. Syverson. How Do Incumbents Respond to the Threat of Entry? Evidence from the Major Airlines. *The Quarterly Journal of Economics*, Vol. 123, No. 4, 2008, pp. 1611–1633.
13. Kwoka, J., K. Hearle, and P. Alepin. From the Fringe to the Forefront: Low Cost Carriers and Airline Price Determination. *Review of Industrial Organization*, Vol. 48, No. 3, 2016, pp. 247–268.
14. Daraban, I. *Low Cost Carrier Entry, Incumbent Responses and Spaital Competition in the U.S. Airline Industry*. PhD dissertation, Florida State University, 2007.
15. Brueckner, J., D. Lee, and E. S. Singer. Airline Competition and Domestic US Airfares: A Comprehensive Reappraisal. *Economics of Transportation*, Vol. 2, No. 1, 2013, pp. 1–17.
16. Evans, W., L. Froeb, and G. Werden. Endogeneity in the Concentration-Price Relationship: Causes, Consequences, and Cures. *Journal of Industrial Economics*, Vol. 41, No. 4, 1993, pp. 431–438.
17. Cao, K. H., B. Krier, C. Liu, B. McNamara, and J. Sharpe. The Nonlinear Effects of Market Structure on Service Quality: Evidence from the U.S. Airline Industry. *Review of Industrial Organization*, Vol. 51, No. 1, 2017, pp. 43–73.

The Standing Committee on Aviation Economics and Forecasting (AV040) peer-reviewed this paper (18-04748).

Article

TRR
JOURNAL OF THE TRANSPORTATION RESEARCH BOARD

Transportation Research Record
2018, Vol. 2672(23) 20–30

DOI: 10.1177/0361198118757995
journals.sagepub.com/home/trr
SAGE

A Comparison of Traffic Organization Methods for Small Unmanned Aircraft Systems

Anthony Battista[1] and Daiheng Ni[1]

Abstract

Small unmanned aircraft systems (sUAS), better known as "drones," have been drastically increasing in ownership and popularity during recent times. While this technology was once restricted to military and hobbyist use, numerous public and private-sector groups have proposed new uses, such as package delivery and firefighting. As the number of sUAS in operation increases, the methods currently utilized to manage sUAS traffic flow are expected to become inadequate; thus, it is necessary to develop an alternate system of sUAS traffic management. This paper proposes a computerized simulation methodology to quantitatively evaluate air traffic management methods based on safety, efficiency, and computational intensity. This methodology is then used to test the current organization method against free competition for available airspace and a freeway-style method of traffic organization known as an "air highway." It was ultimately found that these alternate methods of organization are less foolproof than the current method with regards to safety, but are significantly more practical to implement due to reduced computational intensity; additionally, these methods are sometimes capable of reducing travel time.

Small unmanned aircraft systems (sUAS), more commonly known to the public as "drones," are becoming increasingly prevalent in today's landscape. While their usage was once restricted to hobbyists and military operations, growing segments of the population have begun using them for more innovative purposes. sUAS have seen use in applications such as traffic monitoring and real estate photography. Furthermore, groups have proposed the usage of sUAS for even more complex purposes such as package delivery and firefighting.

As the number of sUAS in deployment increases, it is becoming increasingly important to develop a method to organize their flight paths. Current regulations from the Federal Aviation Administration (FAA) stipulate that sUAS must fly in a specified segment of low airspace located between 61 m and 122 m (200 ft and 400 ft) above ground. As this available segment of airspace is limited, the possibility arises of flight path conflicts between sUAS. Collisions can result in costly damage to sUAS, potential damage to people and property on the ground below, and the risk of debris.

Currently, mass deployments of sUAS are managed with the same methods as commercial aircraft. For commercial aircraft, centralized monitoring stations are utilized to observe the flight paths of vehicles in real time and re-route them to avoid any potential collisions. With the increasing popularity of sUAS, though, this method is becoming more and more difficult to implement. Some sUAS are built of materials that cannot be easily detected by radar. Furthermore, the sheer amount of sUAS that may be deployed at one time may be so large that it would be difficult and computationally intense to monitor them in this way.

With the rise of vehicle-to-vehicle (V2V) communication technology, several researchers have conducted work on sense-and-avoid algorithms to avoid collisions. Some of the studies focus on the technology and operational behavior necessary to prevent collisions between individual sUAS (*1–4*). This paper, however, aims to address the implementation of a large-scale method to prevent collisions by organizing the collective flow of sUAS traffic. Large-scale organizational methods have only recently been studied in literature. Preliminary efforts have been made to safely integrate sUAS into United States (U.S.) airspace, allocating a specific low-altitude area for sUAS operation (*5*). Interested readers are referred to Dalamagkidis

[1]Department of Civil and Environmental Engineering, University of Massachusetts Amherst, Amherst, MA

Corresponding Author:
Address correspondence to Daiheng Ni: ni@engin.umass.edu

et al. (*6*) for a thorough background on the methods and challenges of integration.

Other groups have further proceeded to pursue methods of organization within this low-altitude space. Amazon has proposed general airspace design concepts for its theoretical "delivery drone" system (*7, 8*), but acknowledges that further work must be conducted on specific organizational methods. NASA has conducted work on a system known as UAV traffic management, which has been tested in several U.S. cities (*9*). While results have proven effective, the system treats sUAS flight similarly to commercial air traffic, in which flights are pre-planned and individually controlled. Joulia et al. (*10*) utilize a four-dimensional control system, in which the temporal and spatial locations of individual sUAS are planned in advance using a system of "contracts." As sUAS become more numerous and adopt more varied purposes, it may prove inefficient to attempt to manage the flights of individual vehicles.

Several studies have proposed mechanisms of air traffic management that control traffic as a collective; these mechanisms often resemble freeways. Gharibi et al. (*11*), for example, worked to lay the groundwork for an airspace network control architecture involving a system of "airways." Yakovlev et al. (*12*), utilized geometric constraints to present a method of organizing flight paths. Chen et al. (*13*) also explored the concept of air highways as pre-planned aviation routes. Further studies have expanded upon the concept of the air highway and developed potential designs (*14*). Although multiple researchers have proposed various methods of sUAS organization, the relative safety and efficiency merits of these systems have not yet been studied in detail. To the best of the authors' knowledge, no standardized methodology currently exists to compare proposed methods of organization.

This paper presents a simulation methodology to study and compare different methods of sUAS traffic control. This methodology is applied to produce concrete operational data for four proposed methods of air traffic organization, which are then compared on the basis of their relative safety and efficiency metrics. Four methods of organization are discussed and analyzed: the aviation method, free competition for available airspace, a grid-style "air highway" design, and a roundabout-style air highway design. The computer simulation methodology is proposed to determine quantitative safety and efficiency merits of these organization systems under some simplification assumptions; this base methodology can be modified and adapted to suit the needs of other experiments. The methodology is then applied to each of the four organizational methods using specific values and terms. Preliminary results are presented and discussed; conclusions about the relative merits of each method are drawn. Finally, the results of the study are summarized, and areas of further study are suggested.

Discussion of Proposed Organizational Methods

Three Methods of Control

Three theoretical methods currently exist for managing and organizing sUAS traffic: aviation, free competition, and air highways. The three methods are described as follows.

Aviation. Aviation is the current method of air traffic control used by the FAA for commercial flights; it is the safest possible method of air traffic organization. In this method, air traffic is meticulously planned and controlled to ensure that no two vehicles will occupy the same spatial location at the same time. When applied to a group of sUAS, each traveling from an origin point to a destination point, the departure times and velocities of the sUAS are carefully observed and moderated to ensure that no collisions occur. This may involve delaying scheduled departure times for safety purposes. Such a system will result in the least likelihood of collision, but will create time delays in the process.

While this system is optimal with regards to safety, it presents notable disadvantages that must be taken into account. Such a system may be ultimately impractical to implement for sUAS, which are likely to see large-scale usage soon. The algorithms utilized by this method require a large amount of computer processing power to keep the system in order. These computations could take lengthy amounts of time when large numbers of sUAS flights must be planned; all users are thus burdened with the task of waiting out the planning process. Furthermore, these algorithms may not be able to account for deviations from the plan. An sUAS may be blown off course by a strong wind, or a rogue user may recklessly fly a new sUAS into the airspace without regard for the safety of others. In these cases, other sUAS may have to deviate from their pre-planned flight paths to avoid collision; this could result in a potentially disastrous domino effect of deviations.

Free Competition. The free competition method of air traffic organization falls within the umbrella of highway-style systems. This method is the most minimalistic style possible; in free competition, no structured organizational system exists. An operator can launch an sUAS at any point in time, and all sUAS in operation compete for the available airspace. sUAS simply travel the shortest straight-line path from their origin point to their destination point, with no concerns as to the flight paths of other sUAS. When two sUAS are on a collision course, sense-and-avoid technology will alert the vehicles. One will temporarily ascend to a higher altitude and the other will temporarily descend to avoid collision.

This is the optimal method with regards to minimizing travel time; sUAS will take the most direct possible route, and experience no delays in the process. It is also, by far, the easiest method to implement, as it requires few new

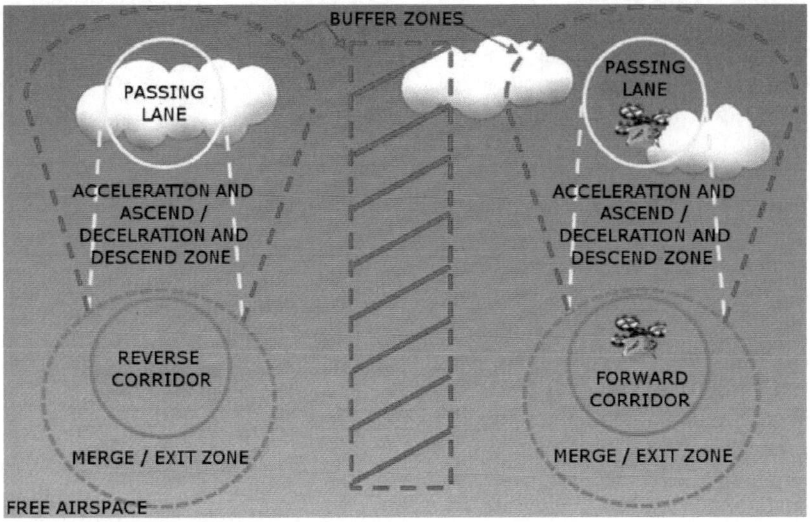

Figure 1. Potential air highway design.

regulations, geofences, or technological modifications. Unfortunately, it also presents the most significant safety challenges. In this method, no measures are taken to prevent flight paths from crossing. Although sUAS can ascend and descent to avoid collisions, this method is not entirely reliable. Thus, the free competition method presents the highest likelihood of collision.

Air Highway. The air highway method of air traffic control is another highway-style organization method. This method of flight control involves geofenced, controlled-access air traffic paths that sUAS are programmed to follow. An air highway is modeled after a land freeway. Each air highway lane is reserved for travel in a specific direction; the lane includes space for sUAS to merge into oncoming traffic, as well as a travel area, a passing area, and a "breakdown lane" for damaged sUAS. An illustration of one possible air highway design is shown in Figure 1. Air highways are carefully planned and constructed to prevent sUAS traveling in different directions from crossing paths. sUAS must follow a special merge procedure to change travel directions.

This system carries a major benefit over free competition for airspace: it creates a system of organization that increases safety and gives order to flight paths. This system does, however, require sUAS to travel a less direct path from their origin point to their destination point. If optimizing travel time is a concern, air highways are not capable of surpassing free competition. No universal standardized air highway design currently exists. While the idea has been proposed, it has not been thoroughly studied or tested; the optimal design is not currently known. Logistically, implementing air highways does require geofencing, and may require sUAS to be modified with additional, though inexpensive, technology. While this system is less foolproof than the aviation method, it is significantly less computationally intensive. No pre-flight delays are implemented, sUAS can simply be started at will and directed to the air highway.

Aviation-Style and Highway-Style

Of the three aforementioned methods of control, aviation falls into the category of aviation-style control, while free competition and air highways fall into the category of highway-style control. Aviation-style control is the method utilized by the FAA to control commercial air flights, while highway-style control is the method utilized by the Federal Highway Administration to control land highway travel.

sUAS travel bears many more similarities to land highway travel than to commercial flight operation:

- In sUAS flight, it is ultimately the operator's responsibility to route the vehicle to avoid collisions; this is similar to highway travel, in which drivers are responsible for the control of their own vehicles. Conversely, in commercial air traffic operations, the safety of users is meticulously ensured by pilots, controllers, and engineers.
- Whereas the aviation method strictly controls every detail of every flight, highway-style methods only loosely control traffic as a collective entity. With the sheer number of sUAS that may be in operation at one time, it is much easier to control their flights collectively than individually.
- The operators of sUAS, much like drivers on a highway, make decisions autonomously. Airline pilots are required to listen to and obey instructions. Furthermore,

Table 1. Input Variables and sUAS Characteristics

Variable	Description
Universal Input Variables	
Time Length	Length of time over which sUAS flight is simulated (s)
Time Step	Time interval over which simulation is updated (s)
Demand	Total number of individual sUAS to generate
Tests	Number of tests to conduct
Radius	Effective radius of each sUAS (m)
Buffer Constant	Multiplier constant; influences center-to-center distance at which two sUAS are considered to be "in collision"
Individual sUAS Characteristics	
Index	Numerical index of the sUAS
Departure Time	Randomized, unique time at which sUAS begins its trip (s); no two sUAS will share the exact same departure time
Free Velocity	Velocity of sUAS when traveling in unorganized airspace (m/s). Randomized within a specified range
Initial Position (x,y)	Randomized origin point of sUAS flight (m)
Destination Position (x,y)	Randomized destination point of sUAS flight (m). Will always be on a different edge of the airspace than the origin point

airline pilots are held to very high standards of competence. Attaining a driver's license is much simpler than attaining a pilot's license; operating an sUAS does not even require special training.

- Airline flights require pre-trip planning, while sUAS operators and motorists do not necessarily need to schedule trips in advance.
- sUAS and highway trips can begin from almost any geographic location, while commercial airline flights must begin at airports.

When the usage of sUAS becomes widespread, their behavior is much more likely to resemble that of highway travel than that of commercial air travel. Thus, the aviation-style methods currently in place are likely to be inefficient in comparison to highway-style methods. It is important to implement a method that takes these differences into account.

Simulation Methodology

Simulation Procedure

In this section, a generic computerized simulation methodology is proposed that can be used to acquire operational data. Test simulations were conducted in MATLAB. The simulation consists of a central script that generates a fleet of sUAS, and an individual function script for each control method used to dictate the behavior of the sUAS. This methodology can be utilized and modified to conduct tests with different methods of air traffic management and different air highway designs.

Several initial simplification assumptions are made to govern the operation of the sUAS fleet. First, it is assumed that all sUAS involved in the simulation are quadrocopter-style vehicles; they are capable of at-will acceleration and deceleration, as well as stopping to hover in place. Under current governmental regulations, sUAS are allowed to travel only within a relatively small, low-altitude segment of airspace. Thus, for the purposes of experimental simulation, the available airspace can be treated as a two-dimensional plane, with segments of airspace reserved above and below for sUAS to temporarily ascend and descend. All simulation tests are conducted in a square-shaped, isolated, 1000 m x 1000 m segment of airspace with clear weather. It is assumed that actual sUAS flights would cover longer distances than those in this segment.

When the simulation is run, the overarching script begins by creating the 1000 m x 1000 m virtual airspace, and then creates a test fleet of sUAS. The simulation utilizes a set of universal input variables that are specified by the user. It then produces a specified number of virtual sUAS, each possessing a set of characteristics. See Table 1.

Once this test fleet is created, copies of it are then passed onto the various simulation functions. Each function corresponds to a different method of sUAS organization. This ensures fair comparisons; in each simulation, each individual sUAS departs from the same initial position at the same time, and travels to the same destination using the same free airspace velocity. Simulation functions are created for the different methods of control: a free competition function, an aviation function, and air highway functions.

Free Competition Function. In the free competition function, at an sUAS's specified departure time, the shortest Euclidean distance from its origin to its destination in two-dimensional space is calculated, and the vehicle begins traveling along the path at its specified *Free Velocity*. No attempts to preemptively prevent collisions are made – if the flight paths of two sUAS overlap, one will temporarily ascend and the other

Table 2. Additional Variables for Air Highways

Variable	Description
Additional Universal Input Variables	
Speed Min.	Minimum speed requirement of air highway (m/s). It is assumed that all sUAS can meet this speed, and will accelerate up to it upon entering the air highway
Merge Delay	Time required for an sUAS to move from one air highway lane to another (s)
Merge Buffer	Multiplier constant; influences center-to-center distance at which a highway lane is considered clear to enter
Additional sUAS Characteristics	
In Point (x,y)	Air highway entrance point closest to sUAS origin point (m)
Out Point (x,y)	Air highway exit point closest to sUAS destination point (m)
Switch Point(s) (x,y)	Point(s) at which sUAS must conduct a lane merge to change directions, if any exist (m)
Air Highway Velocity	Max (Speed Min., Free Velocity) (m/s)

will temporarily descend until the path is clear. An sUAS will come to a stop upon reaching its destination.

Aviation Function. The aviation function utilizes an algorithm to ensure that all sUAS reach their destinations without any collisions. Before any sUAS are launched, a full "theoretical" free competition simulation is conducted. If any collisions occur in this theoretical run, the smaller numerical index of the two offending sUAS is recorded. After the conclusion of the theoretical simulation, a 1-s departure time delay is added to each recorded sUAS; the theoretical free competition simulation is then repeated with the updated departure times. The process of running theoretical simulations and delaying departure times is repeated until a theoretical run is conducted that produces no collisions. Once this point is reached, the fleet can finally be launched. If the departure time of an sUAS is delayed, this delay is added to the total travel time of that sUAS.

Air Highway Function. At the beginning of the air highway function, additional universal variables are created, and each sUAS is given additional characteristics. See Table 2.

As an air highway safety feature, it is assumed that air highway entry points are located at an altitude slightly above traveling altitude, and exit points are located slightly below traveling altitude. This way, inward-bound sUAS are never at risk of colliding with any nearby outward-bound sUAS.

At an sUAS's specified departure time, it steps through the following flight path algorithm:

- **Stage 1 - Origin to Entry Point:** The sUAS ascends and travels to its given *In Point*. The shortest Euclidean distance from origin to *In Point* is calculated, and the sUAS travels toward this point at its *Free Velocity*. If the flight paths of two sUAS overlap, one will temporarily ascend and the other will temporarily descend until the path is clear. (Effectively, it behaves as in free competition.)
- **Stage 2 – Entry Queuing:** Upon reaching the entry point, the sUAS examines the area around it within a radius specified by the *Merge Buffer*. If this radius is clear of any oncoming sUAS, the sUAS will enter the highway. If it is not, the sUAS will hover in place and form or join a first-in, first-out queue; vehicles will wait until the airspace is clear to enter.
- **Stage 3 – Entry Point to Switch Point:** Upon entering the highway, the sUAS adopts its *Air Highway Velocity*, and travels along the path laid out for it. For simplicity, car-following models are neglected; they have not been adequately studied in the context of air highways, and require additional coding complexity for only marginal increases in simulation accuracy. Instead, the *Speed Min.* is set to ensure a simplified, narrow range of speeds; when combined with queuing, this ensures no rear-end collisions and no need for passing.
- **Stage 4 – Switch Point Merge Queuing:** If the sUAS does not need to change lanes to change directions, this stage is bypassed. If a lane merge is required, the sUAS will follow the same merge procedure as in **Stage 2**. If it is assumed that the two lanes are at different altitudes, *Merge Delay* can be set to a nonzero value; this accounts for the time required to ascend or descend. *Merge Delay* is added to the vehicle's total travel time.
- **Stage 5 – Switch Point to Exit Point:** The sUAS travels to its highway *Out Point*, behaving as in **Stage 3**.
- **Stage 6 – Exit Point to Destination:** Upon exiting the air highway, the sUAS returns to its *Free Velocity*. It descends and travels to its final destination, behaving as in **Stage 1**.

Quantitative Operational Metrics

Three quantitative metrics are tracked over the course of each simulation: average total travel time, conflict rate for each sUAS, and total simulation run time. These metrics correspond to the safety, efficiency, and computational intensity of each organizational method, respectively.

Average Total Travel Time. Total travel time is tracked for each individual sUAS. In free competition, this is the time it takes for an individual sUAS to move from its starting point to its destination. In the air highway, this is starting point-to-destination time, including time spent waiting to merge, and time delay from changing lanes. In the aviation method, this is starting point-to-destination time, including any departure time delays put in place to prevent collisions. This metric is the average of total travel time for each sUAS.

Conflict Rate. Conflict rate is a weighted index used to determine the number and severity of flight path conflicts that occur. Two types of path conflicts can occur in these simulations: merging points and collisions. A merge point occurs in an air highway when an sUAS plans to change or enter lanes, but the intended entry point is not clear. The relative danger of a merge situation can be defined as a variable, X. A collision occurs when two sUAS occupy positions within a specified radius of each other at the same point in time; this radius is determined by the *Buffer Constant*. The relative danger of a collision in terms of safety can be defined as the variable Y. In general, it can be assumed that a collision is more dangerous than a merge; thus, Y will generally be a greater number than X.

At the conclusion of each step of a simulation, conflicts are checked for; a running conflict rate count is kept throughout each simulation. When an sUAS in an air highway is forced to stop and wait before merging, the amount of conflicts will increase by X; when two sUAS collide in any of the simulations, the amount of conflicts will increase by Y. At the end of the simulation, the final amount is normalized using the number of sUAS in the simulation and the amount of time covered by the simulation. Conflict rate, in *conflicts/(sUAS/s)*, is expressed in Equation 1:

$$\textbf{Conflict Rate} = t\left(XM + YC\right) / D \qquad (1)$$

where
t = time length of simulation;
X = merge situation danger factor;
M = number of merges;
Y = collision danger factor;
C = number of collisions; and
D = demand.

As the aviation method is specifically designed to prevent conflicts from occurring, every run of the aviation function will produce a conflict rate of 0. As the free competition method does not contain any opportunities for merges, the free competition conflict rate will only include direct collisions. Both air highway designs present opportunities for both collisions and merges; their conflict rates will include both.

Run Time. Run time is the total elapsed time from beginning to end of a simulation function. This value encompasses everything that occurs after the test fleet is generated. Run time is representative of the computational intensity required by an organizational method. High run times indicate that the organizational method requires high amounts of processing power, and will produce additional wait times for end-users.

Example Simulation Study

With the underlying simulation methodology in place, an example study is now conducted to compare four different methods of sUAS control: aviation, free competition, and two possible air highway designs.

Air Highway Designs

For this example, two air highway designs are tested: a grid-based system, and a roundabout-based system. In the grid-based system, four air highway lanes exist: a northbound and southbound lane, located at a slightly lower altitude, and an eastbound and westbound lane, located at a higher altitude. The highway system features eight access points; each lane has one access point used strictly for entry, and one access point used strictly for exit. The system also features four merge points for northbound/eastbound, northbound/westbound, southbound/eastbound, and southbound/westbound interchanges, respectively.

The roundabout-based system utilizes one large, looped lane. This highway utilizes only four access points, each of which functions as both an entry point and an exit point; these four points are located on the north, south, east, and west ends of the air lane. sUAS enter the loop, travel in a uniform counterclockwise direction, and exit upon reaching the access point nearest their destination. Both highways are illustrated in Figure 2.

Input Variables

Comparative tests of the four different organization methods were run using eight different values of *Demand*: 33 sUAS/h, 65 sUAS/h, 131 sUAS/h, 196 sUAS/h, 262 sUAS/h, 524 sUAS/h, 785 sUAS/h, and 1047 sUAS/h. To reduce the effects of random variability, 20 tests were run for each demand level, and results were averaged. For every test run, the values from Table 3 were utilized.

Example Simulation Findings

Simulation Results

Summaries of results from the test runs are displayed in Table 4. All results are compared to those of aviation, the system currently in place, so it is easy to see quantitative advantages or disadvantages caused by replacing the system. Figure 3 shows graphical plots of changes in safety, efficiency, and computational intensity against changes in demand.

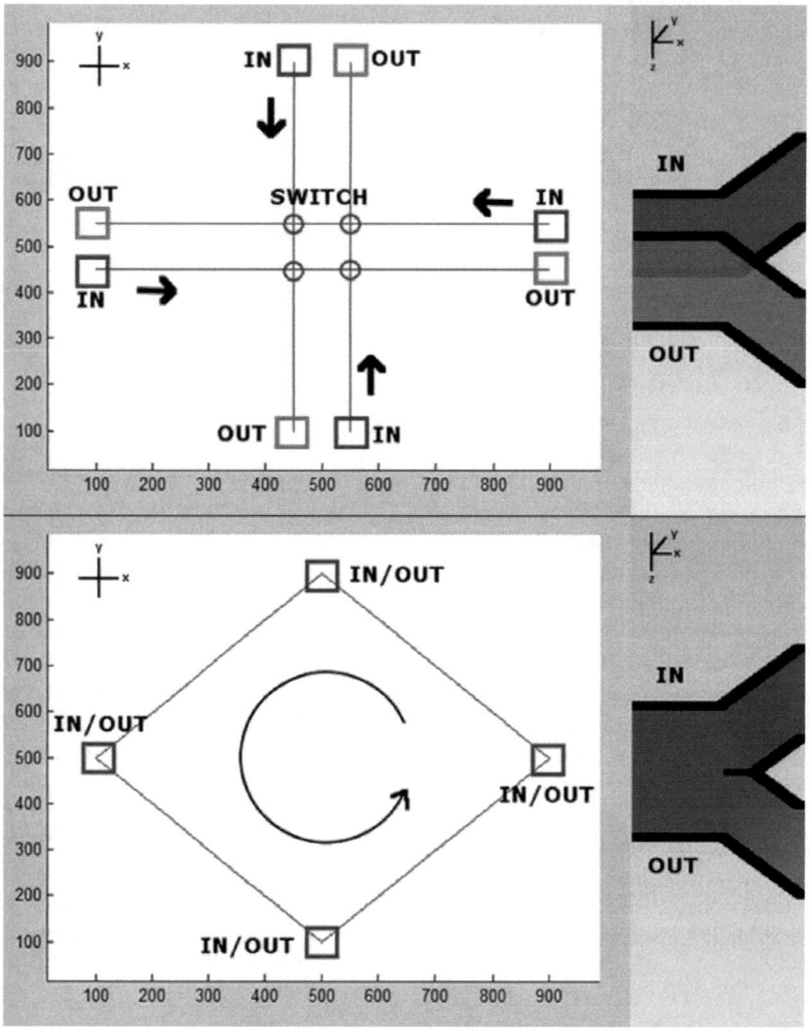

Figure 2. Illustration of grid-based (top) and roundabout-based (bottom) air highways with entry/exit mechanisms shown.

Table 3. Variable Inputs

Variable	Value
Time Length	1,100 s
Time Step	1 s
Free Velocity Range	10 m/s–20 m/s (22.4 mph–44.7 mph) (Common operating speeds of currently available sUAS)
Radius	4 m (13.1 ft)
Buffer Constant	3
Speed Min.	19 m/s (42.5 mph)
Merge Delay	5 s
Merge Buffer	1.5
X (Danger of Merge)	1
Y (Danger of Collision)	3

Discussion of Results

An examination of the resulting figures shows the expected advantages of the aviation method with regards to safety and efficiency. In comparison to aviation, which always produces a conflict rate of 0, free competition and the two air highway designs do increase the conflict rate by a non-negligible amount. Furthermore, both air highway designs create non-negligible increases in travel time, as sUAS are not traveling the most direct possible route. As sUAS do not experience departure time delays in the free competition method, decreases in overall travel time do occur, growing greater in magnitude as volume increases. These decreases, though, are negligibly small: all fall under 4%.

Upon examination of average run times, the weakness of the aviation method becomes apparent: the collision-avoidance

Table 4. Relative Simulation Results Versus Aviation

	Additional avg. conflict rate vs. aviation			Additional avg. travel time vs. aviation (s)			Additional avg. run time vs. aviation (s)		
Avg. flow (SUAS/h)	Free comp.	Grid highway	Round. highway	Free comp.	Grid highway	Round. highway	Free comp.	Grid highway	Round. highway
33	33.00	44.00	22.00	−0.02	23.68	40.41	−0.21	0.50	0.84
65	49.50	11.00	44.00	−0.04	19.20	37.55	−0.57	0.96	1.66
131	66.00	64.63	82.72	−0.07	19.43	36.55	−2.45	0.45	1.77
196	132.00	103.58	111.87	−0.16	20.53	38.04	−6.99	−2.14	−0.14
262	169.28	113.54	177.18	−0.22	19.47	36.44	−12.03	−5.87	−3.23
524	339.50	253.12	274.73	−0.63	18.72	37.03	−76.44	−65.86	−60.95
785	506.75	355.54	440.77	−1.10	18.42	36.04	−233.26	−215.88	−208.88
1047	707.63	513.14	592.33	−2.07	17.39	34.79	−473.69	−454.89	−447.28

	Additional avg. travel time vs. aviation (% change)			Additional avg. run time vs. aviation (% change)			Avg. aviation run time (s)
Avg. flow (SUAS/h)	Free comp.	Grid highway	Round. highway	Free comp.	Grid highway	Round. highway	Aviation method
33	−0.04	40.02	68.29	−75.12	177.12	298.95	0.28
65	−0.06	31.11	60.86	−74.64	126.90	218.89	0.76
131	−0.12	31.01	58.35	−77.34	14.36	55.98	3.16
196	−0.26	33.19	61.52	−78.52	−24.09	−1.57	8.90
262	−0.35	31.08	58.19	−80.80	−39.43	−21.65	14.90
524	−1.01	29.85	59.05	−87.37	−75.28	−69.67	87.49
785	−1.74	29.14	57.03	−89.63	−82.96	−80.27	260.23
1047	−3.19	26.77	53.55	−91.86	−88.22	−86.74	515.65

Note: Avg. = average, Comp. = competition, Round. = roundabout-based

algorithm is an extremely time intensive process. By bypassing this algorithm, the free competition method routinely reduces run time by over 75% at all levels of demand. At low levels (under 200 sUAS/h), the algorithm runs quickly, and thus produces a shorter run time than the air highway methods. Beyond 200 sUAS/h, however, both air highway designs begin to show significantly and increasingly lower run times. At very high levels of demand, all three alternate methods of control continue to show steady percent run time decreases, converging to a near 100% decrease. The aviation algorithm requires very long run times at high levels of demand, consistent with what may be expected in the future; other methods need less time and computational power.

The roundabout-style highway design produces strictly worse quantitative metrics than the grid design in all cases. While it is the simpler air highway design, and thus likely the easier method to implement, it routinely produces higher conflict rates, longer travel times, and higher run times than the grid method. Furthermore, the looped pathway is even less direct than the grid-based system, forcing sUAS to travel longer distances. Thus, the roundabout highway is not a viable design option.

Of the remaining methods, free competition shows the greatest improvements over aviation with regards to travel time and run time. As no measures are taken to prevent collisions from occurring, sUAS experience no delays in starting time; this results in a lower average travel time for all sUAS. As the volume of sUAS flowing through the test area increases, the decrease in average travel time also grows, albeit negligibly. In every test, free competition produces the highest reductions in run time; it is the only method that produces lower run times at any level of demand. However, the free competition method is by far the least safe. Safety results are similar to those of the other methods at low demand levels. Conflict rate roughly increases by a power of two with flow, though, and the safety gap grows even wider in the process. As free competition will never produce optimal safety at higher levels of demand, its usefulness depends on the operator's value of time and the volume of sUAS flowing through an area.

Grid air highways are the safest highway-based traffic management system. In the test simulations, they consistently displayed the lowest conflict rates, which increase roughly linearly with flow. Although they produce lower conflict rates than free competition, they do produce longer average travel times due to their less direct routes: roughly 30% increases from the aviation method. Of note, it can be seen that these added travel times decrease over time when compared with the aviation method: as the volume of sUAS flying through an airspace increases to high levels, the time cost of waiting to merge becomes less disadvantageous in comparison to aviation's departure time delays. The grid air highway also produces run time reductions in all tests conducted at and above 196 sUAS/h. Although free competition produces the greatest percent run time reductions overall, the grid highway run time reductions rapidly

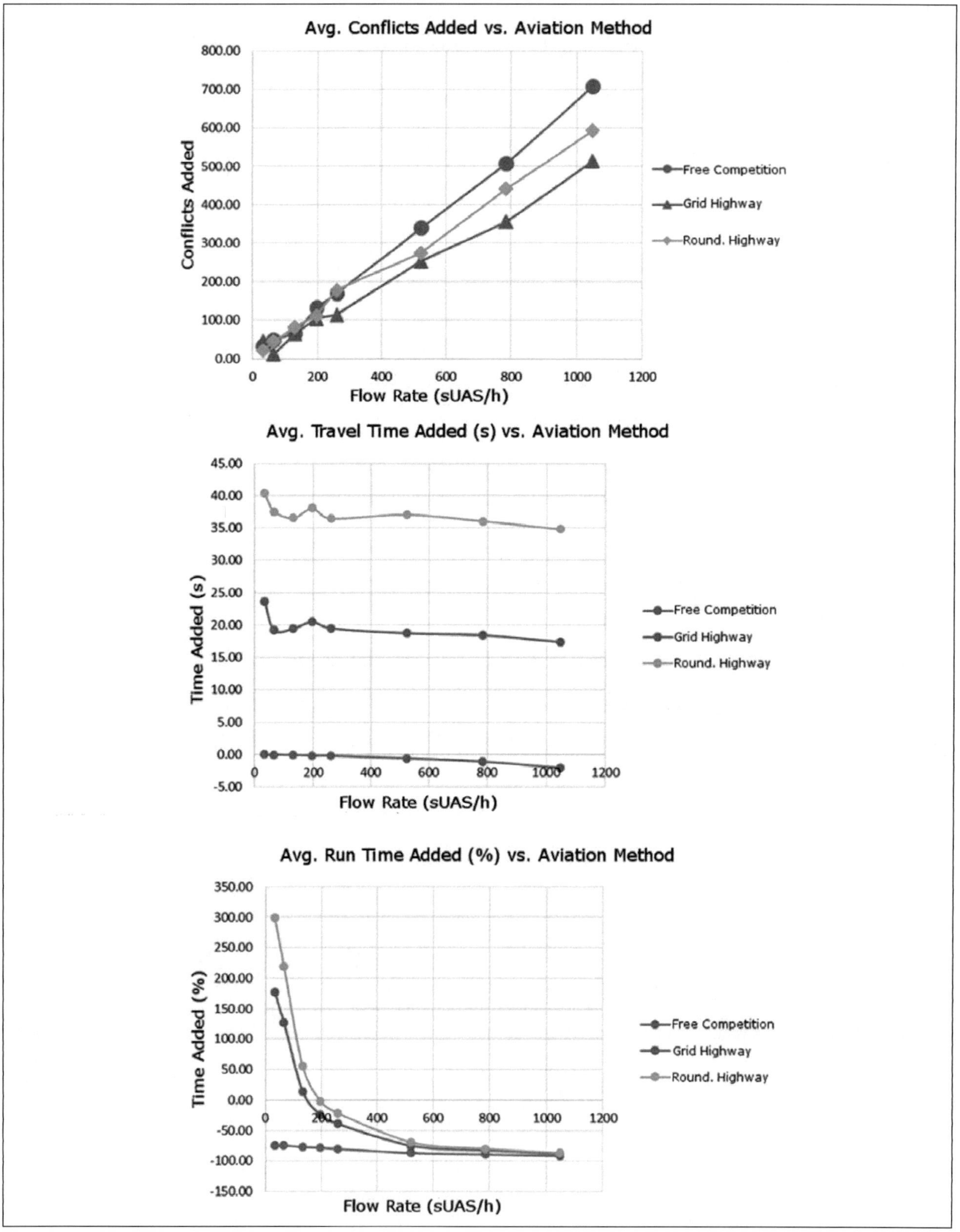

Figure 3. Operational metric changes versus aviation control.

converge toward those of free competition at the highest tested demand levels. The utility of grid-based air highways also depends on the value of time of the operators. If decreasing travel time is the main priority, operators may accept the decrease in safety; if the operators can afford added travel time, though, it may be advantageous to utilize the grid highway to increase safety.

While the aviation method is the best option for safety optimization, it can ultimately be impractical and inefficient to implement as a large-scale control method for sUAS. At very low levels of demand, all alternate methods produce similar safety results, but free competition requires the lowest travel times and least computational intensity. At very low demand levels, simple free competition may be the optimal method. When selecting an alternate method at higher demand levels, the air highway will organize traffic more safely, but at a higher time cost; the appropriateness of this method thus depends on the priorities of the system operators.

Conclusion

In this paper, the relative advantages and disadvantages of four sUAS traffic control methods were discussed and quantified. Two distinct styles of traffic management exist: aviation-style, which controls each individual vehicular trip, and highway-style, which controls the operation of traffic as a whole. In the near future, the operation of sUAS is more likely to resemble highway travel than commercial flight; furthermore, the methods currently used to monitor sUAS flight are computationally intensive, and will likely prove inefficient when dealing with massive volumes of sUAS. It is important to determine a viable alternative.

Two highway-style alternatives were proposed: free competition for available airspace, and an air highway system modeled after freeways. Two versions of the air highway were proposed: a grid of one-way, controlled-access paths, and a roundabout-style system. A computer simulation methodology was then proposed. This methodology can be used and modified to test different forms of sUAS traffic control and determine quantitative safety, efficiency, and computational intensity metrics. The simulation methodology was used in conjunction with the proposed systems to attain quantitative data on the relative advantages and disadvantages of each system.

Ultimately, it was found that the aviation method is ideal with regards to safety, but due to logistical requirements, it is not the ideal method to implement for large-scale sUAS control. The roundabout-style highway was also found to be inefficient. Its results were strictly worse than those of the grid design in all categories. Free competition is less computationally intense at all demand levels tested; the grid-based air highway is less computationally intense at the majority of demand levels tested. Overall, the grid-based highway will produce safer results than free competition, with this difference becoming more marked at increased levels of demand. Free competition, on the other hand, has the unique benefit of reducing travel time. At very low levels of demand with similar safety results, it may be optimal to simply allow free competition; at higher levels of demand, when safety differences become more marked, it is safer to implement air highways, so long as the operators are willing to accept travel time increases.

This study opens doors for significant further work in the field of operational sUAS management. Only one aviation collision-prevention algorithm was tested; a more efficient algorithm may exist. Furthermore, while two air highway designs were utilized in this simulation, a range of possibilities exist that have not been tested. Potential improvements to the model could include additional lanes, changes in flight directions, or alterations to sUAS merging and queuing behavior. Multiple designs should be simulated in order to determine a true optimal model. Finally, these tests were conducted in a simplified, isolated airspace. It is important to test these systems in a more realistic environment, and for more practical applications than simple origin–destination travel.

References

1. Haessig, D. A., R. T. Ogan, and M. Olive. Sense and Avoid – What's required for Aircraft Safety? *IEEE SoutheastCon*, Norfolk, Va., IEEE, New York, 2016, pp. 1–8.
2. Liu, Z., and A. G. Foina. An Autonomous Quadrotor Avoiding a Helicopter in Low-Altitude Flights. *IEEE Aerospace and Electronic Systems Magazine*, Vol. 31, No. 9, 2016, pp. 30–39.
3. Ma, Z., T. Hu, L. Shen, W. Kong, and B. Zhao. A Detection and Relative Direction Estimation Method for UAV in Sense-and-Avoid. *2015 IEEE International Conference on Information and Automation*, Lijiang, China, IEEE, New York, 2015, pp. 2677–2682.
4. Ong, H. Y., and M. J. Kochenderfer. Markov Decision Process-Based Distributed Conflict Resolution for Drone Air Traffic Management. *Journal of Guidance, Control, and Dynamics*, Vol. 40, No. 1, 2017, pp. 69–80.
5. Ogan, R. T.Integration of Manned and Unmanned Aircraft Systems into U.S. Airspace. *IEEE SoutheastCon*, Lexington, Ky., IEEE, New York, 2014, pp. 1–4.
6. Dalamagkidis, K., K. Valavanis, and L. Piegl. On Integrating Unmanned Aircraft Systems into the National Airspace System. *Intelligent Systems, Control and Automation: Science and Engineering*, Vol. 36, 2009, pp. 63–108.
7. Amazon. *Determining Safe Access with a Best-Equipped Best-Served Model for Small Unmanned Aircraft Systems*, July 2015. www.smalluavcoalition.org/wp-content/uploads/2015/07/Amazon_Determining-Safe-Access-with-a-Best-Equipped-Best-Served-Model-f....pdf. Accessed June 30, 2017.
8. Amazon. *Revising the Airspace Model for the Safe Integration of Small Unmanned Aircraft Systems*, July 2015. www.utm.arc.nasa.gov/docs/Amazon_Revising%20the%20Airspace%20Model%20for%20the%20Safe%20Integration%20of%20sUAS%5B6%5D.pdf. Accessed June 30, 2017.

9. Rios, J., D. Mulfinger, J. Homola, and P. Venkatesan. NASA UAS Traffic Management National Campaign: Operations across Six UAS Test Sites. *2016 IEEE/AIAA 35th Digital Avionics Systems Conference (DASC)*, Sacramento, Calif., IEEE, New York, 2016, pp. 1–6.
10. Joulia, A., T. Dubot, and J. Bedouet. *Towards a 4D Traffic Management of Small UAS Operating at Very Low Level.* Presented at 30th Congress of the International Council of the Aeronautical Sciences (ICAS 2016), Daejon, Korea, 2016.
11. Gharibi, M., R. Boutaba, and S. L. Waslander. Internet of Drones. *IEEE Access*, Vol. 4, 2016, pp. 1148–1162.
12. Yakovlev, K. S., D. A. Makarov, and E. S. Baskin. Automatic Path Planning for an Unmanned Drone with Constrained Flight Dynamics. *Scientific and Technical Information Processing*, Vol. 42, No. 5, 2015, pp. 347–358.
13. Chen, M., Q. Hu, J. Fisac, K. Akametalu, C. Mackin, and C. Tomlin. Guaranteeing Safety and Liveness of Unmanned Aerial Vehicle Platoons on Air Highways. *CoRR*, Vol. arXiv:1602.08150, 2016.
14. Geller, J., and D. Ni. *Operational Organization of Small Unmanned Aerial System Physical Airspace*. Presented at the 96th Annual Meeting of the Transportation Research Board, Washington, D.C., 2017.

The Standing Committee on Airfield and Airspace Capacity and Delay (AV060) peer-reviewed this paper (18-02539).

Article

TRR
JOURNAL OF THE TRANSPORTATION RESEARCH BOARD

Transportation Research Record
2018, Vol. 2672(23) 31–39

DOI: 10.1177/0361198118756619
journals.sagepub.com/home/trr
SAGE

Effect of Pavement Structure on the Mechanical Response and Performance of Perpetual Pavements at the National Airport Pavement Test Facility

Carlos E. Cary[1], Zilong Wang[1], Hao Yin[1], Navneet Garg[2], and Ryan Rutter[2]

Abstract

Construction Cycle 7 (CC7) conducted at the FAA National Airport Pavement Test Facility (NAPTF) was aimed to study the effect of hot mix asphalt (HMA) layer thickness and develop perpetual pavement design criterion for airfield flexible pavements. Four fully instrumented perpetual test pavements were designed, constructed, and tested, with all test items trafficked under heavy aircraft loads using a three duals in tandem configuration. Pavement condition was monitored using heavy weight deflectometer tests, distress surveys, and surface profiles. Comprehensive response data analysis revealed that increasing HMA layer thickness significantly reduced the tensile strain at the bottom of the HMA layer, permanent deformation in the unbound layer, and vertical stress at top of the subgrade. Thicker HMA test items exhibited better rutting performance, as evidenced in both multiple depth deflectometer and surface profile data. The interaction between HMA layer temperature and thickness was captured in the pressure cell responses. More tests are planned to determine the threshold HMA strain to prevent fatigue cracking.

Located at the FAA William J. Hughes Technical Center, Atlantic City International Airport, New Jersey, the National Airport Pavement Test Facility (NAPTF) aims at generating full-scale pavement response and performance data for the development and verification of airport pavement design criteria. The NAPTF became operational on April 12, 1999 and consists of a 900 feet long by 60 feet wide test pavement area, embedded pavement instrumentation with dynamic data acquisition system, environmental instrumentation with static data acquisition system, and the NAPTF test vehicle for trafficking the test pavement with configurations of up to eighteen wheels with loads up to 75,000 pounds per wheel (*1, 2*).

Asphalt pavements are usually designed for a 20-year life based on extrapolation far beyond road test or full-scale experiment conditions. One consequence of this extrapolation is ever-increasing thickness with traffic volume, resulting in overly designed asphalt pavements for high-volume roadways (*3*). The Asphalt Pavement Alliance (APA) introduced the concept of perpetual pavements in 2000 (*4*). Perpetual pavement is defined as an asphalt pavement designed and built to perform for 50 years or more without requiring major structural rehabilitation or reconstruction, and needing only periodic surface renewal in response to distresses confined to the top of the pavement. With perpetual pavements, the potential for traditional fatigue cracking is reduced, and pavement distress is typically confined to the upper layer of the structure. This concept is an appealing alternative to airport pavements, where it is desirable to minimize rehabilitation and reconstruction costs as well as minimize closures to traffic. Aimed at studying the effect of hot mix asphalt (HMA) layer thickness and developing perpetual pavement design criteria for airfields, varying thickness pavement test sections were designed using FAARFIELD (FAA Airport Pavement Thickness Design Software) and constructed for Construction Cycle 7 (CC7) at the NAPTF (*5*).

The major goal of this paper is two-fold: to study the effect that HMA layer thickness has on the mechanical response and field performance of perpetual pavement and; provide recommendations for the future development of perpetual pavements design criteria.

[1]Gemini Technologies, Inc., Egg Harbor Twp, NJ
[2]Federal Aviation Administration, William J. Hughes Technical Center, Atlantic City International Airport, NJ

Corresponding Author:
Address correspondence to Carlos E. Cary: carlos.cary@gemitek.com

Background and Methods

Pavement Test Items

Four test pavements were designed using FAARFIELD considering a gear configuration corresponding to a B-777 aircraft with 55,000 lb wheel load. The test item thicknesses were designed to prevent any subgrade failure and induce a fatigue failure mechanism. The test items were denominated LFP1-N, LFP2-N, LFP3-N, and LFP4-N.

Each perpetual item is 40 ft long and 30 ft wide, with a 10 ft transition between test items. The test items were constructed on a specification P-152 (*6*) low-strength subgrade with a target California bearing ratio (CBR) of 5.5. The as-built thicknesses of FAA specification P-401 (*6*) HMA, and FAA specification P-154 subbase (*6*) for different test items are shown in Figure 1*a* (pavement cross-sections).

Instrumentation

Over 150 sensors of various types were installed in the test items to monitor the mechanical response and performance of the pavement, generating extensive dynamic and static data through the duration of traffic test. Some of the dynamic sensors include H-bar asphalt strain gauges (ASGs) near the bottom of the P-401 surface layer, multiple depth deflectometers (MDDs) at various depths, and pressure cells (PC) on top of the P-152 subgrade material. It is worth mentioning that the MDD components capture deflections at multiple depths: E and F at different depths within P-152; B to D at different depths within P-154; and A near the bottom of P-401. A total of six ASGs were installed in each test item: three in a transverse orientation (TSG) and three in a longitudinal orientation (LSG). Thermocouples were embedded within the P-401 surface layer at multiple depths and formed the static data collection system recorded at regular intervals. Figure 1*b* illustrates the approximate location of MDD, PC, and ASG sensors within the test items. Only one half of a test item (i.e., 20 ft) is illustrated in the schematic. Although the P-401 and P-154 thicknesses vary across test items, the relative position and depth of each type of sensor remain constant.

Traffic Tests

Full-scale traffic tests for CC7 began on 09/15/2014 using the NAPTF test vehicle (NAPTV) with a three duals in tandem landing gear configuration (55,000 lb per wheel) at a vehicle speed of 2.5 mph. The wheel load was increased to 65,000 lb after 30,030 vehicle passes on test sections LFP1 and LFP2. The wander pattern consisted of 66 repetitions, 33 traveling east and 33 traveling west. These 66 passes are arranged in nine wheel tracks to simulate a normal distribution with standard deviation of 30.5 in, which is equivalent to a taxiway distribution for design. Pavement distress surveys and surface rut depth measurements were conducted at various intervals during trafficking activities.

The failure criteria used for flexible pavements are: 1-in upheaval outside the traffic lane, signifying structural shear failure in the subgrade or other supporting layers; and surface cracking to the point that the pavement is no longer waterproof, signifying complete structural failure of the surface layer. These are the same criteria as used for the multiple wheel heavy gear load (MWHGL) test series run by the U.S. Army Corps of Engineers Waterways Experiment Station in early 1970s (*7*). Traffic tests were terminated on LFP1-N and LFP2-N after completion of 37,686 passes with no evidence of any fatigue cracks, while the tests were terminated on LFP3-N and LFP4-N after completion of 35,046 and 27,192 passes due to the presence of severe fatigue cracking.

Pavement Mechanical Response and Performance

The mechanical pavement response captured by the dynamic instrumentation is analyzed in this section. The pavement performance was correlated to the mechanical response to draw relevant conclusions. It should be noted that the dynamic data analysis presented is based on the critical responses encountered for each sensor. This response corresponds to the wheel track (transverse position) of the NAPTV that gave the maximum sensor response value. During the wander pattern, the critical wheel track is trafficked multiple times. The responses from only one pass at the critical track per wander pattern were selected. The repetition number within the wander pattern, and travel direction for the selected critical response was consistent for every data point analyzed throughout the entire traffic test (*8*).

Permanent Deformation in the Subbase Material

Given the nature of instrumentation used in this study, initial treatment of the data was found to be a critical step prior to conducting advanced analysis. The raw MDD data underwent extensive adjustment, screening and preliminary evaluation processes, producing data meeting a high rationality and consistency standard available for analysis. Based on the evaluation of data rationality, it was deemed unfeasible to determine the subgrade contribution to permanent deformation as the deeper MDD components E and F, located within the P-152 subgrade layer, were found to produce unreliable data most of the time. Thus, the analysis covers mostly the contribution of P-154 material to the cumulative permanent deformation measured above the subgrade. This was believed to be suitable since prior experiences at the NAPTF led to the speculation that unbound material layers are the major contributor to permanent deformation in the pavement structure (*9, 10*).

Figure 2*a* shows a comparison of the P-154 permanent deformation for the perpetual test items. The increase in P-154 permanent deformation with decreasing P-401 thickness is evident. In addition, a notable increase in the rate of

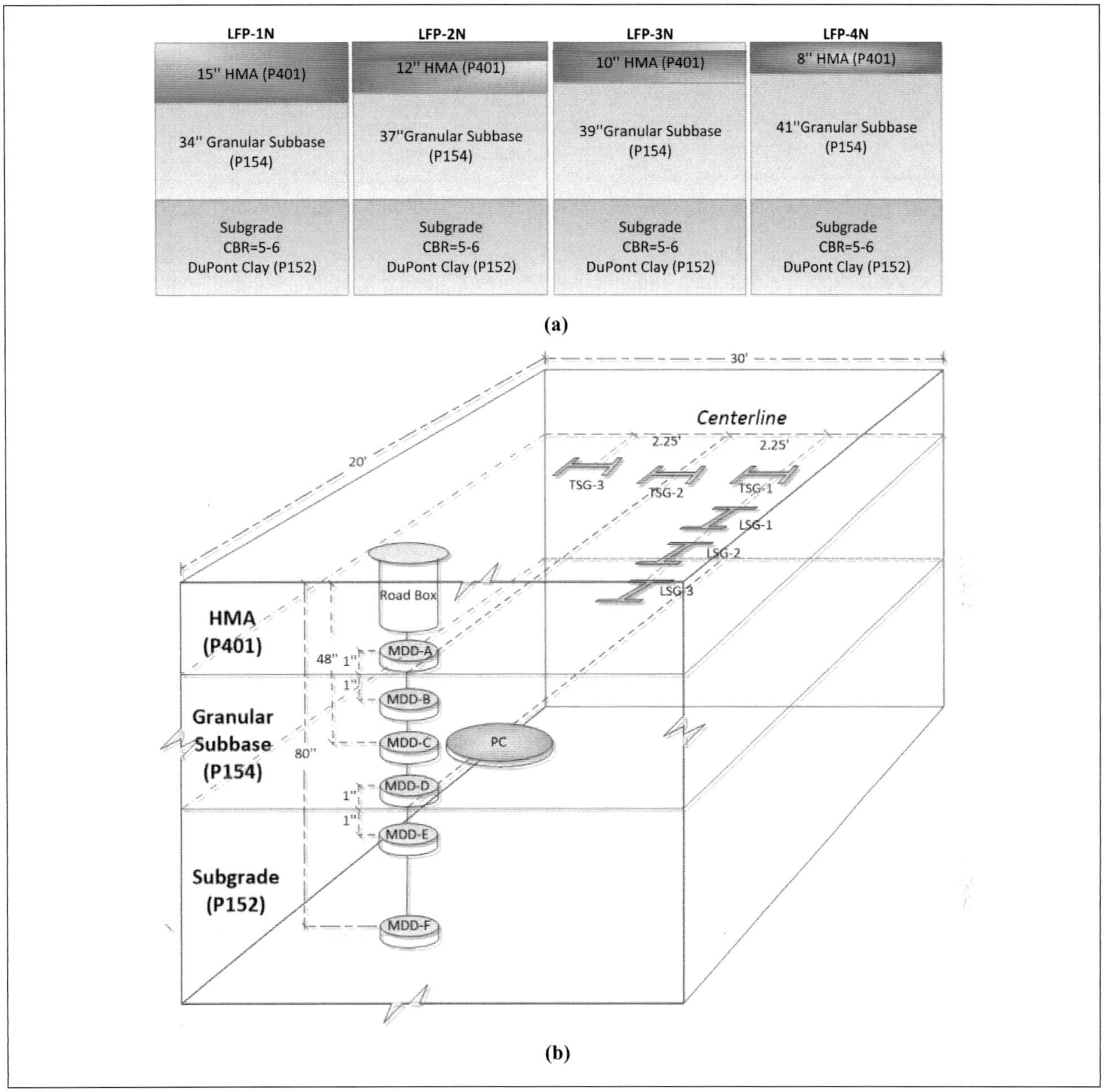

Figure 1. Pavement structure and instrumentation: (*a*) perpetual test item cross-sections; (*b*) generic instrumentation layout.

accumulation of permanent deformation in response to the increase in wheel load from 55,000 to 65,000 lb after 30,030 vehicle passes is readily apparent for perpetual items LFP1-N, LFP2-N, and LFP3-N.

P-154 permanent deformation is plotted versus P-401 thickness at various points along the trafficking timeframe in Figure 3. Note that each line connecting data points is labeled with the corresponding pass number. The correlation is always negative: the thicker the P-401 layer the lower the cumulative permanent deformation regardless of event (pass) number. The decrease in the gap between lines with increasing pass number indicates stabilization of the material under the constant 55,000-lb wheel load. Red lines indicate the effect of the wheel load increase to 65,000 lb after 30,030 vehicle passes generated further increase in the accumulation of P-154 permanent deformation. It is worth indicating that for the loading conditions used in this study, the incidence of increasing the P-401 thickness on reducing P-154 permanent deformation does not seem to be very significant beyond the 12-in benchmark.

The ratio of permanent deformation (PD ratio) contributed by P-154 to the total pavement surface permanent

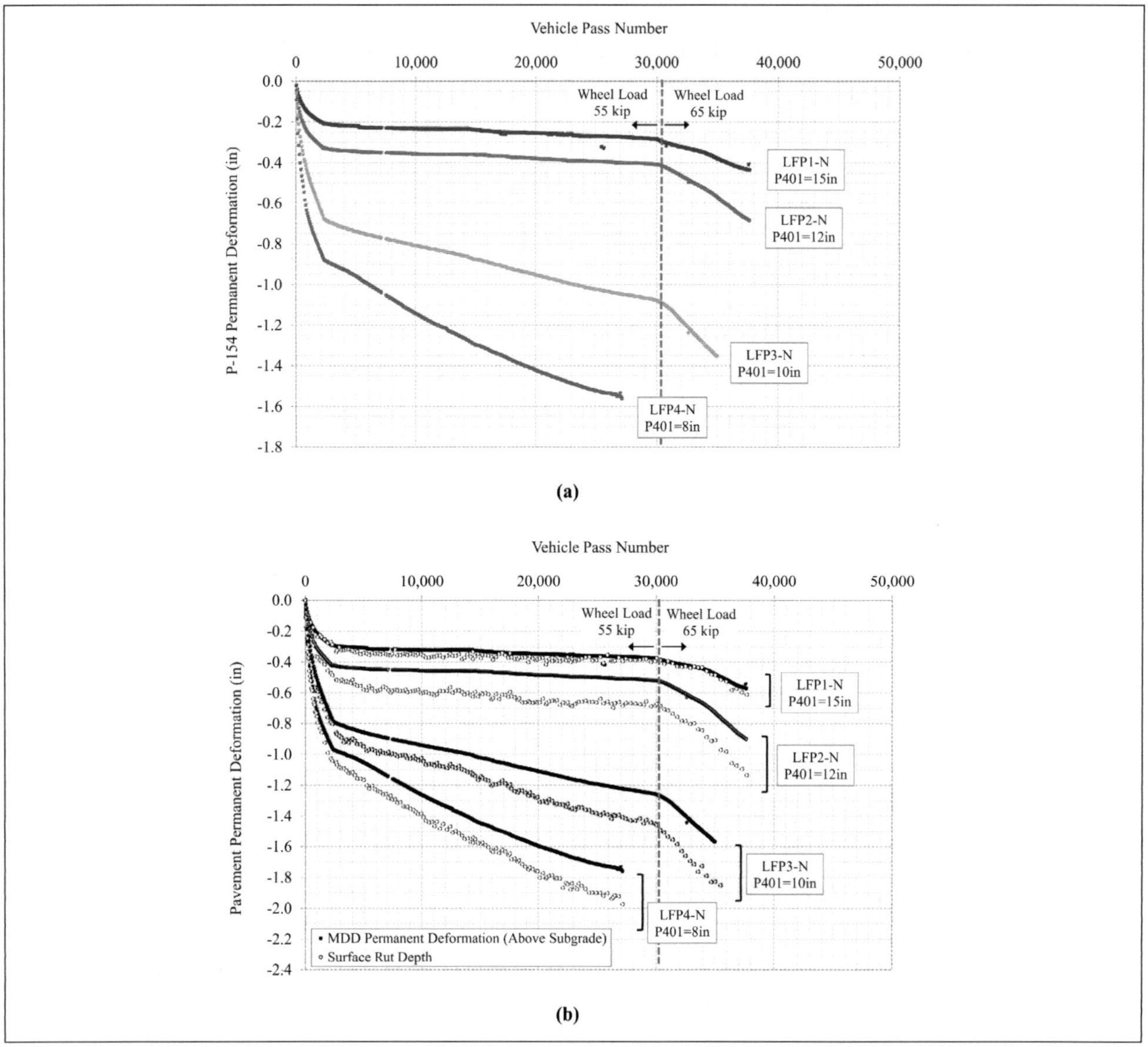

Figure 2. Effect of P401 thickness on the permanent deformation: (*a*) permanent deformation of P154; (*b*) permanent deformation above subgrade versus surface rut depth.

deformation was determined. Any potential contribution of the subgrade to the total permanent deformation was not reflected in the ratios. As no permanent deformation in HMA layer (average temperature during traffic tests was 52°F) was anticipated and confirmed by observations from cores, the difference between MDD measured deformation and surface rutting is caused by permanent deformation in the subgrade. In Figure 3, average ratios for perpetual test items are provided in a tabular form. The average PD ratio ranges from 0.74 to 0.91, depending on the thickness of the P-401 layer. Recognizing the instrumentation limitations to measure permanent deformation of both the P-401 layer and the P-152 subgrade material, such a range of ratios reveals that the permanent deformation underwent by the P-154 subbase material was considerably significant. Most of the permanent deformation in the perpetual test items seems to take place in the subbase layer, so future attempts to enhance the performance of perpetual pavement in terms of permanent deformation should aim at improving the quality of P-154 subbase material.

During the traffic test, pavement surface profiles were measured in all perpetual test items. In Figure 2*b*, the maximum rut depth is compared with the cumulative permanent deformation of all materials above the subgrade. Although discrepancies in magnitude when comparing the two sets of data are evident, the trends in deformation accumulation are very consistent. This validates the methodology applied for the preliminary data treatment and for the determination of

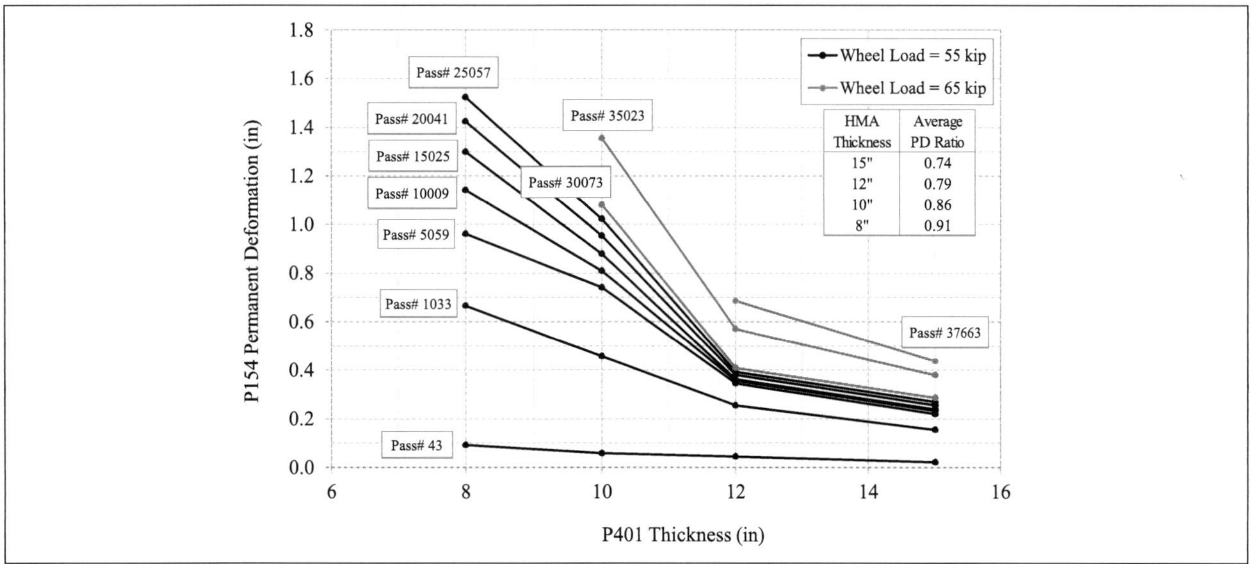

Figure 3. Relative benefit of increasing P401 thickness on P154 permanent deformation.

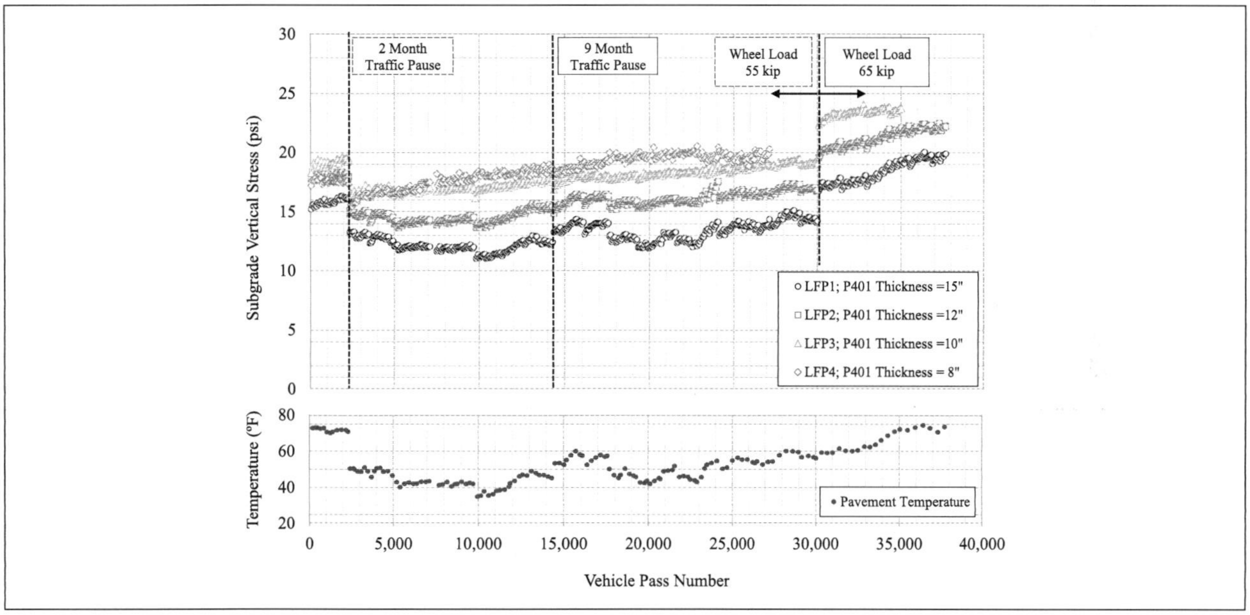

Figure 4. Peak subgrade vertical stresses measured in perpetual test items.

permanent deformation from MDD data. Assuming that the methodology adopted to estimate rutting is effective in isolating rutting from upheaval, any discrepancies between the measured surface profiles and the permanent deformation above subgrade estimated from MDD data can be attributed to the contribution of P-152 subgrade to the total permanent deformation. Note that the discrepancy between surface rut depth and MDD permanent deformation for LFP1-N is minimal. This indicates that the 15 in. of P-401 layer was very effective at protecting the subgrade from the traffic-induced vertical stresses. As the P-401 thickness decreases, the discrepancy between data sets increases indicating increasing subgrade exposure to vertical stresses and hence, permanent deformation damage.

Vertical Stresses at the Top of Subgrade

The analysis was aimed at evaluating the effect that varying thicknesses of P-401 layer have on the vertical stresses measured by the PCs at the top of the P-152 subgrade. The peak vertical stresses considered in the analysis correspond to the critical traffic path only (*8*). In Figure 4, all PC

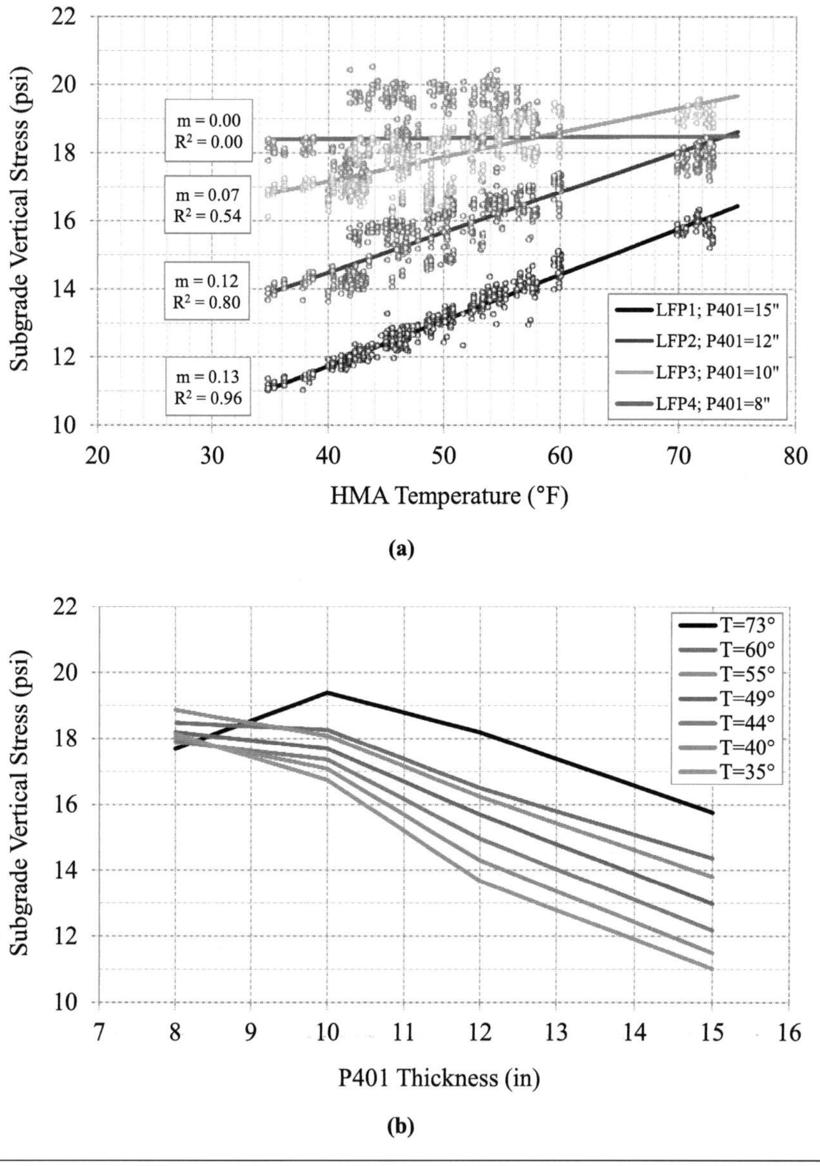

Figure 5. Relationship between subgrade vertical stress and pavement temperature: (*a*) effect of P401 thickness; (*b*) effect of pavement temperature.

response data are summarized. The average temperature captured by thermocouples embedded at different depths within the P-401 surface layer is included in the bottom part of the Figure. The effect of the thickness of the P-401 layer on the magnitude of vertical stress at the top of subgrade is evident. It is apparent that, in general, as the thickness of the P-401 layer increased, the vertical stress measured on top of the subgrade decreased, due to lower subgrade stress ratios.

Some anomalies observed in the data are explained by the events marked in dashed black lines. The effect of traffic pauses is mainly associated with seasonal changes in the pavement temperature. Increments in the vertical stresses are observed as the wheel load increased from 55,000 to 65,000 lb. When correlating the PC responses to the pavement temperature, it is evident that LFP1-N was the most susceptible, likely due to the stiffness of a 15 in. P-401 layer being more prone to seasonal temperature changes than thinner layers of P-401 materials. The correlation weakened as the P-401 thickness decreased. Although no signs of severe traffic-induced damage were observed in test item LFP1-N, it was observed that the vertical stress fluctuation with temperature changes can be up to approximately ±20% of the average stress value.

Additional evidence on the correlation between vertical stress and pavement temperature is presented in Figure 5. In Figure 5*a*, the direct correlation between pavement temperature and PC responses is illustrated for each test item. The high slope (*m*) shown by LFP1-N relative to other perpetual items, clustered data along the trend line, and relatively high coefficient of determination (R^2) indicate that the PC responses were highly correlated to pavement temperature. As the thickness of the P-401 layer decreased for LFP2-N and LFP3-N, it can be observed that *m* and R^2 decreased, and the data became more scattered around the trend line. For LFP4-N, the trend line turned flat and the data became highly scattered, indicating a poor correlation between the PC responses and pavement temperature.

To consolidate the observations regarding the effect of the thickness of the P-401 layer on the PC responses, the relationship between vertical stress at the top of subgrade and layer thickness is illustrated in Figure 5*b* for a wide range of pavement temperatures covering the timeframe of trafficking. Again, it is generally observed that regardless of pavement temperature, increasing P401 thickness resulted in a decrease of the vertical stresses on top of the subgrade.

Distresses in Perpetual Pavements

Traffic tests were terminated on LFP4-N after completion of 27,192 passes. After the completion of 30,030 passes, traffic tests were continued on LFP1-N, LFP2-N, and LFP3-N at an increased wheel load of 65,000 lb. Traffic tests were terminated on LFP3-N after completion of 35,046 passes due to the presence of severe fatigue cracking. Traffic tests were terminated on LFP1-N and LFP2-N after completion of 37,686 passes, with no evidence of any fatigue cracks.

Figure 6 demonstrates the development of cracking in linear feet for each perpetual test item over the course of trafficking. LFP4-N data are only represented up to the point of failure. LFP1-N and LFP2-N did not experience significant traffic-associated distresses; in fact, these two test items showed zero fatigue area and nearly zero linear cracking as evidenced in Figure 6. The distress mapping revealed that longitudinal cracks were generally the first to develop. As all perpetual items were subjected to the same traffic load, a sound explanation for the difference in performance is the varied thickness of the P-401 layer. The apparent thickness of the P-401 layer to limit fatigue cracking (under loading conditions used) is higher than 10 in. and at most 12 in.

Given the scattered pattern encountered in the ASG output, the data were first subjected to a thorough filtering process to improve the data quality. A strong correlation was found between the tensile strain and temperature at the bottom of the P-401 layer. The ASG data were then adjusted to a reference temperature of 52°F. Advanced analysis of the strain pulse indicated that tensile strains at the bottom of the P-401 layer decreased with increased thickness (*8*). It is known that as consequence of increasing the surface layer thickness, the traffic load spreads over a larger area in the

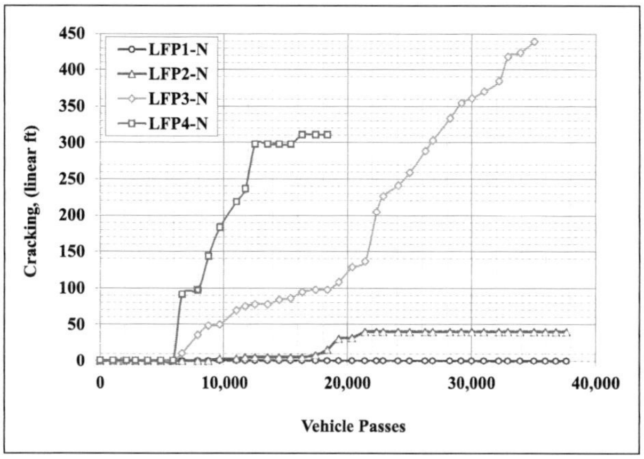

Figure 6. Crack quantization of linear cracks for perpetual test items.

pavement structure. Figure 7 shows the loading pulse duration (loading time) from representative longitudinal and transverse ASGs of all test items, at various event numbers along trafficking. Figure 7, *a* and *b* show an obvious increasing trend in loading time with increasing thickness of the P-401 layer. This is consistent with a decreasing tensile strain magnitude as the thickness of the P-401 layer increased.

Among the main findings from ASG data analysis, it was encountered that tensile strains measured in the transverse direction were higher than those in the longitudinal direction. Figure 8 shows a comparison of temperature adjusted responses between representative ASGs in both directions for test item LFP1-N. Significant unrecoverable strains are observed in the case of transverse strains which result in development of longitudinal crack. Unrecoverable strain magnitude in the case of longitudinal strain gauges is almost negligible. Similar to the findings reported by other researchers (*2, 11, 12*), this observation explains the dominant longitudinal cracks in perpetual test items.

Additional full-scale accelerated pavement tests are planned as part of Construction Cycle 9 (CC9). The combined data from CC7 and CC9 will be used to determine the HMA strain threshold to prevent fatigue cracking.

Conclusions

This paper presents the effects that varying the thickness of the P-401 HMA layer has on both the performance and mechanical response of perpetual pavements. The following main conclusions can be made:

- Increasing the thickness of the P-401 layer significantly reduced the magnitude of permanent deformation in P-154, vertical stress at top of subgrade, and tensile strain at the bottom of P-401 layer.
- The perpetual test item LFP1-N appears to virtually eliminate the subgrade damage, by providing 15 in. of

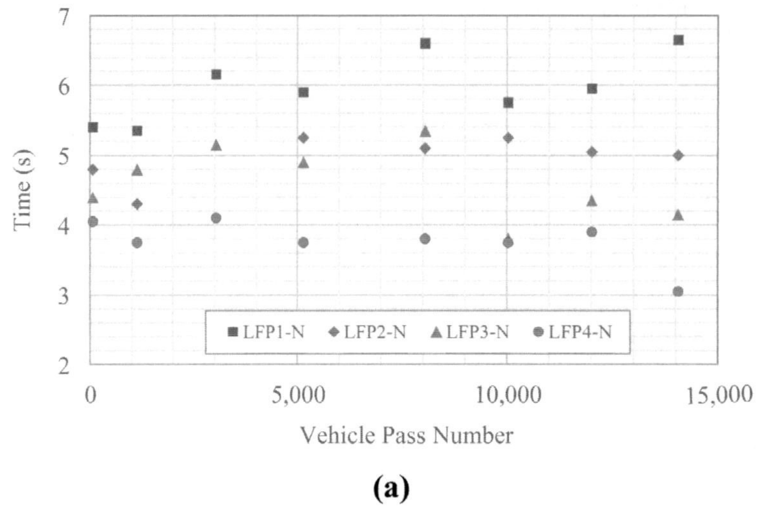

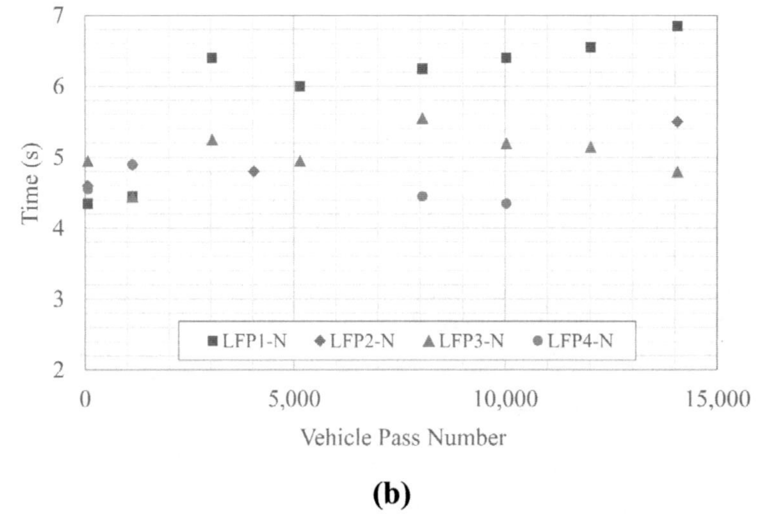

Figure 7. Effect of P-401 thickness on the strain pulse loading time: (*a*) longitudinal strain gauges; (b) transverse strain gauges.

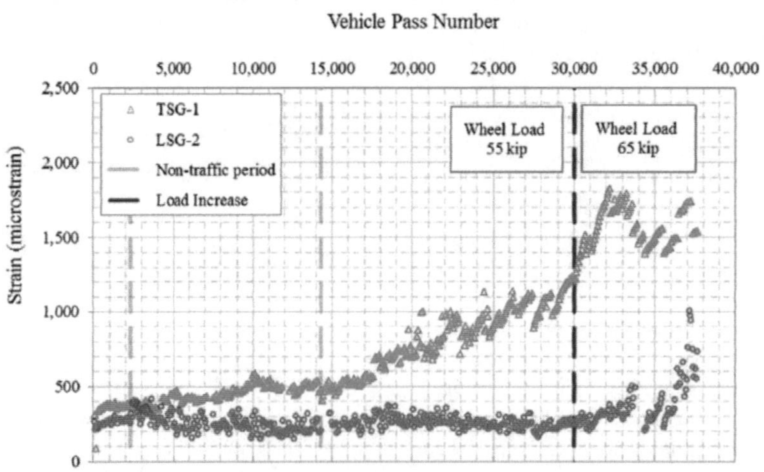

Figure 8. Comparison of strain gauge response in LFP1-N: longitudinal versus transverse orientation.

P-401 layer protection. Increasing from 12 to 15 in. of P-401 layer does not translate into an important reduction of P-154 permanent deformation, but does appear to completely remove the potential for damage in the subgrade.

- Increasing the thickness of the P-401 layer reduced the vertical stress measured at the top of the subgrade as the overall quality of the protecting layers was improved by increasing the thickness of the surface HMA layer. The correlation between vertical stresses at the top of the subgrade and seasonal HMA layer temperature changes further strengthened with increasing the thickness of the P-401 layer.
- The difference in pavement performance is explained by the varied thickness of the P-401 layer. The apparent P-401 thickness to limit fatigue cracking is around 12 in.
- Analysis of ASG data revealed that transverse strains were higher than the longitudinal strains at the bottom of HMA layer. Distress mapping validated such findings, as longitudinal cracks were the first to develop in the process of cracking inception.

Acknowledgment

The work described in this paper was supported by the FAA Airport Technology Research and Development Branch.

References

1. Hayhoe, G. F., R. Cornwell, and N. Garg. Slow-Rolling Response Tests on the Test Pavements at the National Airport Pavement Test Facility (NAPTF). *Advancing Airfield Pavements*. ASCE, 2001, pp. 30-58.
2. Garg, N., and G. F. Hayhoe. Asphalt Concrete Strain Responses at High Loads and Low Speeds at the National Airport Pavement Test Facility (NAPTF). *Advancing Airfield Pavements*. ASCE, 2001, pp. 1-14.
3. Asphalt Pavement Alliance. *Perpetual Pavements: A Synthesis*. APA 101, Lanham, Maryland, 2002.
4. Newcomb, D. E., R. Willis, and D. Timm. *Perpetual Asphalt Pavements: A Synthesis*, APA IM-40, Lanham, Maryland, 2010.
5. Federal Aviation Administration. *Airport Pavement Design and Evaluation*. AC 150/5320-6F, FAA, Office of Airport and Standards, 2016.
6. Federal Aviation Administration. *Standards for Specifying Construction of Airports*. AC 150/5370-10G, FAA, 2016.
7. Ahlvin, R. G., W. J. Turnbull, J. P. Sale, and A. A. Maxwell. *Multiple-Wheel Heavy Gear Load Pavement Tests, Basic Report*. Technical Report S-71-17 (AFWL-TR-70-113, Vol I), 1971.
8. Yin, H., C. E. Cary, Z. Wang, Q. Li, S. Augustyn, and J. Davis. Construction Cycle 7 Test Report – Perpetual Test, 2017.
9. Garg, N., and G. F. Hayhoe. Permanent Deformation Behavior of the Granular Layers Tested at the FAA's National Airport Pavement Test Facility. Paper Submitted for Presentation and Publication to the 3rd International Conference on Accelerated Pavement Testing, October 1–3, 2008, Madrid, Spain.
10. Hayhoe, G. F., N. Garg, and M. Dong. Permanent Deformations During Traffic Tests on Flexible Pavements at the National Airport Pavement Test Facility. Airfield Pavements Specialty Conference, 2003.
11. Islam, M. R., and R. A. Tarefder. Evaluating the Longitudinal and the Transverse Horizontal Strains at the Bottom of Hot Mix Asphalt. *International Journal of Scientific and Engineering Research*, Vol. 4, No. 3, 2013, pp. 1-6.

12Gopalakrishnan, K., and M.R. Thompson. Effect of Dynamic Aircraft Gear Loads on Asphalt Concrete Strain Responses. *Journal of ASTM International*, Vol. 3, No. 8, 2006, pp. 1-16.

The Standing Committee on Aircraft/Airport Compatibility (AV070) peer-reviewed this paper (18-05157).

Article

TRR
JOURNAL OF THE TRANSPORTATION RESEARCH BOARD

Transportation Research Record
2018, Vol. 2672(23) 40–51

DOI: 10.1177/0361198118787361
journals.sagepub.com/home/trr
SAGE

The Global Potential for CO_2 Emissions Reduction from Jet Engine Passenger Aircraft

Lynnette M. Dray[1], Andreas W. Schäfer[1], and Kinan Al Zayat[1]

Abstract

We analyze the costs of CO_2 emissions mitigation measures available to aviation using a global aviation systems model. In that context, we discuss the relationship between mitigation potential and scenario characteristics, and how these interact with policy measures that increase the effective price of fuel, for example, ICAO's CORSIA emissions offset scheme. We find that global fuel lifecycle CO_2 emissions per revenue passenger km could be reduced by 1.9% to 3.0% per year on average by the use of a combination of cost-effective measures, for oil prices which reach \$75 to \$185 per barrel by 2050. Smaller additional emissions reductions, of the order of 0.1% per year, are possible if carbon prices of \$50 to \$150/tCO_2 are assumed by 2050. These outcomes strongly depend on assumptions about biofuels, which account for about half of the reduction potential by 2050. Absolute reductions in emissions are limited by the relative lack of mitigation options for long-haul flights, coupled with strong demand growth.

Currently, aviation contributes around 2.5% of global fuel combustion-related CO_2 (*1*, *2*). Aviation emissions have risen by 3.6% per year since 1980 (*2*). This increase is expected to continue, even with ongoing reductions in emissions per revenue passenger-kilometer (RPK), due to the sector's rapid growth. Increases of 4% to 5% per year in RPK are projected over the next 20 years (*3*, *4*). Consequently, the sector is a target for local, regional, and global emission-reduction policy efforts—most notably ICAO's Carbon Offsetting and Reduction Scheme for International Aviation (CORSIA) (*5*), which aims to offset aviation CO_2 growth after 2020 via the purchase of emissions allowances from other sectors. This approach does not require emissions reductions within the aviation sector, although airlines may choose to reduce emissions if allowance costs become sufficiently high. This is a cause for concern because aviation also has substantial non-CO_2 climate impacts (*6*), which are not covered by CORSIA. It may also represent an over-pessimistic approach to cost-effective within-sector reductions. Recent research has shown that a combination of technological and operational measures which are cost-effective at oil prices of \$50 to \$100 per barrel could reduce CO_2 per RPK for North American narrowbody aircraft by 2% per year to 2050 (*7*). Given projected regional RPK growth rates of around 3% per year (e.g., *4*) these strategies could make a significant contribution to carbon-neutral growth at zero marginal cost. More generally, reductions in aviation fuel use will also help address the sector's non-CO_2 climate and health impacts.

Many of the measures outlined in Schäfer et al. (*7*) are replicable on a global scale. New aircraft technologies and retrofits may be applied anywhere, although different regional operating costs may affect their cost-effectiveness. The main obstacle in achieving carbon-neutral growth in other world regions is that their projected RPK growth rates are typically higher than in North America (e.g., *4*, *5*). Other studies have also investigated the cost-effectiveness of mitigation measures applied to other aircraft size classes, albeit with limited geographic scope and/or limited communication of the underlying assumptions (e.g., *8*–*14*). However, there remains limited research assessing cost-effective mitigation potential at a global level.

A further challenge in assessing the potential of within-sector mitigation measures is the complexity of interactions between passengers, airlines, airports, regulators, and other stakeholders. For example, airlines may respond to the introduction of fuel-saving technologies by lowering ticket prices, leading to demand increases and lower-than-expected reductions in overall emissions (e.g., *15*, *16*). The mitigation potential of new technologies at any given time is determined by the rate of fleet turnover (*17*). The potential of new aircraft models may also be limited by existing infrastructure.

[1]Air Transportation Systems Lab, UCL Energy Institute, London, UK

Corresponding Author:
Address correspondence to Lynnette M. Dray: l.dray@ucl.ac.uk

These issues are typically neglected in the literature on marginal abatement costs.

In this paper, we address these gaps by extending the analysis of Schäfer et al. (*7*) to cover passenger transport by regional jets and widebody aircraft on a global scale. We use this analysis as input to an aviation systems model (AIM2015) (*18*) to assess how system-level interactions and carbon pricing affect outcomes. The second section discusses our assumptions on technology characteristics, systems modeling, and uncertainty. The third and fourth sections discuss modeling outcomes and conclusions.

Methodology

Mitigation Options

CO_2 emissions per RPK can be reduced by reducing fuel life cycle carbon content, by increasing engine and/or aerodynamic efficiency, by decreasing structural weight, or by adding passengers (e.g., *7*). In addition, overall emissions can be lowered by reducing inefficiencies that lead to longer flight or taxi distances. Because relatively small benefits are available in each domain, a combination of measures is required. We consider 25 mitigation measures, comprising all the measures analyzed in Schäfer et al. (*7*), except better matching of aircraft size to mission, plus blended-wing body and advanced turboprop aircraft. Broadly, the measures can be divided into: new aircraft models, retrofits to existing aircraft, operational changes, and alternative fuels. These areas are discussed individually below. We omit measures aimed primarily at influencing demand, for example, increased high-speed rail and telepresence, and measures which may result in a significantly different service being offered to passengers (for example, reduced baggage limits). For each measure we require fuel use, emissions, cost characteristics, the extent to which it can be (or is already) introduced, and any interactions with other measures. We use estimates from (*7*), updated and/or supplemented from academic studies, industry sources, and our own calculations.

We divide the fleet into nine size classes, including: regional jet (RJ), single aisle (SA), twin aisle (TA) and very large aircraft (VLA) classifications. For each size class we use an existing aircraft model to measure relative benefits against the CRJ700. The aircraft used are: Embraer 190; Airbus A319; Airbus A320; Boeing 737-800; Boeing 787-800; Airbus A330-300; Boeing 777-300ER; and Airbus A380-800. These are chosen based on the proportion of global scheduled aircraft-km performed by each aircraft within its size class in 2015 (*19*).

New Aircraft Models. Even if conventional aircraft technologies dominate into the future, we would still expect CO_2 per RPK to reduce. Fleet turnover will lead to older, less fuel-efficient aircraft being replaced (*17*), and new models of conventional-technology aircraft will become available. Over the period from 2015 to 2025 these include the Airbus A350, A320neo and A330neo, Boeing 737MAX and 777-X, and Embraer E-Jet E2 families. Therefore we also need to estimate the fuel burn and costs of evolutionary improvements to existing technology.

Between 1959 and 1995, the cruise-specific fuel consumption of new aircraft engines decreased by 40% (*20*), roughly 1.5% per year. Figure 1 shows the lowest representative mission fuel use per RPK for in-production jet aircraft over time, using the PIANO-X performance model (*21*). Representative missions, including passenger load factor (LF) are derived from global flight schedule and demand data (*19*), and typical seating configurations and production dates from historical fleet data (*22*). Within the nine size categories, Figure 1 indicates that major improvements in technology occur roughly every 15 to 20 years. Following initial large reductions in fuel burn per RPK between 1960 and 1990, the reduction rate has slowed. Between 1990 and 2016, fuel burn per RPK declined on average between 1.1% and 0.3% per year by size class, an average of 0.7% per year. For example, the A320neo has fuel burn per RPK 20% lower than the original A320 for a 750-km mission with 76% LF (*21*), a reduction equivalent to around 1% per year.

For the upcoming generation of aircraft models, we use published estimates of fuel use, cost, and emissions (e.g., *23–29*), assuming a 30% discount from list price where appropriate. For future generations, we assume for our central case a 20-year gap between new models in each size class and 0.7% per year fuel burn reduction for comparable missions. We anticipate this reduction will be achieved by greater use of composite materials, higher bypass ratio engines, and a greater lift/drag ratio. For costs, we use data from Schäfer et al. (*7*) for narrowbodies; for other aircraft types we assume costs will remain roughly consistent with the 2015–2025 aircraft generation in real terms. These characteristics are summarized in Table 1. Yearly cost changes are primarily maintenance-related.

We also model upper and lower scenarios for each variable. For the upcoming aircraft generation, these are based on the range of variation in different models and/or configurations within aircraft classes. For capital costs we assume 25% to 35% discount on list price if data on only one model is available. For future generations, the gap between generations is assumed to be 15 to 25 years, and the average improvement per year to be between 0.5% and 1%.

As well as updates to existing technology, alternative engine and airframe technologies have been proposed which could have a larger impact on emissions. Different technologies may be appropriate for different size aircraft (*30*). We select concepts that are relatively established and for which cost estimates are available, including an optimized counter-rotating propeller (CRP) engine aircraft for narrowbodies (*7*, *26*), an advanced turboprop for smaller size classes (*26*), and blended-wing body (BWB) designs for widebodies (*30*). The assumed characteristics of these aircraft are summarized in

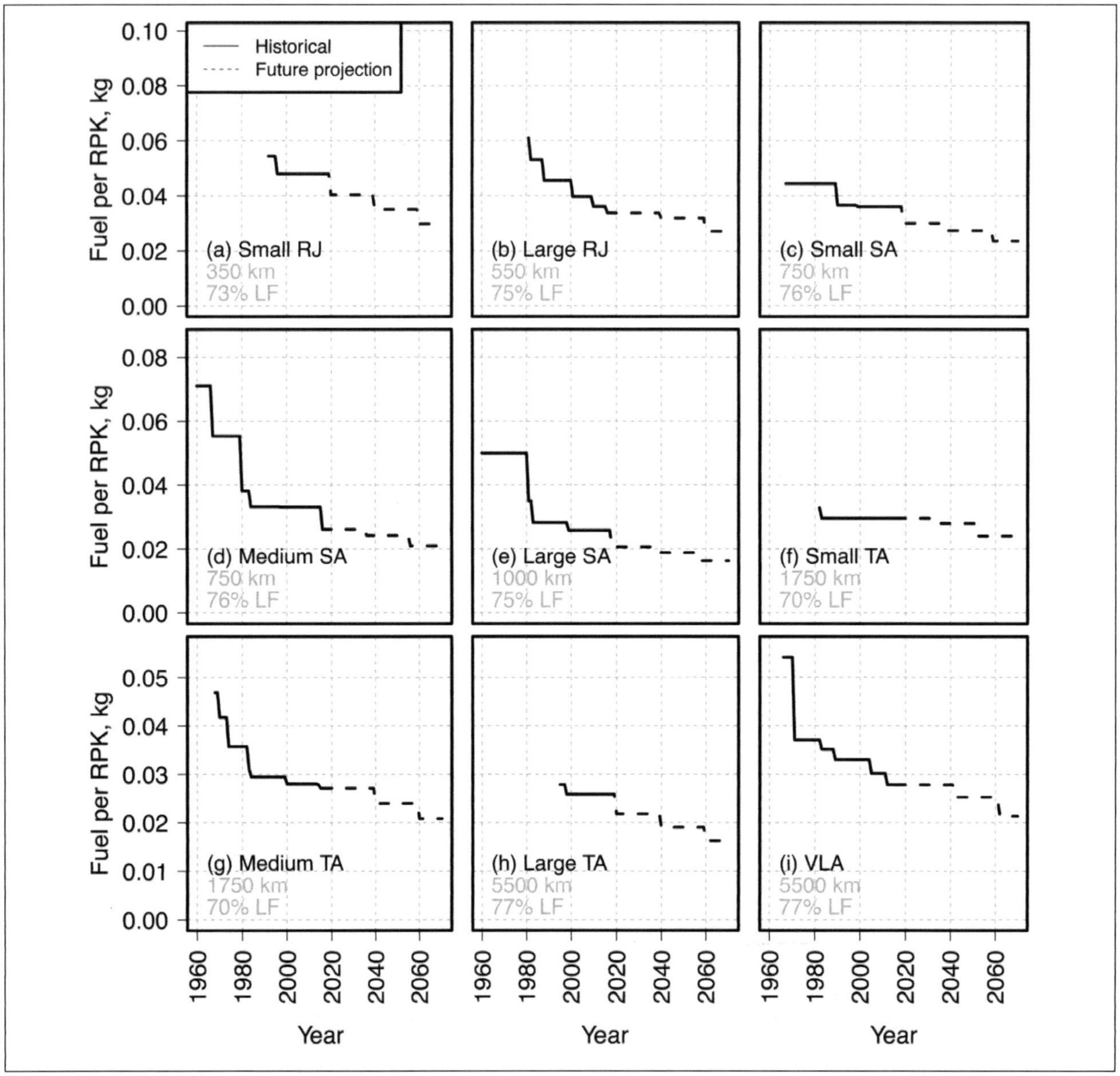

Figure 1. Lowest modeled sample mission fuel burn per RPK available from in-production regional jet (RJ), single aisle (SA), twin aisle (TA), and very large aircraft (VLA) models over time by increasing aircraft size class: (*a*)–(*i*).

Table 2. We assume capital and maintenance costs scale between size classes similarly to those for the present-day reference aircraft when no other scaling information is available. We also assume global availability of new aircraft models, and that operators in each region will decide which technology to purchase based only on the associated costs.

Many other aircraft concepts exist, including the NASA N+3 designs (*31*), and hybrid or battery electric aircraft (e.g., *25*). We exclude these because they are more speculative than the options currently considered, would require extensive infrastructure changes, and/or lack cost estimates.

Retrofits. More immediate impacts can be obtained by applying retrofits to currently-operational aircraft. Retrofit measures include blended winglets for aircraft types which do not already have them, re-engining (if a suitable engine exists), carbon brakes, and cabin weight reduction ("lightweighting"). The impact of these measures can be complex; for example, electric taxi equipment increases weight, slightly, increasing fuel use in non-taxi flight phases. Table 3 gives a summary of the assumed characteristics of retrofit measures. Many retrofit measures have already been applied to part of the fleet, limiting future applicability. We follow

Table 1. Assumed Characteristics of Updates to Conventional Aircraft Technologies Compared with Existing Reference Technologies

Technology	Size class	Available from	Capital cost, million US$ (2015)	Change in non-fuel yearly cost, million US$ (2015)	Change in block fuel use, %	References
Next generation conventional	Small RJ	2020 (2018–2025)[a]	40.9 (35.7–46.1)	−0.35 (−0.3 – −0.47)	16 (15–21)	*24, 25*
	Large RJ	2020 (2018–2025)	53.6 (46.8–60.4)	−0.4 (−0.35 – −0.55)	16 (15–21)	*24, 25*
	Small SA	2019 (2018–2020)	69.6 (64.7–74.6)	—[b]	20 (15–22)	*23, 7, 26*
	Med SA	2016	75.8 (70.4–81.3)	—	20 (15–22	*23, 7, 26*
	Large SA	2018 (2017–2019)	88.9 (82.5–95.2)	—	20 (15–22)	*23, 7, 26*
	Small TA	No update; reference aircraft is already based on the 787–800				
	Med TA	2020 (2018–2022)	211 (189 – 233)	−0.026	12 (10–14)	*23, 27*
	Large TA	2020 (2018–2022)	251 (233–270)	−0.35 (0 – −0.07)	21 (17.5–23.7)	*23, 28*
	VLA	2020 (2017–2022)	305 (284–323)	−0.2 (0 – −0.4)	4	*29*
Subsequent generation conventional	Small RJ	2040 (2033–2050)	41 (36–46)	−0.35 (−0.3 – −0.47)	28 (25–32)	—
	Large RJ	2040 (2033–2050)	54 (47–60)	−0.4 (−0.35 – −0.55)	28 (25–32)	—
	Small SA	2039 (2031–2045)	75 (68–82)	—	30 (26–34)	*7, 24*
	Med SA	2036 (2031–2041)	83 (75–90)	—	30 (26–34)	*7, 24*
	Large SA	2038 (2032–2044)	97 (87–106)	—	30 (26–34)	*7, 24*
	Small TA	2032 (2027–2037)	123 (114–132)	—	14 (12–14)	—
	Med TA	2040 (2033–2047)	211 (188–233)	−0.026	24 (22–24)	—
	Large TA	2040 (2032–2047)	251 (233–270)	−0.35 (0 – −0.7)	31 (29–33)	—
	VLA	2042 (2039–2045)	306 (284–324)	−0.2 (0 – −0.4)	17 (15–17)	—

[a]Values in brackets indicate the estimated range of potential values for each variable, which are later used to define the optimistic and pessimistic lenses used in the section "Future scenarios and uncertainty" below.
[b]A dash in cost data indicates no cost change relative to reference aircraft; a dash in the references column indicates new calculations for this paper.

Table 2. Assumed Characteristics of New Aircraft Technologies

Technology	Size class	Available from	Capital cost, million US$ (2015)	Change in non-fuel yearly cost, million US$ (2015)	Change in block fuel use, %	References
Advanced Turboprop	Small RJ	2030 (2025–2035)	22 (19–24)	1.7 (0.9–2.6)	43 (37–46)	*26*
	Large RJ	2030 (2025–2035)	28 (24–31)	1.7 (0.9–2.6)	43 (37–46)	*26*
Optimized CRP	Small SA	2035 (2030–2040)	73 (61–85)	0.4 (0.2–0.5)	41 (40–45)	*7, 26*
	Med SA	2035 (2030–2040)	98 (82–115)	0.4 (0.2–0.6)	41 (40–45)	*7, 26*
	Large SA	2035 (2030–2040)	99 (83–116)	0.4 (0.2–0.6)	41 (40–45)	*7, 26*
Blended- Wing Body	Small TA	2040 (2035–2045)	217 (180–289)	−0.3 (−0.2 – −0.5)	30 (15–40)	*30, 26*
	Med TA	2040 (2035–2045)	233 (194–310)	−0.3 (−0.2 – −0.5)	30 (15–40)	*30, 26*
	Large TA	2040 (2035–2045)	249 (207–332)	−0.3 (−0.2 – −0.5)	30 (15–40)	*30, 26*
	VLA	2040 (2035–2045)	364 (303–485)	−0.3 (−0.2 – −0.5)	30 (15–40)	*30, 26*

Schäfer et al. (*7*) in limiting uptake to take account of this based on current fleet composition.

Operational Measures. Operational strategies include measures to decrease routing inefficiency and/or congestion at airports; increased maintenance to reduce performance deterioration; and changes in airline behavior, such as reducing tankering. The costs and benefits of these measures can be difficult to evaluate, and can have different impacts on different system stakeholders. Additionally, some measures have different regional impacts, for example, depending on existing inefficiency levels or fuel price differentials between origin and destination. For this paper, we use a global aviation systems model which includes regional estimates of airline costs, fleet composition, and system inefficiency (e.g., *18*, *32*; section "Aviation Systems Modeling"). We assume similar percentage reductions in system inefficiencies are available by region worldwide. This likely underestimates the benefits achievable in regions with less developed or more congested aviation systems. We also omit benefits which may be available on specific routes (for example, meteorology-optimized routing). Similarly, we assume tankering behavior worldwide is similar to that in the U.S.A. Table 4 shows the assumed characteristics of operational

Table 3. Assumed Characteristics of Retrofit Measures

Technology	Size class	Available from	Capital cost, million US$(2015)	Change in non-fuel yearly cost, million US$ (2015)	Change in fuel use, %	References
Blended winglets	Small SA–Med TA	2015	0.85–1.9[a]	—[b]	3 (2–4)[c]	7, 12
Surface polish	Small RJ–Med TA	2015	0.03–0.13	0.03–0.16	1 (0.5–1.5)	7, 12
Carbon brakes	Small RJ–VLA	2015	—	0.015–0.045	0.15 (0.1–0.2)	7
Engine upgrade kit	Small RJ–Med TA	2015	0.5–1.8	—	1 (0.5–1.5)	7, 12
Re-engining	Small RJ–Med TA	2015[d]	7.1–16.6	—	12.5 (10–15)	7, 12
Electric taxi	Small RJ–VLA	2018	0.3–4	—	2.8 (1.8–3.8)[e]	7
Cabin weight reduction	Small RJ–VLA	2015	0.2–2.3	—	1.2 (1.2–2.1)	7, 12

[a]Where characteristics vary by size class, the range over all size classes is shown.
[b]Dash indicates no change from reference aircraft.
[c]Cruise only; reduced benefits for other phases (*12*).
[d]Assumed only applicable to aircraft over 20 years old.
[e]Average impact on block fuel.

Table 4. Assumed Characteristics of Operational Measures

Measure	Size class	Available from	Cost, million US$ (2015)	Change in fuel use, %	References
Surface congestion management	Small RJ–VLA	2015	0.015–0.06	15 (10–20)[a]	7, 33
Single engine taxi	Small RJ–VLA	2015	0–0.06	30 (20–40)[a]	7
Optimize departures	Small RJ–VLA	2015	0.2–0.6	20 (10–30)[b]	7, 33
Reduce cruise inefficiency	Small RJ–VLA	2015	0.07–0.13	5.5 (2.8–8)[c]	7, 33
Optimize approach	Small RJ–VLA	2015	0.2–0.6	40 (15–50)[d]	7, 33
Reduced fuel reserves	Small RJ–VLA	2015	0–0.5	0.01–0.4	7
Reduced tankering	Small RJ–Large SA	2015	0[e]	0.26 (0.34–0.27)	7, 12
Increased engine maintenance	Small RJ–VLA	2015	0.001–0.002	2.4 (1–4)[f]	8
Increased aerodynamic maintenance	Small RJ–VLA	2015	0.001–0.002	1 (0.2–1.5)[f]	8
Engine wash	Small RJ–VLA	2015	–0.1–0.09	0.75 (0.25–1)	7, 12
Increased LF / reduced frequency	Small RJ–Large SA	2015	0.2–7.6	0[g]	7
Increased turboprop use	Small RJ–Large RJ	2015	2.6	30 (25–32)	7

[a]Taxi phases only.
[b]Takeoff and climb only.
[c]Cruise only.
[d]Approach only.
[e]14% increase in regional fuel costs assumed.
[f]Age-dependent, maximum benefit for 30-year-old aircraft shown.
[g]Per flight; impact of 5%–10% increased LF on frequency and performance is modeled in the systems model, below.

measures for this paper. Further discussion of what is included in each measure is given in Schäfer et al. (*7*) and Marais et al. (*33*); for example, surface congestion management includes collaborative decision-making-type measures. Only airline costs are shown. We do not model costs to airports or air navigation service providers, that is, we assume no pass-through to airlines for these costs, which are assessed in the supplementary material to (*7*).

Alternative Fuels. Several alternative fuels have been proposed and/or trialed in aircraft (e.g., *34*), with different life-cycle emissions, cost, and fleet compatibility characteristics. For simplicity, we choose a single alternative fuel option. Drop-in biofuels can be used directly in existing engines and so have much faster potential for market entry than fuels that require major design modifications. Although algae-based fuels are potentially promising, they are associated with high uncertainty and potentially high cost (*35*). Cellulosic biomass is a relatively abundant feedstock which has low impact on food production and favorable cost and scalability characteristics. Therefore we model drop-in synthetic Jet A from cellulosic biomass feedstock, as in Schäfer et al. (*7*). Given potential demands for biomass from other sectors, it is uncertain whether enough fuel can be produced to supply global demand. Using data from (*7*) and from the U.S. Department of Energy (*36*), we estimate that U.S. biomass production potential is comparable to Jet A demand, and that costs of \$3.0 to \$3.6 per gallon are feasible for commercial-scale production beginning in 2020, with a reduction of 80% to 85% in lifecycle CO_2 compared with fossil-derived Jet A. For this paper, we use a cost curve model to estimate biofuel availability and price. Global biomass cost curves are taken from Hoogwijk et al. (*37*), linearly adjusted according to Searle and Malins (*38*), and we assume the aviation industry will have priority access to biomass before other sectors, based on higher willingness to pay due to the relative difficulty of reducing aviation emissions by other means. We assume transport, plant investment, and operating and maintenance costs add an additional $3.6 per gallon in 2020, falling to $1.8 ($1.3–$2.3) per gallon in 2050, based on a range of different assumptions about economies of scale and conversion plant sizes (*7*). Given current certification limitations, we limit the maximum biofuel blend to 50% (*39*).

Interaction between Measures. Many of the measures considered above interact. Care must be taken to avoid double-counting and/or the adoption of incompatible measures. We assume retrofit-type measures which are applicable to new aircraft are already included in the benefits from future aircraft models, unless there is evidence to the contrary. Similarly, we assume that measures which target the same fuel burn reduction opportunity are incompatible, for example, single engine taxi and electric taxi. For compatible measures, we apply percentage reductions in emissions cumulatively and assume that applied measures do not alter the percentage benefits achievable from other measures.

Aviation Systems Modeling

To assess the cumulative benefits achievable from the measures discussed above, we use them as input to the open-source aviation systems model AIM2015 (*18*). AIM2015 models interactions between passengers, airlines, airports, and other stakeholders into the future for a set of 878 global cities, containing 1,169 airports and representing around 95% of global scheduled RPK. Passenger demand between cities is projected using a gravity model based on city population, income, and fare, and distributed between airports and routes using an itinerary choice model. An aircraft size choice model projects which aircraft are used on each route; historical LFs are then used to project flight schedules and demand for aircraft, which in turn are used to project airport-level delay. The segment costs to airlines of flying this schedule with the given fleet are estimated and input to a fare model, which feeds back into the demand and itinerary choice calculations. When equilibrium is reached between supply and demand, output metrics including CO_2, noise, and local emissions can be calculated. The structure and validation of these sub-models and of the overall model is discussed extensively in Dray et al. (*18*). Within this structure, technology adoption affects airline costs, which in turn affect fare, demand, and scheduling.

For this paper, we adapt the fleet and technology choice routines of AIM2015 to use the technology characteristics from the above section "Mitigation Options." These routines have previously been used to assess the uptake of limited combinations of technology measures (e.g., *40*) but are here updated to consider many more technologies, to model their impacts more fully, and to include uncertainty. AIM2015 projects the number of new aircraft required by year, region, and size class. We model technology purchase decisions based on net present value (NPV), comparing existing reference aircraft in each size class against available alternative models with costs appropriate for each region (e.g., *18*, *25*, *40*). We assume a discount rate of 10% and an evaluation timeframe of seven years across all regions (*41*). For retrofits, alternative fuels, and operational measures, a simpler model is used in which airlines adopt a measure if the payback period is less than three years. The impact of changing these parameters depends on the balance between initial and ongoing costs for each measure, with shorter timeframes and higher discount rates favoring measures with low initial costs over those with longer-term benefits. However, since capital costs are amortized in the NPV model, the overall impact is relatively small. Operational measures and drop-in alternative fuels are assumed to be reversible if no longer cost-effective. The uptake of these measures is evaluated by world region, size class, and year of aircraft age. Sales of older aircraft between regions are captured by a model based on regional GDP, as in Dray et al. (*40*).

Analyses of historical technology adoption show uptake is not instant even when measures are cost-effective, but rather follows an S-curve dynamic over time (*42*). This reflects several underlying effects. For example, some retrofits can only be installed at major maintenance checks; production lines have limited capacity; airlines differ in their willingness to try new technology; and many aircraft are leased rather than owned by operators. Based on the timeframes discussed in Kar et al. (*42*), we implement initial limits on adoption for each technology to simulate this behavior.

Future Scenarios and Uncertainty. The adoption and impact of future technologies depends on many uncertain variables. Total emissions and technology penetration depend on demand growth, which depends on future developments in population, income, and fuel price, amongst other variables. Additionally, technology characteristics are uncertain. We model ranges for all technology variables. Since the run time of AIM2015 is too long for Monte Carlo modeling, we incorporate uncertain variables using a lens approach (e.g., *43*). Each lens is a linked set of input values chosen from the central, upper, and lower cases described above. For example, we combine all central case variables into a central lens to explore the most likely outcome. An optimistic lens combines variables associated with early availability, lower fuel use, lower costs, and other favorable outcomes, and a pessimistic lens similarly combines variables associated with late availability, higher fuel use, and higher costs. Running the model with these three lenses allows an idea of the potential variation in outcomes due to variation in technology characteristics.

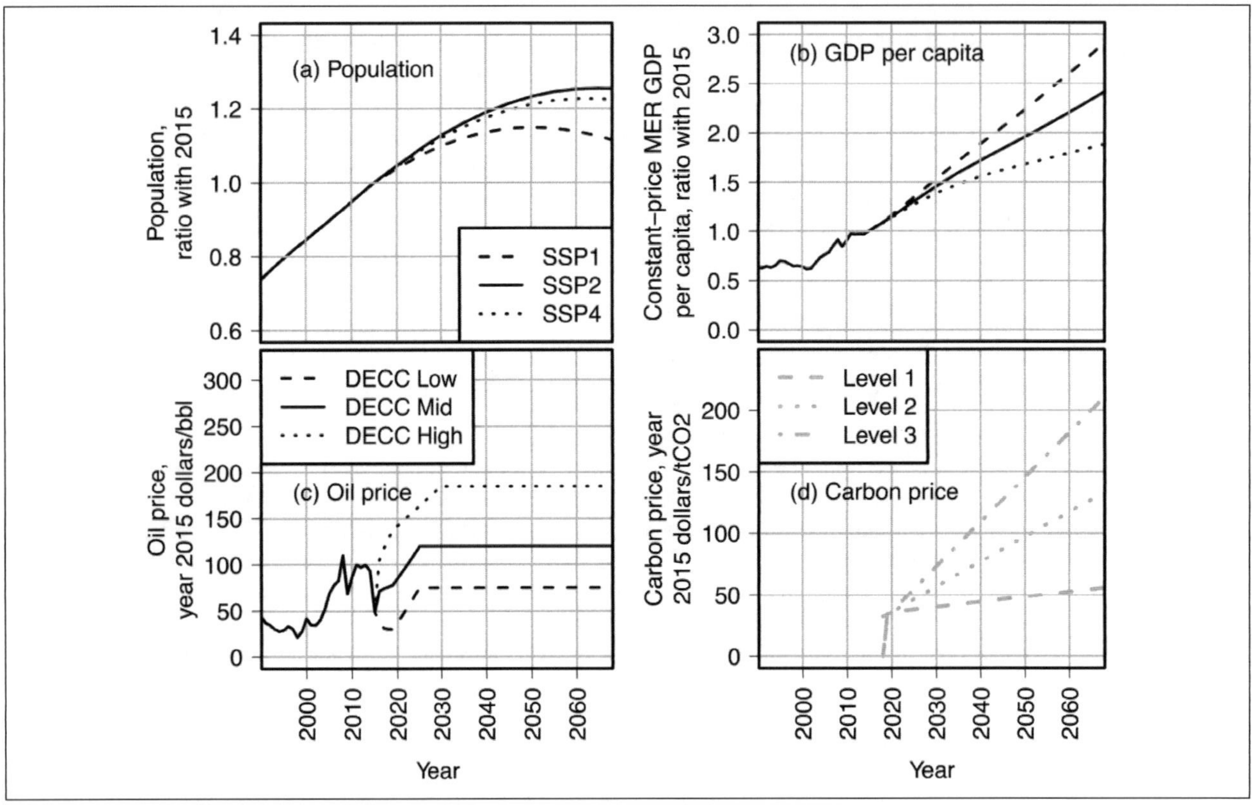

Figure 2. Future scenario characteristics used in this paper: (*a*) population, (*b*) GDP per capita, (*c*) oil price, and (*d*) carbon price.

To model demand growth uncertainty, we use different scenarios for the country-level development of population, income, and oil prices. Following Dray et al. (*18*), we use country-level population and GDP per capita scenarios from O'Neill et al. (*44*), and oil price scenarios from the UK government (*45*). These are summarized in Figure 2. We combine high oil prices with the low GDP growth SSP4 scenario, and low oil prices with the high GDP growth SSP1 scenario, to construct scenarios with slow and rapid demand growth respectively. The central SSP2 scenario uses central oil price projections. Each scenario is run with each lens, giving a total of nine combinations. Additionally we consider a range of carbon prices (Figure 2*d*).

Results

Aviation RPK growth, CO_2, and fleet size strongly depend on the socioeconomic scenario assumed (*18*). Global fleet composition by scenario/lens is shown in Figure 3. For 2050, global RPK varies between 16 and 34 trillion, and global fuel lifecycle CO_2 between 620 and 1690 megatonnes (Mt). Similarly, total fleet varies by a factor of two across scenarios. In the pessimistic lens, fleets are primarily evolutionary updates to current technology. The central lens has uptake of CRP and BWB aircraft, which is increased in the optimistic lens. These and other measures contribute to differences in carbon intensity between lenses. In the SSP2 central lens, around 39 g CO_2 is emitted per RPK in 2050 on a fuel lifecycle basis. This value is 9% higher in the pessimistic lens and 10% lower in the optimistic lens. Biofuels account for somewhat over half of reductions in carbon intensity to 2050. If these scenario/lens combinations are run using only evolutionary updates to conventional aircraft, RPK varies between 16 and 32 trillion, and fuel lifecycle CO_2 between 1240 and 2740 Mt.

The central SSP2 scenario has average RPK growth of 4.4% per year from 2015 to 2035, comparable to projections from Airbus and Boeing over the same time period of 4.6% (*3*) and 4.8% (*4*). For North American narrowbody aircraft, the closest scope modeled here to the U.S. domestic narrowbody fleet considered in Schäfer et al. (*7*), we see broadly similar outcomes once different biofuel assumptions are accounted for. In the SSP2 central lens, we see a 3.2% per year reduction in fuel lifecycle CO_2/RPK to 2050, compared with 2% in (*7*). This is the result of higher biofuel use arising from the cost curve model used in this paper, which reaches 39% of fuel use by 2050; if biofuels are limited to the level assumed in (*7*), we see a yearly reduction of 1.7%, consistent with later assumed introduction years for new aircraft models. All measures found to be cost-effective in (*7*) see some

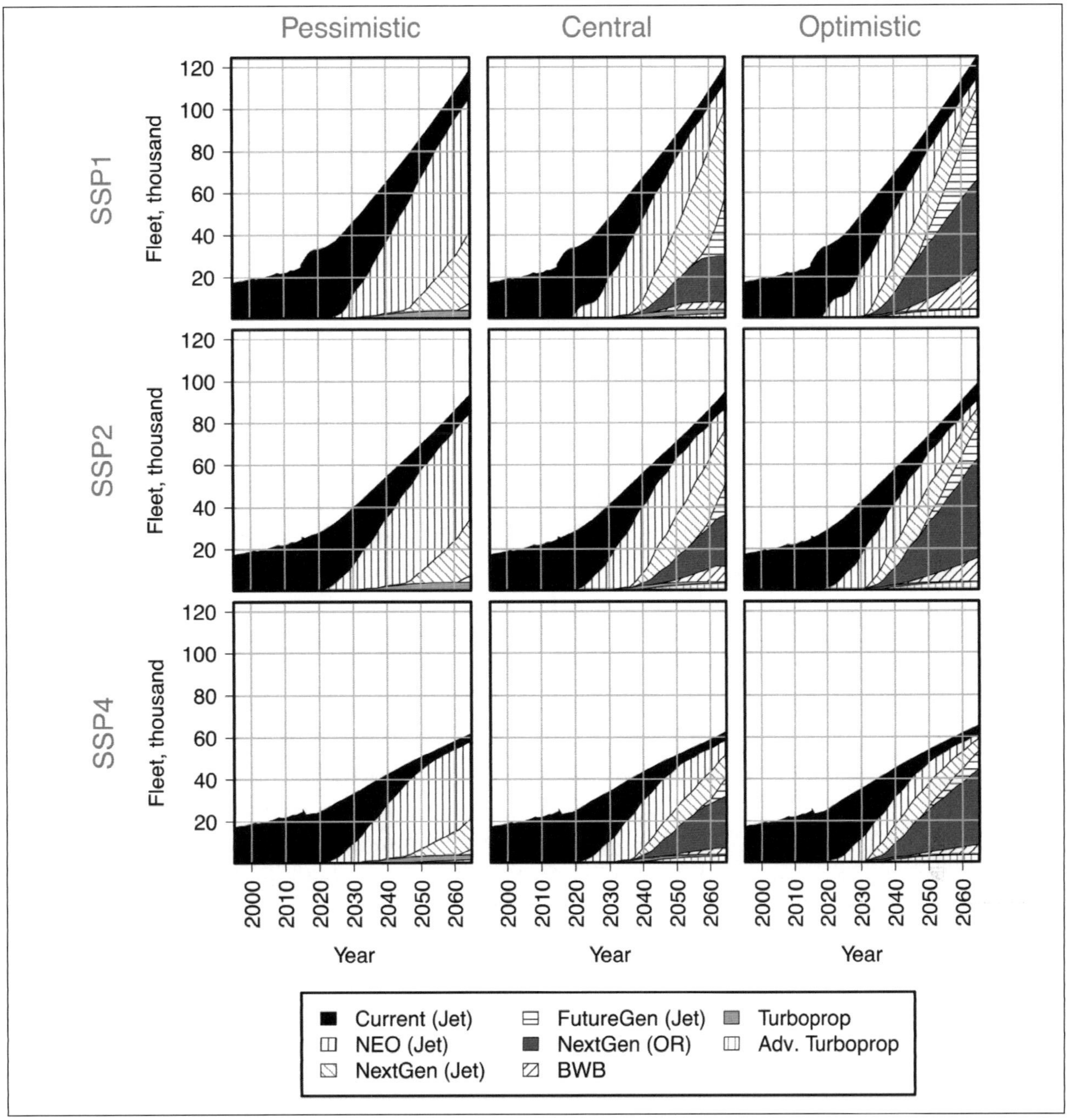

Figure 3. New aircraft technology uptake by lens and scenario in the model runs with no carbon price.

level of adoption. As the fleet considered is more diverse (in terms of geographic scope, size classes, and age cohorts) we also see low uptake for measures which were not cost-effective in that study, for example, surface polish and carbon brakes. No or minimal uptake is seen across scenarios for re-engining, engine upgrades, and reducing tankering.

Globally, we find a smaller potential for cost-effective mitigation, ranging between 2.7% to 2.9% per year reductions in fuel lifecycle CO_2/RPK to 2050 for the central lens (1.2% to 1.4% if biofuel is limited to the level assumed in (*7*)), and 1.9% to 2.7% and 2.9% to 3.0% per year for the pessimistic and optimistic lenses respectively. This reflects the importance of biofuels in mitigating long-haul emissions, where high demand growth is expected (*3*, *4*). In comparison, reference scenarios featuring just evolutionary updates to existing technology demonstrate a 0.5% to 0.9% per year reduction in lifecycle CO_2/RPK. Global RPK and CO_2 trajectories are shown in the left-hand panels of Figure 4. Past

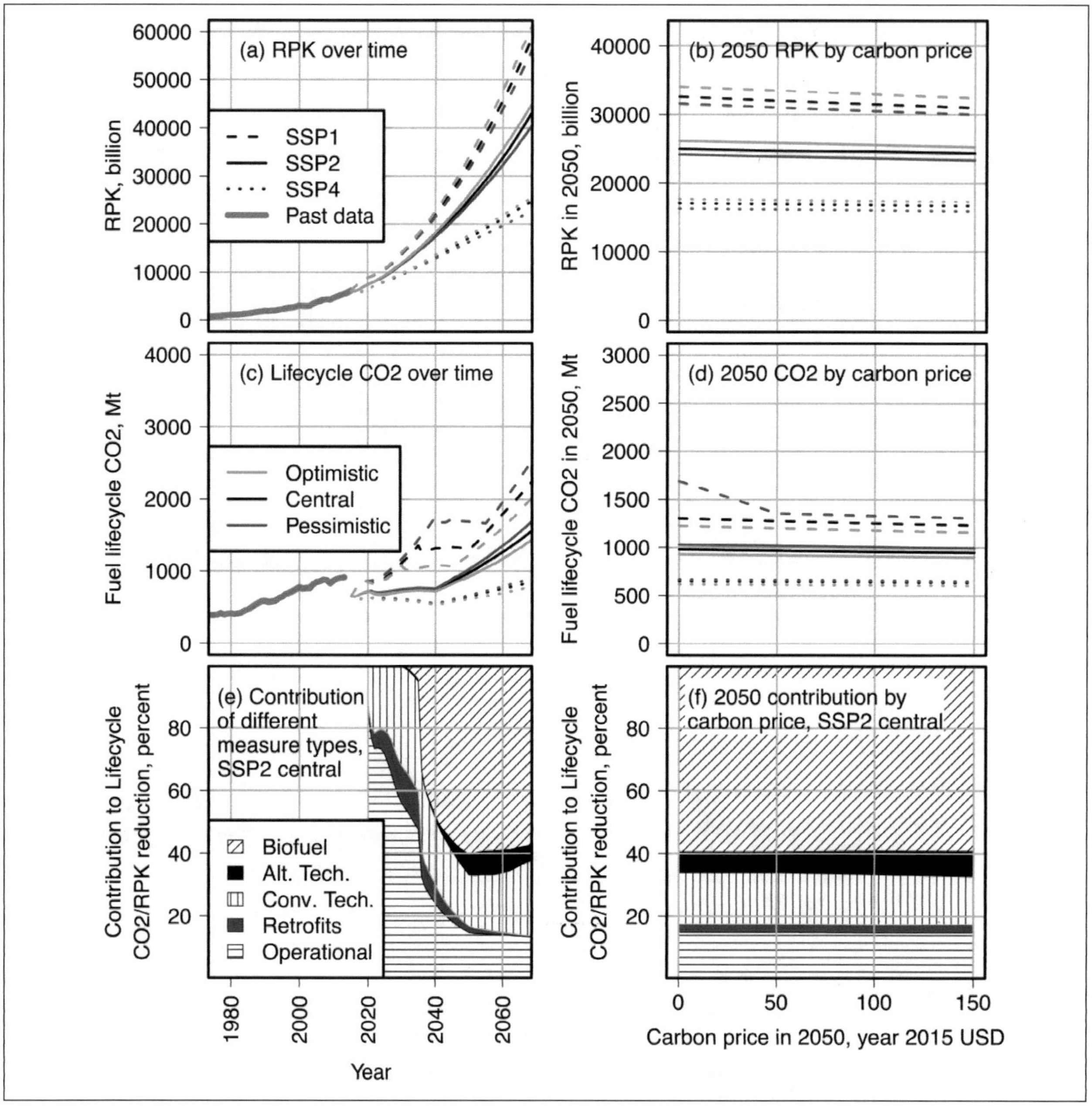

Figure 4. Relationship between RPK, global fuel lifecycle CO_2 and carbon price: (*a*) total RPK, (*b*) RPK in 2050 by carbon price, (*c*) total fuel lifecycle CO_2, (*d*) fuel lifecycle CO_2 in 2050 by carbon price, (e) contribution to reduction in fuel lifecycle CO_2/RPK by measure type, SSP2 central lens, from 2020, (*f*) contribution to reduction in fuel lifecycle CO_2/RPK by measure type in 2050 by carbon price, SSP2 central lens.

data on RPK and CO_2 is from the World Bank (*46*) and IEA (*47*); note that past CO_2 data includes all aviation sources including freight, military, and unscheduled flights, which account for around an extra 30% of emissions (*18*) and are not modeled here. Although the differences in carbon intensity between lenses are substantial, their impact on total CO_2 is much smaller than the impact of different rates of scenario demand growth. Figure 4*e* shows contribution by measure type over time in the SSP2 central lens. Rapidly applicable operational and retrofit measures are important initially, but by 2050 longer-timescale measures such as new aircraft models and biofuels dominate.

We also see a small rebound effect associated with better technology. In 2050, compared with the central lens, optimistic lens runs have approximately a 4% increase in RPK traveled; scenarios using the pessimistic lens have a 3% to

4% decrease. These changes reflect overall changes in airline costs and fare per RPK of 2% to 4% between the different lenses.

Projected energy intensity reductions increase as fuel-related costs increase, making more measures cost-effective. If a carbon price is applied to aviation, as in the EU Emissions Trading Scheme (ETS) and CORSIA, we would expect within-sector emissions reductions as well as reductions outside the sector funded by aviation carbon allowances. A full analysis of CORSIA is outside the scope of this paper, but the EU ETS price has never risen above the equivalent of US$40 (year 2015) and is currently much lower than this, suggesting CORSIA initial carbon prices will be similarly low (*48*) and its impact will be small.

To test system responsiveness to increased carbon price, we run each scenario/lens combination with different carbon prices, as shown in Figure 2*d*. The carbon price in 2050 in these model runs is 50, 100, and 150 $/tCO_2$ respectively. This is much higher than current EU ETS prices; it represents an additional 47 cents to 142 cents on the price of a gallon of fuel. For comparison, the SSP2 fuel price is US$3.7 per gallon in 2050.

The right-hand panels of Figure 4 show how RPK, CO_2, and the contribution of measures to CO_2/RPK reduction vary in 2050 by carbon price. Due to fleet turnover timescales, emissions in 2050 also depend on historical carbon price trajectories. The impact on RPK and relative contribution by measure type is small. However, increased overall uptake of mitigation measures is seen with increased carbon price. For a carbon price of $50/t$CO_2$ in 2050, emissions are reduced by 1% to 20% across the range of scenario/lens combinations from the case with no carbon price. This is due to earlier and greater uptake of biofuels and other mitigation measures, plus a 1% to 2% reduction in RPK from increased fares. Measures with little or no uptake in scenarios without carbon prices are typically not cost-effective at the carbon prices used here. We see diminishing returns for greater carbon prices. For a carbon price of $150/t$CO_2$ in 2050, RPK is reduced by 2% to 5% from scenarios with no carbon price and fuel lifecycle CO_2 by 3% to 23%. This corresponds to a global reduction in aviation fuel lifecycle CO_2 per year of 2.5% to 3.1% per year to 2050 (2.6% to 3.0% per year in the central lens).

Although not directly modeled, we also expect reductions in aviation non-CO_2 emissions and other externalities. This arises from several sources: the overall reduction in fuel use; a reduction in excess distance flown, which will reduce contrails; and a reduction in NOx emissions indices associated with some technologies, for example, open rotor engines.

Conclusions and Future Developments

In this paper we significantly extend an analysis of the costs and benefits of new technologies which could be used to reduce aviation carbon intensity, and use the characteristics of these technologies as input to an aviation systems model to project likely technology uptake and use. This modeling suggests that cost-effective mitigation options exist which could be used to reduce global fuel lifecycle CO_2 per RPK by around 2.7% to 2.9% per year on average to 2050, assuming central values for all input variables. Considering more pessimistic or optimistic assumptions, this value may range from 1.9% to 3.0% per year. However, these values are strongly dependent on biofuel availability, which typically accounts for over half the total reductions and dominates outcomes for long-haul flights. As noted by Schäfer et al. (*7*), more diverse opportunities exist for smaller aircraft. Introducing a carbon price led to reductions in carbon intensity, but these were typically small. The overall impact of changes in aviation carbon intensity on CO_2 emissions also remains smaller than that of the socioeconomic scenario used to project demand. Using a range of plausible future scenarios (*44*), we project demand in 2050 of between 16 and 34 trillion RPK. Although world regions with slower projected demand increases may experience carbon-neutral growth using the technologies considered here (*7*), this is unlikely to happen globally for more than a brief period, even in the lowest demand growth scenario considered.

Given that CORSIA is intended to reduce CO_2 emissions outside the sector to offset growth in aviation emissions, the rate at which emissions can be reduced within-sector may seem unimportant. However, this ignores the climate impact of aviation non-CO_2 emissions (*6*), as well as the increasing difficulty over time of further reducing emissions in other sectors. One avenue for further research would be to model more explicitly the impact of technology adoption on aviation non-CO_2 impacts such as NOx and contrails. Another would be to investigate more radical new technologies, for example, fully or hybrid electric aircraft, which have the potential to reduce externalities radically at the cost of what may be substantial changes in infrastructure and aircraft capabilities.

Author Contributions

The authors confirm contribution to the paper as follows: study conception and design: LMD, AWS; data collection: KZ, LMD, AWS; analysis and interpretation of results: LMD, AWS; draft manuscript preparation: LMD. All authors reviewed the results and approved the final version of the manuscript.

References

1. International Energy Agency. *CO_2 Emissions from Fuel Consumption: Detailed Estimates*. IEA, Paris, 2014.
2. Energy Information Administration. *International Energy Statistics*, Washington, 2015. http://www.eia.gov/cfapps/ipdbproject/IEDIndex3.cfm. Accessed July 18, 2017.
3. Airbus. *Global Market Forecast 2017-2036*, 2017. http://www.aircraft.airbus.com/market/global-market-forecast-2017-2036/. Accessed July 18, 2017.
4. Boeing. *Current Market Outlook 2017-2036*, 2017. http://www.boeing.com/commercial/market/current-market-outlook-2017/. Accessed July 18, 2017.

5. International Civil Aviation Organization. *Assembly Resolution A39-3*. ICAO, Montreal, 2016.
6. Lee, D. S., D. W. Fahey, P. M. Forster, P. J. Newton, R. C. N. Wit, L. L. Lim, B. Owen, and R. Sausen. Aviation and Global Climate Change in the 21st Century. *Atmospheric Environment*. Vol. 43, 2009, pp. 3520–3537.
7. Schäfer, A., A. D. Evans, T. Reynolds, and L. M. Dray. Costs of Mitigating CO_2 Emissions from Passenger Aircraft. *Nature Climate Change*, Vol. 6, 2016, pp. 412–417.
8. Henderson, J. *Controlling Carbon Dioxide Emissions from the Aviation Sector*. Stratus Consulting, Syracuse, NY, 2005.
9. McKinsey and Company. *Pathways to a Low-Carbon Economy: Version 2 of the Global GHG Abatement Cost Curve*. McKinsey and Company, New York, NY, 2009.
10. Pearce, B. *Commercial Airline CO_2 and Mitigation Potential*. IATA, Montreal, 2009.
11. Holland, M., M. Mann, M. Ralph, B. Owen, D. Lee, G. Horton, N. Dickson, and S. Kollamthodi. *A Marginal Abatement Cost Curve Mode for the UK Aviation Sector*. Department for Transport, London, 2011.
12. Morris, J., A. Rowbotham, A. Angus, M. Mann, and I. Poll. *A Framework for Estimating the Marginal Costs of Environmental Abatement for the Aviation Sector*. Omega Report, London, 2009.
13. Raper, D., and J. Morris. *UK Aviation: Carbon Reduction Futures*. Department for Transport, London, 2009.
14. Jesse, E., P. van Aart, and J. Kos. *Cost-Benefit Studies of Possible Future Retrofit Programmes*. WP/Task No. D4.2 EC-FP7 RETROFIT Project. Fokker Services, Redditch, UK, 2012.
15. Timmis, A. J., A. Hodzic, L. Koh, M. Bonner, C. Soutis, A. Schäfer, and L. M. Dray. Environmental Impact Assessment of Aviation Emissions Reduction Through the Implementation of Composite Materials. *The International Journal of Life Cycle Assessment*, Vol. 20, No. 2, 2015, pp. 233–243.
16. Schafer, A., and A. Evans. The Rebound Effect in the Aviation Sector. *Energy Economics*, Vol. 36, 2013, pp. 158–165.
17. Dray, L. M. Time Constants in Aviation Infrastructure. *Transport Policy*, Vol. 34, 2014, pp. 29–35.
18. Dray, L. M., P. Krammer, K. Doyme, B. Wang, K. Al Zayat, A. O'Sullivan, and A. Schäfer. *AIM2015: Validation and Initial Results from an Open-Source Aviation Systems Model*. 21st ATRS Conference, Antwerp, 2017.
19. Sabre. *Market Intelligence Database*, 2017. https://www.sabreairlinesolutions.com/home/software_solutions/product/market_competitive_intelligence/. Accessed July 17, 2017.
20. Lee, J., S. P. Lukachko, I. A. Waitz, and A. Schäfer. Historical and Future Trends in Aircraft Performance, Cost and Emissions. *Annual Review of Energy and the Environment*, Vol. 26, 2001, pp. 167–200.
21. Lissys. *Piano-X Aircraft Performance Model*, 2017. http://www.lissys.demon.co.uk/PianoX.html. Accessed July 17, 2017.
22. FlightGlobal. *Global Fleet Database*. https://www.flightglobal.com/products/fleets-analyzer/. Accessed July 17, 2017.
23. Airbus. *2017 Average List Prices*. http://www.airbus.com/content/dam/corporate-topics/publications/backgrounders/Airbus_commercial_aircraft_price_list_Jan17.pdf. Accessed July 18, 2017.
24. Embraer. *Investor Relations*, 2016. https://www.slideshare.net/embraerri/embraer-march-2016-book-ri. Accessed July 18, 2017.
25. Al Zayat, K., and A. Schäfer. A Comparative Analysis of Operating Cost between Future Jet Engine and Battery-Electric Aircraft. *21st ATRS Conference*, Antwerp, 2017.
26. Vera-Morales, M., W. Graham, C. Hall, and A. Schäfer. *Techno-Economic Analysis of Aircraft*. TOSCA Project WP2 Report, https://www.yumpu.com/en/document/view/7313209/techno-economic-analysis-of-aircraft-tosca-project. Accessed July 17, 2017.
27. Leahy, J. *The A330neo: Powering into the Future*, 2013. https://www.ausbt.com.au/files/A330neo%20Launch%20Presentation.pdf. Accessed July 18, 2017.
28. Reuters. *Elbows Fly in Airbus and Boeing Battle Over Mini-Jumbos*, 2013. https://www.reuters.com/article/us-airbus-boeing-minijumbo-insight-idUSBRE9620TJ20130703. Accessed July 18, 2017.
29. Airbus. *Airbus Presents the A380plus*, 2017. http://www.airbus.com/newsroom/press-releases/en/2017/06/airbus-presents-the-a380plus.html. Accessed July 18, 2017.
30. Liebeck, R. H. Design of the Blended Wing Body Subsonic Transport. *Journal of Aircraft*, Vol. 41, No. 1, 2004, pp. 10–25.
31. Greitzer, E. M., P. A. Bonnefoy, E. DelaRosaBlanco, C. S. Dorbian, M. Drela, D. K. Hall, R. J. Hansman, J. L. Hileman, R. H. Liebeck, J. Levegren, and P. Mody. *N+3 Aircraft Concept Designs and Trade Studies*. NASA/CR-2010-216794/VOL1, 2010. https://ntrs.nasa.gov/archive/nasa/casi.ntrs.nasa.gov/20100042401.pdf. Accessed July 20, 2017.
32. Reynolds, T. G. Development of Flight Inefficiency Metrics for Environmental Assessment of ATM. *8th USA/Europe ATM R&D Seminar*, Napa, CA, 2009.
33. Marais, K., T. G. Reynolds, P. Uday, D. Muller, J. Lovegren, J.-M. Dumont, and R. J. Hansman. Evaluation of Potential Near-Term Operational Changes to Mitigate Environmental Impacts of Aviation. *Journal of Aerospace Engineering*, Vol. 227, No. 8, 2013, pp. 1277–1299.
34. Hileman, J. I., and R. W. Stratton. Alternative Jet Fuel Feasibility. *Transport Policy*, Vol. 34, 2014, pp. 52–62.
35. Quinn, J. C., and R. Davis. 2015. The Potentials and Challenges of Algae Based Biofuels. *Bioresource Technology*, Vol. 184, pp. 444–452.
36. Department of Energy. *US Billion-Ton Update*. ORNL/TM-2011/224. Oak Ridge National Laboratory, Oak Ridge, TN, 2011.
37. Hoogwijk, M., A. Faaij, B. de Vries, and W. Turkemburg. Exploration of Regional and Global Cost-supply Curves of Biomass Energy from Short-rotation Crops and Abandoned Cropland and Rest Land Under Four IPCC SRES Land-use Scenarios. *Biomass and Bioenergy*, Vol. 33, No. 1, 2009, pp. 26–43.
38. Searle, S., and C. Malins. A Reassessment of Global Bioenergy Potential in 2050. *GCB Bioenergy*, Vol. 7, No. 2, 2015, pp. 328–336.
39. Zschocke, A. *High Biofuel Blends in Aviation*. ENER/C2/2012/420-1, 2012. http://www.hbba.eu/study/index.html. Accessed November 13, 2017.
40. Dray, L. M., A. Evans, T. G. Reynolds, A. Schäfer, M. Vera-Morales, and >W. Bosbach. Airline Fleet Replacement Funded by a Carbon Tax: An Integrated Assessment. *Transport Policy*, Vol. 34, 2014, pp. 75–84.

41.Morrell, P., and L. Dray. *Environmental Aspects of Fleet Turnover, Retirement and Life Cycle*. Final report for the Omega Consortium, 2009. http://bullfinch.arct.cam.ac.uk/documents/FleetTurnover_CranfieldCambridge.pdf . Accessed November 13, 2017.

42Kar, R., P. Bonnefoy, and J. Hansman. Dynamics of Implementation of Mitigating Measures to Reduce Aviation's Environmental Impacts. *9th AIAA ATIO Conference*, Hilton Head, SC, 2009.

43. Allaire, D., G. Noel, K. Willcox, and R. Cointin. Uncertainty Quantification of an Aviation Environmental Toolsuite. *Reliability Engineering and System Safety*, Vol. 126, 2014, pp. 14–24.

44. O'Neill, B. C., E. Kriegler, K. Riahi, K. L. Ebi, S. Hallegatte, T. R. Carter, R. Mathur, and D. P. van Vuuren. A New Scenario Framework for Climate Change Research. *Climatic Change*, Vol. 122, No. 3, 2013, pp. 387–400.

45. DECC. *Fossil Fuel Price Projections: 2015*, 2015. Department of Energy and Climate Change, London. https://www.gov.uk/government/publications/fossil-fuel-price-projections-2015 Accessed April 2017.

46. World Bank. *Global Air Transport Data*, 2016. http://data.worldbank.org/indicator/IS.AIR.PSGR. Accessed July 28, 2017.

47. International Energy Agency. *World Energy Statistics*, 2016. http://www.iea.org/statistics/. Accessed July 28, 2017.

48. European Climate Exchange. *Carbon Prices on the European Climate Exchange, 2017*. http://www.ecx.eu/

The Standing Committee on Environmental Impacts of Aviation (AV030) peer-reviewed this paper (18-04002).

Article

TRR
JOURNAL OF THE TRANSPORTATION RESEARCH BOARD

Transportation Research Record
2018, Vol. 2672(23) 52–58

DOI: 10.1177/0361198118783845
journals.sagepub.com/home/trr
SAGE

Impacts on Vehicle Occupancy and Airport Curb Congestion of Transportation Network Companies at Airports

Karina Hermawan[1] and Amelia C. Regan[2]

Abstract

How does the growth of transportation network companies (TNCs) at airports affect the use of shared modes and congestion? Using data from the 2015 passenger survey from Los Angeles International Airport (LAX), San Francisco International Airport (SFO), and Oakland International Airport (OAK), this research analyzes TNCs' relationship with shared modes (modes that typically have higher vehicle-occupancy and include public transit such as buses and light rail, shared vans or shuttles) and the demand for their shared vs. standard service at the airport. Because TNCs both replace shared rides and make them possible, the research also measured the net effects at these airports. The results suggest that in 2015, TNCs caused 215,000 and 25,000 passengers to switch from shared to private modes at SFO and OAK, respectively. By 2020, the increase is expected to be about 840,000 and 107,000 passengers per year, respectively.

Congestion at airport terminals has tremendous negative impacts on airport patrons and the surrounding communities; perhaps the most familiar and direct impact is the rise in passenger disutility resulting from the need to arrive early to prevent catastrophic schedule delays, or the impacts of those delays when passengers elect not to add in a buffer and then miss their flights. Other effects include increases in opportunity costs, and in fuel consumption and pollution. Los Angeles International Airport (LAX) for example, experiences severe congestion, "the upper level curbside demand exceeds the available curbside length during the peak hour by about 60%, while the lower level demand exceeds capacity by nearly 50%" (LAX Economic Impact Review, 2009).

With the advent of Uber and Lyft at airports, a major concern is that these on-demand, app-based ride services (also known as transportation network companies [TNCs]), which are mostly private-use instead of pooled-use, will increase traffic. TNC operations at airports are highly controversial. For example, Uber and Lyft, founded in 2009 and 2012, respectively, have only recently been permitted to pick up passengers at LAX (since January 2016); at Long Beach Airport, it was even later, in April 2017. Data from the 2015 passenger surveys from LAX, San Francisco International Airport (SFO), and Oakland International Airport (OAK) provide information on the share of pooled vs. standard TNC services, as well as how much they are they replacing and complementing shared vehicles (shared vehicles are public transit such as buses and light rail, shared vans or shuttles, which typically have higher vehicle-occupancy). The objective is that by understanding these relationships, one can also set some modal shift targets (shifts to vehicles with higher occupancy) and reduce congestion.

Literature Review

There are several studies that have investigated the growing demand for TNCs and their effects on vehicle-miles-traveled (VMT) and congestion. Unfortunately, there does not seem to be consensus regarding whether in aggregate, TNCs add or abate existing high levels of traffic congestion and VMT. Studies specifically on the demand for TNCs at airports are more limited. Research by Hermawan and Regan found that a mode choice model of ground transportation options to LAX, which included TNCs as one of the alternatives, provided insights into the demand for TNCs as an airport access mode (*1*). In the study, they evaluated travel times and costs to the airport by TNCs and by all other modes (from various locations and with different party sizes) using Google Maps Directions and other sources. Their estimated nested logit model implies that if TNC fares were to increase to match the

[1]Institute of Transportation Studies, University of California, Irvine, CA
[2]Department of Computer Science, Bren School of Information and Computer Science, University of California, Irvine, CA

Corresponding Author:
Address correspondence to Karina Hermawan: hermawak@uci.edu

cost of taxis, the demand for TNCs would fall from 16.8% to 13.9% (20.9% decline relative to initial TNC shares) for business passengers, and from 19.8% to 16.0% (23.3% decline relative to initial TNC shares) for leisure passengers. Meanwhile, if TNC fares were reduced by 50% and travel times increased by 10 min (this represents the difference that would be imposed by pooled TNC services) the change in demand would be negligible. Although Hermawan and Regan did not examine specifically the effects on congestion or VMT, their analysis of passenger sensitivity to changes in TNC travel time and cost could be useful for identifying effective tools to encourage modal shifts to higher-occupancy vehicles.

In a study by Rayle et al., a survey of TNC users in San Francisco when ride-hailing services were relatively new, suggested that most modal shifts to TNCs comes from taxis and buses, and that some users were using Uber or Lyft as a first/last mile service to go to or from the transit stations (*2*). According to the survey, 39% of the TNC users would have taken a taxi if TNCs were not available, whereas 24% would have taken the bus. Moreover, 4% of users either got picked up or dropped off at a public transit station; the average distance of TNC trips was about 3.1 mi. Presumably, these riders were using TNCs to reach public transit stations that were slightly too far to reach on foot. Lastly, Ralyle et al. also found TNC users drive less because of the availability of the service. Forty percent of users reported that they had reduced their amount of driving with the availability of TNC options.

Another paper examined data from a travel and residential survey that was conducted in seven major U.S. cities; this research was also conducted when ride-hailing services were relatively new (*3*). They found that prevalence of ride-hailing services decreases car ownership (9% of ride-hailing users reported disposing of their personal vehicle), as well as personal driving (26% of ride-hailing users reported driving their personal car less). With regards to how ride-hailing services relate to transit, the demand for these services complements the demand for rail but replaces the demand for buses. In addition, significant differences in level of transit use among ride-hailing adopters and non-adopters were found, however the differences were minimal.

A more recent study by Schaller, claims that TNCs are "unsustainable" because they heighten traffic congestion in New York City (NYC). Schaller analyzed electronic trip logs of TNCs and taxis, other trip logs of for-hire vehicles (which comprise taxis, limousines, and TNCs) from the Taxi and Limousine Commission, monthly taxi trip volumes data, current licenses, and vehicle mileage data for NYC (*4*). They found the demand for ride services (yellow and green cabs, and for-hire vehicles including TNCs) rose every year from 2012 to 2016, whereas the demand for subway and bus ridership dropped significantly in the four consecutive years. Taking into account the declines in mileage for taxis, black cars, and car services, declines in personal vehicle use by some TNC drivers and passengers, as well as declines in mileage due to some TNC passengers "pooling" or sharing the ride with other passengers, they estimate the net increase in driving from TNCs to be about 600 million miles from 2013 to 2016.

Finally, despite findings that current TNCs, including their pooled services, increase VMT, researchers at Massachusetts Institute of Technology's Computer Science and Artificial Intelligence Laboratory see great potential in shared rides, especially those by pooled TNC services such as Uber Pool or Lyft Line (*5*). According to this research, with the help of GPS and linear programming algorithms, these services have the technology to run more efficiently than taxis, and reduce the number of taxis on the road by 75% without significantly affecting travel time. MIT's researchers claim that with the right application, "95% of demand would be covered by just 2,000 ten-person vehicles, compared to the nearly 14,000 taxis that currently operate in New York City" (*5*).

Data

The data used in this research are from the 2015 LAX Passenger Survey dataset (*6*) and the 2015 SFO-OAK Customer Survey dataset (*7*). In the study, the focus is exclusively on passengers departing from and not connecting at LAX, SFO, or OAK. The 2020 projections are also based on the 2014, 2015, and 2016 SFO Customer Survey datasets (*8*).

2015 LAX Passenger Survey

The 2015 LAX passenger survey is a one-on-one survey administered using electronic tablets at LAX boarding gates. It was commissioned by Los Angeles World Airports and is typically conducted every 5 years. With nearly a hundred multiple choice and open-ended questions, the survey is extensive. Among the many questions, the survey inquired whether the departing passenger(s) were traveling for business or leisure, the duration of the trip, and their primary mode of transportation to the airport. Based on the responses, there were different follow-up questions. For example, respondents who specified that their primary form of transportation was a private vehicle were asked more specific questions regarding their chosen mode of ground transportation, such as whether they were dropped off, parked at the airport, or parked off site, and how many people were in the same vehicle. Both visiting and Southern California residents, who originated from or were connecting at LAX, were interviewed over the course of two nonconsecutive weeks, April 13–19 and July 13–19, 2015; over 13,400 survey responses were collected. The 2015 survey was the first time that TNC was included as an alternative mode.

2015 SFO-OAK Passenger Survey

In the 2015 SFO-OAK Passenger Survey, different departing passengers (who were originating or connecting) were interviewed every month between May 2014 and May 2015 and asked over a hundred questions. The survey was administered at OAK and SFO either by tablet, paper, or online, and its sample size is over 27,000 travel parties. Many of the questions were similar to that in the LAX survey; for example, they asked what the passenger's primary mode to the airport was, and those using private modes were further asked how many people in the travel party were in the same vehicle. In contrast to the LAX survey, the OAK-SFO survey asked several in-depth questions related to TNCs. Some of those questions included the name of the TNC service that they took (Uber, Lyft, Sidecar, Uber Pool, and so on), their preferred airport access mode prior to the availability of TNC service, and whether they used TNCs to get to the station if they took the bus, train, or rail service.

2014–2016 SFO Customer Survey

These customer surveys are administered annually at SFO at the boarding gates, by mail, or online. They include both passengers originating from and connecting at SFO. The surveys share many common questions to those in the LAX and SFO-OAK surveys. For example, the customer surveys inquired about the passenger's primary mode to the airport, with TNCs as an alternative. Another common question was whether those who drove a private vehicle parked it at the airport. In contrast to the other surveys, it does not ask much more about TNCs or about access modes. It also provides the data year over year. The sample size is about 3,000 individuals for each year.

Results

One of the leading causes of congestion at airports is the high use of private modes, which tend to be much lower in vehicle occupancy. Since missing a flight can be very costly to air travelers, most of them prefer private modes because they are much faster and more reliable. Data from the 2015 LAX Passenger Survey and the 2015 SFO-OAK Passenger Survey, suggest that this is true at LAX, SFO, and OAK. Here private modes include passenger cars, taxis, limousines, private TNC rides, and private vans and shuttles. Shared modes comprise public transit such as buses and light rail, shared vans or shuttles, hotel courtesy shuttles, and rental cars (passengers use shared shuttles because rentals are returned or picked up off site).

At least half of the departing passengers at LAX, SFO, and OAK fly alone. Table 1 lists how they travel to the airport. 78%, 63%, and 70% used private modes at LAX, SFO, and OAK respectively. Bay Area airports had slightly fewer solo travelers who used a private mode compared with LAX. This may be because shared modes (including rail and other public transit services) in the Bay Area are more extensive and thus more convenient to use relative to that in Los Angeles.

Table 1. Of Passengers Flying Alone, Use of Private vs. Shared Modes, 2015

Mode	LAX	SFO	OAK
Private	78%	63%	70%
Shared	22%	37%	30%

Note: LAX = Los Angeles International Airport; SFO = San Francisco International Airport; OAK = Oakland International Airport.

Table 2 shows vehicle occupancy by private modes. In the LAX survey, passengers were asked how many people (including them) were in the vehicle; this might have resulted in respondents including their drivers as well. Assuming that drivers were also counted, nearly 71% of private vehicle users at LAX had no more than two passengers in the vehicle. At SFO and OAK, passengers were asked how many people in their travel party came in the same vehicle. If it is assumed that drivers (not in the travel party) were excluded, almost 90% of private vehicle users had no more than two passengers in the vehicle.

Currently TNCs can operate at various airports, including SFO and OAK. Tables 3 and 4 provide some insights into how the inception of TNC services at airports affects the use of private vs. shared modes. Most of the identified modal shift to TNCs from shared modes at SFO and OAK came from Bay Area Rapid Transit (BART), which is the regional rail service in the San Francisco Bay area. Fifteen percent of current TNC users at SFO and 21% at OAK typically used BART to go to the airport before TNCs were available (1.2% and 0.63% of all departing passengers at the respective airports). Twenty-one percent (SFO) and 29% (OAK) of air passengers who arrived at the airport by TNCs claim to have switched from shared modes. Meanwhile, the share of departing passengers who used TNC's is 8% (SFO) and 3% (OAK). This means that about 1.7% (SFO) and 0.9% (OAK) of all departing passengers shifted from vehicles with a typically higher occupancy to TNCs. However, among shifts from all modes, many passengers claimed to have switched from taxis.

Table 4 shows how passengers arriving to SFO or OAK by BART get to the BART station. At both airports, about 11% of all airport passengers take the BART and nearly 3% of those passengers used TNCs as a secondary mode. Thus, TNCs are complementing mass transit somewhat for a very small share of airport patrons (about 0.33%). These passengers may be just a little too far from the station and a private mode such as a TNC is perfect for their short ride to the station. The share using TNCs as a secondary mode is just a little short of those using taxis as a secondary mode, which is surprising because TNCs are usually much less expensive than taxis. Additionally, the most common secondary modes are "dropped off" and "walk" which suggests that most people who take the BART live near the stations.

Table 2. Vehicle Occupancy at LAX, SFO, and OAK Airport by Private Vehicles, 2015

No. of people in vehicle	LAX			SFO			OAK		
	Freq.	Percent	Cum.	Freq.	Percent	Cum.	Freq.	Percent	Cum.
1	734	12.63	12.63	7,362	67.27	67.27	4,309	67.25	67.25
2	2,032	34.96	47.59	2,413	22.05	89.32	1,431	22.33	89.59
3	1,345	23.14	70.73	562	5.14	94.45	335	5.23	94.82
4	696	11.98	82.71	371	3.39	97.84	182	2.84	97.66
5	420	7.23	89.93	100	0.91	98.76	75	1.17	98.83
6 or more	585	10.07	100	136	1.24	100	75	1.17	100

Note: LAX = Los Angeles International Airport; SFO = San Francisco International Airport; OAK = Oakland International Airport.

Table 3. Of Passengers Arriving by TNCs, Preferred Mode to SFO and OAK Prior to TNCs, 2015

Mode	Shared?	SFO			OAK		
		Freq.	Percent	Cum.	Freq.	Percent	Cum.
BART	Yes	215	14.91	14.91	51	20.56	20.56
Bus	Yes	15	1.04	15.95	1	0.4	20.96
Shared vans/ shuttles	Yes	46	3.19	19.14	8	3.23	24.19
Rental	Yes	26	1.8	20.94	13	5.24	29.43
Limousines	No	35	2.43	23.37	5	2.02	31.45
Private car	No	308	21.36	44.73	80	32.26	63.71
Taxi	No	797	55.27	100	90	36.29	100

Note: SFO = San Francisco International Airport; OAK = Oakland International Airport.

Table 6 summarizes all of the estimates of the number of passengers departing from and not connecting at these airport (these estimates include the share using TNCs, the share who substituted or complemented shared modes with TNCs, and the share of standard vs. pooled TNC services). There were nearly 25 million, 20 million, and 5 million departing passengers who started their trip at LAX, SFO, and OAK respectively. At LAX, approximately 2 million of the 25 million passengers used a TNC as the primary access mode; At SFO and OAK, it was about 1.6 million and 134,000 passengers respectively. Data on shifts and complementarities were only available for SFO and OAK. An estimated 340,000 and 40,000 passengers at the respective airports substituted shared modes with TNCs. These numbers are much greater than the number of passengers who complemented shared modes with TNCs: about 72,000 and 15,000 passengers from SFO and OAK, respectively. Again, most of the shift from shared modes to TNCs came from BART.

Approximately 240,000 and 28,000 passengers at SFO and OAK, respectively, used to take BART before shifting to TNCs, whereas those who complemented BART with TNCs numbered approximately 66,000 and 15,000 passengers respectively. Because TNCs replace shared rides (or just BART) more than they complement them (it), the net effect is that TNCs add to congestion as they increase the number of low-occupant vehicles at the airport. This is still true given that some passengers share TNC rides by using Uber Pool or Lyft Line. Assuming that passengers who use TNCs to get to transit stations and those who use pooled TNCs are mutually exclusive, TNCs helped approximately 125,000 and 15,000 passengers use shared rides at SFO and OAK respectively (by getting them to the station or offering a pooled TNC service). These numbers are still much smaller than the number of passengers who switched from shared rides to TNCs. The 215,000 and 25,000 more passengers who used private modes at SFO and OAK respectively, will result in many more additional private vehicles circling the airport curbside terminals.

According to the survey data from LAX, SFO, and OAK (*10–12*), the number of passengers originating at the airports increased by 6.3%, 6.3%, and 8.4% respectively in 2014–2015. Meanwhile, TNC shares at SFO rose 176% in 2014–2015 and 62% in 2015–2016 (*8*). It was assumed that the number of departing passengers would rise constantly each year at those rates, and that TNC shares would also grow, but at a slower pace each year (its growth will decelerate by about 65% each year, as in 2014–2016). It was further assumed that travel times, costs, reliability of the modes, and all other inputs of interest stay constant (ceteris paribus). Figure 1 shows the projected total number of departing passengers who started their trip at LAX, SFO, and OAK and, of those passengers, the number who are expected to use TNCs in 2015–2020. The projections suggest that over 5 million

Table 4. Of Passengers Arriving at SFO and OAK by BART, Chosen Mode to BART Station, 2015

SFO			OAK		
Mode	Freq.	Percent	Mode	Freq.	Percent
ACE	1	0.05	ACE	1	0.1
SamTrans	1	0.05	Biked	1	0.1
VTA	1	0.05	SamTrans	1	0.1
Emery go round	2	0.1	Golden gate transit	3	0.31
Caltrain	4	0.21	VTA	3	0.31
Golden gate transit	7	0.36	Emery go round	4	0.41
Amtrak	9	0.46	Country connection	5	0.51
AC transit	26	1.34	Caltrain	8	0.82
TNC	49	2.53	TNC	28	2.86
Taxi/cab	53	2.73	Taxi/cab	41	4.19
Drove and parked	65	3.35	AC transit	46	4.7
Muni	177	9.12	Drove and parked	48	4.9
Dropped off	437	22.53	Muni	57	5.82
Walked	1,108	57.11	Dropped off	235	24
			Walked	498	50.87

Note: SFO = San Francisco International Airport; OAK = Oakland International Airport.

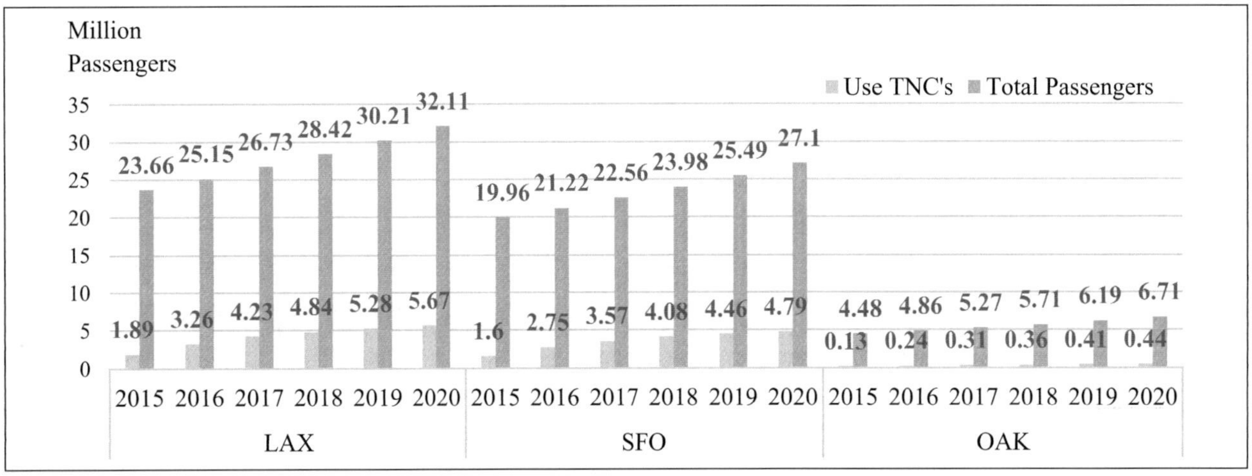

Figure 1. Projected volumes of departing (and non-connecting) passengers and those who will use TNCs.

passengers will take a TNC to go to LAX by 2020; at SFO, projections suggest it will be about 5 million passengers. Although similar growth trends (in the demand for TNCs) at the two airports are expected, LAX will have more TNC users because it serves more passengers than SFO. Of the three airports, OAK serves the fewest number of passengers, it also has the smallest TNC share. In 2020, about 440,000 passengers are expected to take a TNC ride to go to OAK.

According to the most recent data of actual TNC use at SFO (*13*), about 1.5 million, 2.6 million, and 1.7 million trips to the airport were by TNCs in 2015, 2016, and the first half of 2017 (January–June) respectively. Assuming that TNCs have an average of 1.28 passengers per vehicle (*7*), there were about 1.92 million, 3.33 million and 2.17 million passengers who traveled to SFO by TNCs in this period. The projections in Figure 1, which were estimated using data of TNC shares in 2015 and corrected for annual growth in passenger volumes and TNC shares, are a bit lower but still comparable to actual volumes.

Finally, this study also projects future shared modes' shifts to and complementarities with TNCs, demand for pooled TNC services, and the net effects of TNCs on

Table 5. TNC Services: Standard vs. Pooled, 2015

Description	SFO	OAK
Standard TNCs (UberX, Lyft, Sidecar and Wingz)	76.66%	100%
Pooled TNC (Uber Pool and Lyft Line)	3.34%	0%

Note: SFO = San Francisco International Airport; OAK = Oakland International Airport.

Table 6. Estimates of Volumes of Passengers Departing From but Not Connecting at LAX, SFO, and OAK, 2015

Description	LAX	SFO	OAK
Total	23,656,528	19,964,015	4,482,834
Used TNCs	1,892,522	1,597,121	134,485
Substituted shared modes with TNCs	NA	339,388	40,345
Complemented shared modes with TNCs	NA	71,740	15,069
Substituted BART with TNCs	NA	239,568	28,241
Complemented BART with TNCs	NA	65,881	14,793
Used pooled TNCs	NA	53,344	0

Note: NA = not available; LAX = Los Angeles International Airport; SFO = San Francisco International Airport; OAK = Oakland International Airport.

Table 7. Projected Estimates of Volumes of Departing (and Non-Connecting) Passengers

Airport	Year	Substituted shared modes with TNCs (a) (millions)	Complemented shared modes with TNCs (b)	Used carpool versions of TNCs (c)	Net effects (a) – [(b) + (c)]
SFO	2015	0.34	71,740	53,344	214,916
	2016	0.58	76,260	56,705	447,035
	2017	0.75	81,064	60,277	608,659
	2018	0.86	86,171	64,074	709,755
	2019	0.94	91,600	68,111	780,289
	2020	1.01	97,371	72,402	840,227
OAK	2015	0.04	15,069	0	24,931
	2016	0.07	16,335	0	53,665
	2017	0.09	17,708	0	72,292
	2018	0.11	19,195	0	90,805
	2019	0.12	20,807	0	99,193
	2020	0.13	22,555	0	107,445

Note: SFO = San Francisco International Airport; OAK = Oakland International Airport.

replacing shared rides and making shared rides possible. In these projections, which are listed in Table 7, it was assumed that all other shares are held constant (shares of passengers who shifted from shared modes to TNCs, shares of passengers who use shared modes, and the shares of passengers who use pooled TNC services). In 2015, TNCs increased the number of passengers who use private vehicles at SFO and OAK by about 215,000 and 25,000 respectively. In 2020, the increase in the number of passengers who use private vehicles at the two airports will be approximately 840,000 and 107,000 respectively.

Conclusion

This study explored the demand for TNCs, as well as TNCs' relationship with shared modes (shifts and complementarities), and the demand for their shared vs. standard service at the airport. Because TNCs both replace shared rides and make them possible, it also measured the net effects at these airports. The data are from the 2015 LAX Passenger Survey, the 2015 SFO-OAK Customer Survey, and the 2014, 2015, and 2016 SFO Customer Survey. Results suggest that TNCs replace shared rides more than they complement them, and the net effect is that TNCs add to congestion as they increase the number of low-occupancy vehicles at the airport. This is still true given that some passengers share TNC rides by using Uber Pool or Lyft Line. In 2015, TNCs caused 215,000 and 25,000 passengers to switch from shared to private modes at SFO and OAK respectively. By 2020, the shift is expected to be about 840,000 and 107,000 passengers per year, respectively.

Acknowledgments

The authors thank Professors Wilfred Recker, Michael McNally, Jean-Daniel Saphores, and Yaming Yu who provided many useful comments. They also thank Patrick Tomcheck of L.A. World Airports, Humphrey Loe and Angus Davol of SFO, and Sharon Sarmiento of Unison for assisting with the data. Karina Hermawan was partially supported by a UCCONNECT fellowship during this period of research. Amelia Regan was partially supported by the Reuben Smeed Professorial Fellowship at University College London.

Author Contributions

The authors confirm contribution to the paper as follows: study conception and design: KH, AR; data collection: KH; analysis and interpretation of results: KH, AR; draft manuscript preparation:

KH, AR. All authors reviewed the results and approved the final version of the manuscript.

References

1. Hermawan, K., and C. A. Regan. On-Demand, App-Based Ride Services at Los Angeles International Airport. Presented at 96th Annual Meeting of the Transportation Research Board, Washington, D.C., 2017.
2. Rayle, L., S. Shaheen, N. Chan, D. Dai, and R. Cervero. *App-Based, On-Demand Ride Services: Comparing Taxi and Ridesourcing Trips and User Characteristics in San Francisco*. University of California Transportation Center, 2014. https://www.heartland.org/_template-assets/documents/publications/uctc-fr-2014-08.pdf. Accessed January 15, 2017.
3. Clewlow, R., and G. Mishra. Shared Mobility: Current Adoption, Use, and Potential Impacts on Travel Behavior. Presented at 96th Annual Meeting of the Transportation Research Board, Washington, D.C., 2017.
4. Schaller, B. *Unsustainable? The Growth of App-Based Ride Services and Traffic, Travel and the Future of New York City*. Schaller Consulting, 2017. http://schallerconsult.com/rideservices/unsustainable.pdf. Accessed August 11, 2017.
5. Conner-Simons, A. *Study: Carpooling Apps Could Reduce Taxi Traffic 75%*. MIT Computer Science and Artificial Intelligence Laboratory, January 1, 2016. https://www.csail.mit.edu/ridesharing_reduces_traffic_300_percent. Accessed September 5, 2017.
6. Los Angeles International Airport 2015 Passenger Survey. *Los Angeles World Airports* (Requested Directly from the Airport).
7. SFO-OAK Survey. *flysfo Landside Department* (Requested Directly from the Airport).
8. SFO Customer Survey Data. *flysfo*, 2014, 2015, 2016. http://www.flysfo.com/media/customer-survey-data. Accessed August 11, 2017.
9. Alex. Announcing Uber Pool. *Uber Blog*. https://www.uber.com/blog/san-francisco/announcing-uberpool/. Accessed August 13, 2017.
10. Ten Year Summary of Passengers. *Los Angeles World Airports*. http://www.lawa.org/welcome_LAX.aspx?id=800. Accessed August 13, 2017.
11. Analysis of Scheduled Airline Traffic Comparative Traffic Report. *flysfo*, December 2015. http://media.flysfo.com/media/sfo/media/air-traffic/as201512.pdf. Accessed August 13, 2017.
12. Year-end Airport Statistics Summary. *Oakland International Airport*. http://www.oaklandairport.com/news/statistics/year-summary/. Accessed August 13, 2017.
13. Commercial Ground Transportation Historical Data. *flysfo* (Requested Directly from the Airport), 2014–2017.

The Standing Committee on Airport Terminals and Ground Access (AV050) peer-reviewed this paper (18-00777).

Any errors or omissions are those of the authors alone.

Article

TRR
JOURNAL OF THE TRANSPORTATION RESEARCH BOARD

Transportation Research Record
2018, Vol. 2672(23) 59–70

DOI: 10.1177/0361198118781386
journals.sagepub.com/home/trr
SAGE

A Non-Dominated Sorting Genetic Algorithm Approach for Optimization of Multi-Objective Airport Gate Assignment Problem

Seyedmirsajad Mokhtarimousavi[1], Danial Talebi[2], and Hamidreza Asgari[1]

Abstract

Gate assignment problems (GAP) are one of the most substantial issues in airport operation. The ever-increasing demand producing high occupancy rates of gates, the potential financial loss from imbalances between supply and demand in congested airports, and the limited scope for expanding facilities present challenges that require an advanced methodology for optimal supply allocation. In principle, tackling GAP involves seeking to maintain an airport's maximum capacity through the best possible allocation of resources (gates). There are a wide range of dependent and independent resources and limitations involved in the problem, adding to the complexity of GAP from both theoretical and practical perspectives. In this study, GAP is extended and mathematically formulated as a three-objective problem, taking into account all resources and restrictions, which can be directly linked to airport authorities' multiple criteria decision-making processes. The preliminary goal of multi-objective formulation is to consider a wider scope, in which a higher number of objectives are simultaneously optimized, and thus to increase the practical efficiency of the final solution. The problem is solved by applying the second version of Non-dominated Sorting Genetic Algorithm (NSGA-II) as a parallel evolutionary optimization algorithm. Results illustrate that the proposed mathematical model could address most of the major criteria in the decision-making process in airport management in terms of passenger walking distances, robustness, and traditional costs. Moreover, the proposed solution approach shows promise in finding acceptable and plausible solutions compared with other multi-objective algorithms (BAT, PSO, ACO, and ABC).

The flight-to-gate assignment problem has gained attention in recent years, particularly in the context of the need to address capacity shortcomings at the airports, resulting from the global increase in air travel. The growth in the size of fleets and the volume of commercial services together with lower fares offered by low-cost carriers have boosted global demand for air travel in recent years. According to statistical reports, air travel demand, in both domestic and international scheduled flights, grew from 2.628 billion passengers in 2010 to 3.464 billion passengers in 2015 (growth of 31%) and continued to 3.696 billion in 2016 (a growth of 0.06%) (*1*). As demand for air travel has continued this apparently inexorable growth over the last decade, problems associated with limited airport capacity have been revealed. In fact, the need for accurate operational and management functions has become even more acute as has the need to maximize operational efficiency maximization and achieve align resources more effectively to the traffic.

Gate scheduling and assignment is a key activity in airport operations and air transportation systems (ATS). A wide range of resource modules need to be taken into account. These are generally divided into three major categories: airlines, air traffic management (ATM), and airport resources (*2*). In the struggle to optimize the use of resources, each single problem, such as aircraft assignment, crew scheduling, runway sequencing, airspace capacity, and so forth, has a huge number of highly interdependent restrictions and limitations (for example, number of flights, available terminal space, baggage claims),

[1]Department of Civil and Environmental Engineering, Florida International University, Miami, FL
[2]Structural Engineering Research Center, International Institute of Earthquake Engineering and Seismology, Farmanieh, Tehran, Iran

Corresponding Author:
Address correspondence to Seyedmirsajad Mokhtarimousavi: smokh005@fiu.edu

where many conflicting objectives have to be considered. This interdependency between resources makes resource optimization a major challenge for airports and raises difficult issues in decision-making procedures which affect airport performance.

From a theoretical perspective, the Gate Assignment Problem (GAP) at airport terminals refers to the task of assigning a given set of flights to a set of available terminal gates (which are limited in number) in a way that both meets the operational requirements as well as passengers' convenience, and minimizes the operating costs of airports and airlines.

In this context, a "gate" is not only a facility through which passengers pass to board and alight from aircraft, but it is the location at which an aircraft is serviced and prepared for its next flight. Such station operations account for a small proportion of an airline's costs, but they have a major impact on the flight schedules and on the levels of passenger satisfaction achieved (*3*).

A set of strict rules and constraints should be adhered to when seeking to achieve a well-constructed schedule (*4*):

- one gate can only process one aircraft at any time
- service requirements
- space restrictions with respect to adjacent gates
- minimizing the ground time of the aircraft
- ensuring minimum time between aircraft

Regarding the NP-hard characteristic of the GAP problem (*5*), seeking to increase services amid growing constraints will only make the problem more intractable. Issues such as the number of flights, the availability of gates, times spent at gates, hold-ups with taxiing, and so forth, all have the potential to exacerbate the problem (*6, 7*).

The intention of this paper is to address a greater number of objectives in tackling the GAP problem by presenting a three-objective mathematical model. A modified well-known meta-heuristic evolutionary algorithm is also proposed to provide a simultaneous solution technique for finding a series of feasible solutions to the problem which still allows operational requirements to be met.

Literature Review

Airport supply allocation has been a hot topic in aviation research for a long time (*8–13*), and the techniques adopted to address the issue have evolved from simple mathematical formulations to sophisticated hybrid algorithms. The ultimate aim of this ongoing improvement is to prepare a sophisticated yet simple-to-use methodology that can be applied through commercial optimization software in airport applications.

Technically speaking, the airport gate assignment problem is modeled with either unit-weighted or evolutionary methods, reflecting the competing objectives that need to addressed. The objectives are usually described as minimizing the total passenger walking distance (or, in the same way, connection times), the number of ungated flights, and towing procedures. In this regard, current GAP literature can be viewed and classified either based on the objectives and constraints being taken into account, or based on the varieties of techniques proposed as solutions.

Passenger satisfaction, usually defined as the minimization of the total walking distance, is probably the most popular objective considered in the literature (*14–27*). By nature, the total passenger walking distance is based on the passenger transfer volume between every pair of aircrafts and the distance between every pair of gates. Therefore, the problem of assigning gates to arriving and departing flights at an airport is a Quadratic Assignment Problem (QAP) (*7*). Minimizing walking distances directly affects airline/airport customer satisfaction and also decreases the odds of passengers missing flights. Different methods such as branch-and-bound algorithms, integer programming, linear programming, expert systems, heuristic methods, tabu search algorithms, and various hybrid methods have been employed (*3*).

The idea of ungated flights, considered as a way of minimizing the number of flights assigned to remote terminal gates (apron stands), or equivalently, maximizing the number of flights assigned to fixed gates is recognized as the second most prevalent objective in the literature (*14, 16, 17, 23, 27*). This issue is discussed predominantly in the context of avoiding mismatches between flight sizes and gate types, as well as satisfying safety requirements in (*27*).

Besides the two main objectives above, some effort has focused on taking cost into account. In fact, minimizing cost is modeled as an objective (*22, 28, 29*). Yu and Lau (2014) and Yu et al. (2016) (*22, 29*), look as minimizing cost in terms of the personnel cost resulting from towing aircraft, conflict cost from the robustness of the schedule, and the cost of passenger transfers. Delays from ramp conflicts are studied in Kim et al. (*30*). Cost was a factor in penalizing the assignment of two flights to the same gate, as examined by Neuman and Atkin (2013) (*28*).

A wide variety of constraints have been considered in the literature. Fundamentally, the GAP has two basic constraints: first, one gate can only accommodate one single aircraft at a time (restricted gate), and second, two flights cannot be assigned to the same gate if they overlap in time (single gate per activity) (*4*). Other issues are:

- airlines' gate preferences
- adjacent gates cannot handle certain types of aircraft simultaneously (shadow restrictions)
- pushback restrictions (integrating gate allocation with the ground movement to limit the number of aircraft that are expected to block each other while maneuvering in the area close to the gates)
- calculating the idle time of each gate before each flight
- guaranteeing that each arrival and departure has one and only one gate assignment

Recent research on the topic has focused on multi-objective solution techniques:

A hybrid heuristics algorithm guided by Simulated Annealing (SA) heuristics accompanying greedy heuristics and tabu search heuristics was applied by Drexl and Nikulin (2008) (*19*). Their proposed integer program worked with multiple objectives including quadratic and quadratic constraints and quadratic assignment formulation solved by Pareto simulated annealing to obtain a representative approximation for the Pareto front. Although computational experiments have produced good results, relatively limited data was used to test the proposed algorithm.

Characteristics of stochastic and heuristic approaches were combined in a hybrid model to generate an assignment order for the all planes corresponding to assignment priority in Genç et al. (*31*). A proposed Single Leap Big Bang Big Crunch (SL-BBBC) evolutionary model worked effectively with both test and real world data waiving the need for detailed constraints or cost calculation.

Yu and Lau (*22*) proposed a hybrid meta-heuristic algorithm which applied the genetic algorithm (GA) capability of exploration as a population-based artificial intelligence method, and the exploitation capability of large neighborhood search (LNS) methods to each individual in the population. Comparison of results between CPLEX and GALNS revealed that the proposed method can deal better with realistically large problems in terms of their size and computation time. Results from Kaliszewski et al. (*32*) also demonstrated inappropriate application of CPLEX in large problems compared with an Evolutionary Multi-objective Optimization (EMO) algorithm.

A multi-island parallel evolution framework (PEA) with a left-right probability migration topology was applied by Guan el al. (*33*) to solve a multi-objective large-scale flight assignment with consideration of the reduction in airspace congestion and flight delay using real traffic data from the China air route network and daily flight plans. Comparing the results from the proposed approach to some well-known multi-objective evolutionary algorithms based on decomposition demonstrated the better solution quality of the PEA.

A probabilistic tabu search was proposed by Aktel et al. (*34*) and the results were compared with SA algorithm performance considering different problem sizes. This method is also applied in (*30*) to minimize taxi time/delay as well as passenger transit time. Based on different airport terminal layouts, the results are then compared with a genetic algorithm and a quadratic integer program (QIP). Using greedy algorithm as a benchmark showed different performances based on problem size. In Deng et al. (*35*), an improved adaptive particle swarm optimization (DOADAPO) algorithm was used to solve multi-objective GAP. Verifying the results by test data and comparing with twelve basic algorithms showed an improved performance by DOADAPO in slow convergence, search precision, and local optimality. However, the proposed method had time complexity in solving the problem.

In a comprehensive methodological study published by Zhang and Klabjan (*36*), diving and variable rolling horizon algorithms are used to minimize total flight delays, the number of gate reassignment operations, and missed passenger connections, all examples of frequent airport operation disruptions.

Although multi-objective evolutionary methods have shown superior capability as a solution technique for GAP in recent efforts, they have not received sufficient attention in the literature.

Problem Definition and Model Structure

The primary purpose of flight-gate assignment is to assign aircrafts to appropriate gates (terminal gates) in a way that meets both operational and safety requirements while minimizing inconvenience to passengers. As discussed in previous sections, the gate assignment problem is multi-objective in nature. A multi-objective mathematical modeling approach, with an appropriate solving method, usually results in a trade-off between several conflicting objectives. Solving the multi-objective problem will consequently result in a compromise between several goals that can positively influence passenger satisfaction and save extra money for airport authority/operator and airlines.

The proposed GPA framework in this paper focuses on two of the most popular objectives: minimizing passenger walking distances and minimizing taxi conflicts. Existing overlap times between consecutive flights assigned to the same gate represents the conflict duration and is used to measure the robustness of a gate assignment plan. High robustness is achieved when the expected conflict duration is low among aircraft (*37*). Considering that aircraft movements will cause additional costs and inconvenience to the landside facilities, staff, and other aircraft, the tow frequency also needs to be minimized. In this research, the reducing cost imposed by towing aircraft and assigning flights to remote (or unfavorable) gates is also considered as the third objective. The proposed model is expected to address all three objectives simultaneously.

Considered objective functions are as follows:

- The first term in the model (*38*) includes the objective of minimizing the total passenger walking distance, which could be divided into the distance a passenger walks 1) to departure gates, 2) to baggage reclaim areas and 3) between connecting flights (for those flights with connections in the flight plan).

$$min \sum_{i=1}^{N}\sum_{k=1}^{M} F_{o,i}.W_{o,k}.Y_{i,k} + \sum_{i=1}^{N}\sum_{k=1}^{M}\sum_{bc=1}^{B} F_{i,bc}.WBC_{i,bc}.Y_{i,k} + \sum_{i=1}^{N}\sum_{j=1}^{N}\sum_{K=1}^{M}\sum_{l=1}^{M} F_{i,j}.W_{k,l}.Y_{i,k}.Y_{j,l}$$

The following notations are used in the first model:

N: number of flights arriving at and/or departing from the airport

M: number of gates, including remote gates

B: number of baggage claims

$F_{o,i}$: number of passengers from check-in area to flight i

$F_{i,bc}$: number of passengers from flight i to the baggage claim area

$F_{i,j}$: number of passenger transferring from flight i to flight j

$W_{o,k}$: walking distance between check-in area and gate k

$WBC_{i,bc}$: walking distance between gate i and baggage reclaim area

$W_{k,l}$: walking distance between gate k and gate l

$Y_{i,k}$: a binary value representing the association of flight i to gate k

$Y_{j,l}$: a binary value representing the association of flight j to gate l

- The second term is intended to address the objective of minimizing taxiway conflicts (push-back blocking, taxi blocking) and measures the assignment robustness as it prevents rescheduling of the original assignment.

$$min \sum_{i=1}^{N}\sum_{j=1}^{N} \frac{y_{i,j}}{a_i + d_j + 2b}$$

The following notations are used in the second model:

$y_{i,j} = 1$ if $Y_{i,k} = Y_{j,k}$, 0 otherwise

a_i: arrival time of flight i

d_j: departure time of flight j, $di > a_i \forall_i$

b: buffer time (to enlarge the interval between any two adjacent aircraft assigned to the same gate, which is set to 30 min)

- The third term includes the objective of minimizing costs from towing and assignments to unfavorable gates.

$$min \sum_{i=1}^{N}\sum_{k=1}^{M} C_{t(i)}.C_{ufg(i)}.Y_{i,k}$$

The following notations are used in the third model:

$C_{t(i)}$: The actual cost of towing an aircraft

$C_{ufg(i)}$: The actual cost of being assigned an unfavorable gate

$Y_{i,k}$: a binary value representing the association of gate k to flight i

This optimization problem is subject to the following constraints:

1) Restricted gates: assures that every flight must be assigned to exactly one gate including the apron (remote gate).

$$\sum_{k=1}^{M} Y_{i,k} = 1, 1 \le i \le n$$

n: total number of flights, i.e. $n = |N|$

2) Activity overlap restriction: prohibits schedule overlapping of two flights assigned to the same gate.

$$Y_{i,k}.Y_{j,k}(d_j - a_i)(d_i - a_j) \le 0$$
$$1 \le i,j \le n, k \ne RG$$

Remote Gate (RG) represents the apron gate where flights arrive when no terminal gates are available.

3) Dummy variables: each gate can only be occupied with one aircraft.

$$Y_{i,k} \in \{0,1\}, 1 \le i \le N, 1 \le k \le M$$
$$Y_{j,l} \in \{0,1\}, 1 \le j \le N, 1 \le l \le M$$

4) Mismatch restriction: avoid mismatch between flights and gates.

$$(AT_i - GT_k)Y_{i,k} \le 0, i \in N, k \in M$$

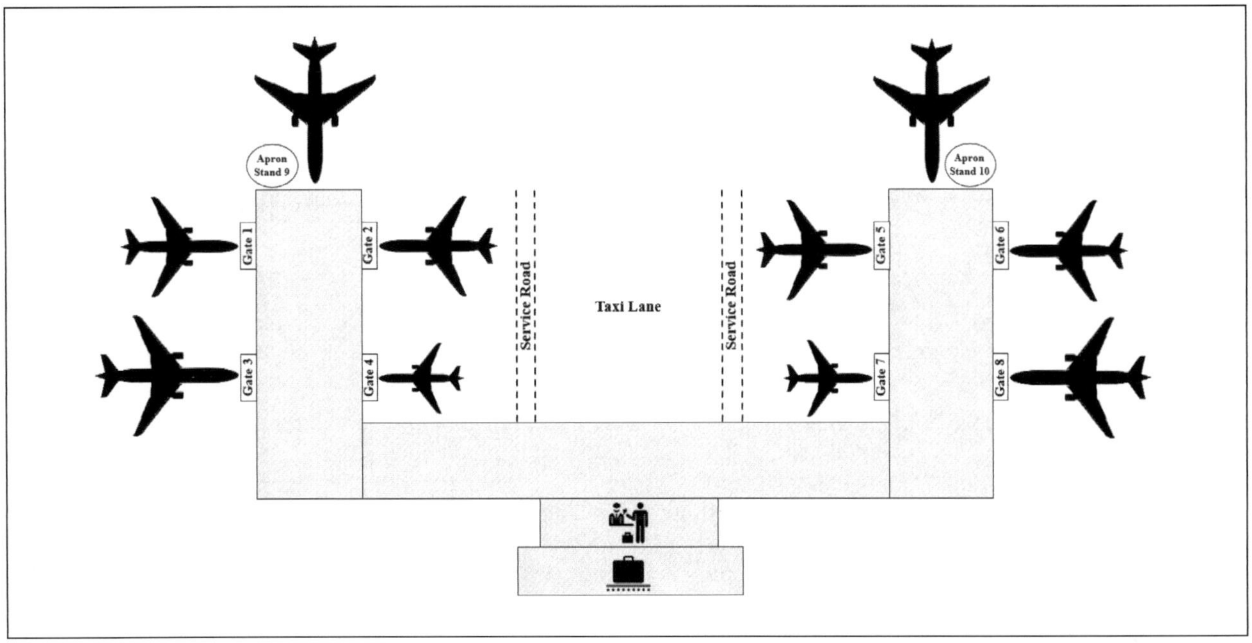

Figure 1. Pier terminal configuration.

AT_i: parameter indicating the aircraft's type of flight i. $AT_i = 1$ if flight i is a large aircraft, else $AT_i = 0$.

GT_k: parameter indicating the type of gate k. $GT_k = 1$ if gate k is a large one, else $GT_k = 0$.

Problem Implementation

It is obvious that different terminal layouts have different impacts on the cost-efficiency of daily airport operations. In our study, one terminal with a pier concourse configuration is considered, the terminal is assumed to have eight gates with two remote/apron stands, as illustrated in Figure 1.

It is difficult to gain access to real world data for GAP. Basically, the data should contain a wide variety of information such as; type of aircraft, type of flight (origin/destination), gate preference (provided by airline/airport), number of passengers who check in at the airport and those whose final destination is the airport, the move from the passenger terminal to a gate or vice versa. Moreover, passengers who have connections at the airport go from one gate to another. Airlines regard these data as private and so do not publish them. As a result, fewer studies use real world data and so do not contain all the passenger information mentioned above. Certain assumptions are made and missing data generated by statistical analysis of the information provided or randomly simulated based on the understanding of the problem (*28*, *39*, 40).

In this study, we investigated the problem using three simulated datasets including sets of 20, 40, and 100 flights assigning to a set of 10 available gates to numerically test the proposed model and solution approach for a defined terminal layout. Inputs include: flight arrival times, corresponding transfer flights, aircraft/gate sizes (small, medium, large) and corresponding capacities, costs associated with mishandling assignments, towing costs, gate-to-gate distances, and gate-to-check-in/baggage reclaim distances.

Arrival flights with corresponding arrive time (minute), aircraft size and transfer flights for the set of 20 aircrafts is illustrated below:

1	2	3	4	5	6	7	8	9	10	11	12	13	14	15	16	17	18	19	20
s	m	l	m	m	l	m	l	s	l	s	m	l	m	m	l	m	l	s	l
0	10	15	20	30	35	40	45	50	60	125	135	140	145	155	160	165	170	175	185
*	*	*	*	*	1	2	3	4	*	*	*	*	*	*	6	7	8	9	*

Considered gate to gate walking distance (Wk,l) is as follows:

	1	2	3	4	5	6	7	8	9	10
1	0	6	8	10	36	45	31	40	5	43
2	6	0	10	8	28	36	23	31	5	43
3	8	10	0	6	31	40	26	35	13	38
4	10	8	6	0	23	31	18	26	13	38
5	36	28	31	23	0	6	8	10	40	5
6	45	36	40	31	6	0	10	8	48	5
7	31	23	26	18	8	10	0	6	35	13
8	40	31	35	26	10	8	6	0	43	13
9	5	5	13	13	40	48	35	43	0	51
10	43	43	38	38	5	5	13	13	51	0

In our simulated datasets, three types of aircraft, those with a capacity of 100, 200, and 300 passengers are considered and listed as small, medium, and large, respectively. In this example Fo,i and Fi,bc are assumed to be equal. In terms of determining transfer passengers, we considered 10% of the capacity of the small aircraft, 15% of the capacity of the medium aircraft, and 20% of the capacity of the large aircraft, i.e. 10, 30, 60 passengers (Fi,j). Aircraft numbers 6 to 9 have connections with aircraft 1 to 4, and aircraft 16 to 19 have connections with aircraft 6 to 9 in this example. This trend is followed to simulate the other considered sequence of aircraft. All departure times (di) are considered to be 1 hour after the arrival time.

Solving Approach

NSGA-II Algorithm

The comprehensive literature review on the GAP revealed that several methods, algorithms and heuristics, and hybrid methods could be used to solve the gate assignment problem. In particular, a Genetic algorithm (GA) has been widely used to solve ATM, airline and airport operation problems such as GAP (*3*, *24*, *41*), and Aircraft Landing Problems (ALP) (*42–44*), fleet/crew scheduling (*45*, *46*), and so forth.

Multi-objective evolutionary algorithms (MOEAs) have attracted the attention of many researchers in the last 20 years, and it is still a challenging topic in the field of optimization research. Following the broad and successful applications of GAs in solving optimization problems in numerous domains, a non-dominated sorting genetic algorithm (NSGA-II) is employed in this study to produce a Pareto optimal set, in which the three aforementioned objectives can be considered simultaneously, resulting in higher levels of flexibility for decision makers.

The Non-dominated Sorting Genetic Algorithm (NSGA) was one of the first such evolutionary algorithms introduced for multi-objective optimization problems (*47*). However, over the years the following shortcomings in its application have been revealed as mentioned in Deb et al. (2002) (*48*):

- High computational complexity of non-dominated sorting

The currently used non-dominated sorting algorithm has a computational complexity of $O\ (MN^2)$, where "M" is the number of objective functions and "N" represents the population size, while, NSGA has a computational complexity of $O\ (MN^3)$. This complexity comes from the non-dominated sorting procedure in each and every generation.

- Lack of elitism

The elitism can speed up the performance of the GA significantly (*47*). The elitism feature of NSGA-II is keeping the best individuals from the parent and child population, so the best individual in terms of objective function values have never been lost once they are found.

- Need to specify the sharing parameter

The crowding distance mechanism in NSGA-II has satisfied diversity in the population. As a result, it no longer requires the specification of a sharing parameter.

The operational function of NSGA-II could be elaborated as follows. At the beginning, the population is initialized as usual GA. With respect to the values of objective functions, the initialized population is consequently sorted based on non-domination into each front. The first front is totally non-dominant set in the current population while the second front is being dominated by

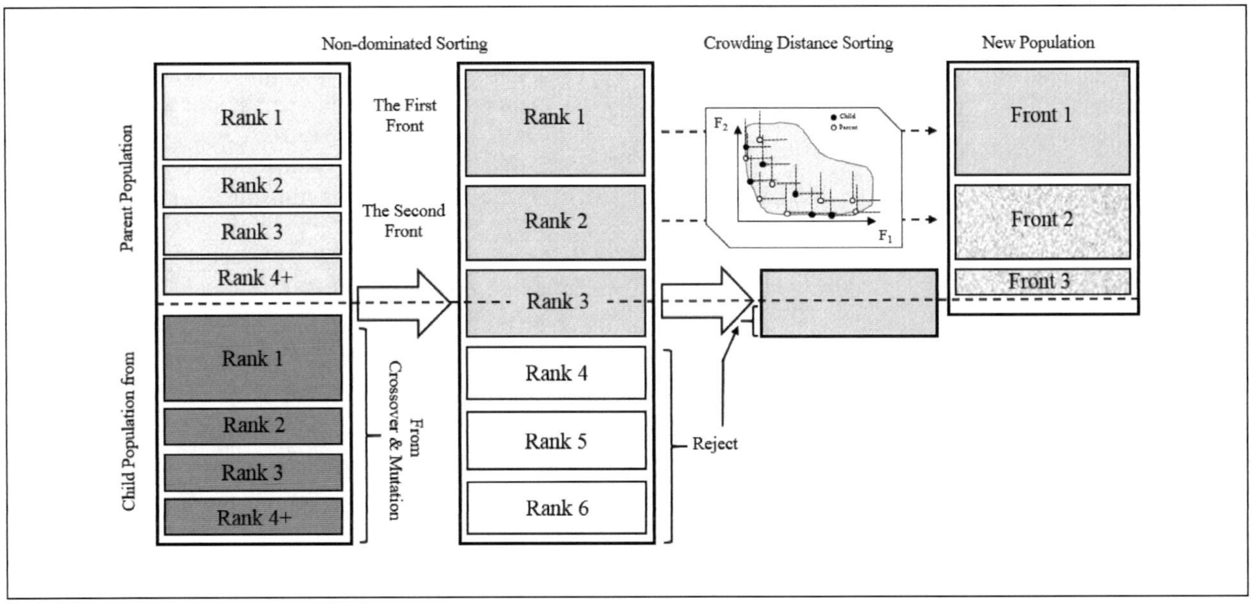

Figure 2. Elitism process conducted in each NSGA-II iteration.

the individuals in the first front only and other fronts are determined in the same way. The crowding distance indicates how close an individual is to its neighbors. Large average crowding distances suggest better diversity in the population. An individual selected in the rank is lesser than the others or, if the crowding distance is greater than the others, this is the main difference between the conventional GA and NSGA-II which is actually the elitism feature previously discussed. Parents are selected from the population by using a roulette-wheel selection based on the rank and crowding distance. The selected population generates offspring from crossover and mutation operators. The population then consists of the current population and the current offspring is sorted again based on non-domination and only the best individuals are selected based on the rank and crowding distances. Once the number of the calculated generation P is larger than the critical number $Pset$, the search will be terminated. The final purpose of this procedure is to get the Pareto-optimal front, that is, the curve with Rank = 1 (*49*).

In general, in NSGA-II, total tasks are conducted in the sections of population production, crossover and mutation as merging and then non-dominated sorting is performed based on the rank and crowding distance; and finally additional parts are truncated. The steps and elitism process conducted in each iteration of the NSGA-II algorithm is shown in Figure 2.

The implementation of the proposed NSGA-II approach for the multi-objective optimization of GAP problem is discussed below.

Implementing NSGA-II Operators and Related Parameters for GAP

Here, some modifications have been done on the algorithm operators explained as follows.

- Selection method

Pairs of parent chromosomes are selected from the current population according to the selection probability and the roulette-wheel (which is also known as “fitness proportionate selection”) approach is then applied to generate the offspring.

- Crossover

Arithmetic crossover is applied in this study. Its operator linearly combines two parent vectors $x^1 = (x_1^1, x_2^1, \ldots, x_n^1)$ and $x^2 = (x_1^2, x_2^2, \ldots, x_n^2)$ selected by applying the roulette-wheel selection method to produce two new offspring $y^1 = (y_1^1, y_2^1, \ldots, y_n^1)$ and $y^2 = (y_1^2, y_2^2, \ldots, y_n^2)$. For our integer decision variables, it is coded as:

$$y_i^1 = (\alpha^* x_i^1) + ((1-\alpha)^* x_i^2)$$
$$y_i^2 = (\alpha^* x_i^2 + ((1-\alpha)^* x_i^1)$$

Where α is a random weighting factor chosen before each crossover operation generates as:

$$\alpha = rand([x^1])$$

- Mutation

A continuous mutation method was applied in which a member is selected using the roulette-wheel selection method and then a new offspring generated as follows:

$$y = 0.1unifrnd\left(-1, 1, \left(\left[x^{1}\right]\right)\right)*(upperbound - lowerbound)$$

- Algorithm parameters

Before generating the initial population, we initialized the population size together with other algorithm parameters. In this study, we came up with the maximum number of algorithm iterations for a series of 20, 40, and 100 flights with 10 available gates as 40, 60, and 120 respectively. The number of initial populations are set to 20, 40, and 100. The percentage of the population taking part in crossover is equal to 0.6 and the index of mutation is equal to 0.4.

The parallel computing of the objective function with three different objectives simultaneously will be explained here: the flights are first sorted according to their arrival times and the gates are sorted according to their corresponding number and size. Then, for each new flight, we start by testing if the first gate can receive the considered flight in comparison with each pair of flight and gate, so if a flight cannot be assigned to gate "1" for instance, the next gate being checked would be "2". The process is repeated for all the flights and gates. Meanwhile, the walking distance between flights and baggage reclaim, flights and check-in area, and between gates for those flights which have connections, are taken into account based on aircraft size which represents the number of passengers and the percentage that are transfer passengers. The flights are also evaluated and sorted in the same way as in the previous approach, based on corresponding towing and mishandling costs. Then, we use the previous heuristics with a random sorting of gates. Thus, new possible solutions can be reached at each new sorting of gates.

Results

Using MATLAB R2016a software, we checked the performance of the proposed method with a simulated dataset including sequences of 20, 40, and 100 aircraft arriving in a three time intervals; 0 to 185, 0–385, and 0–1,140 minutes respectively to be assigned to one of 10 gates on a specific terminal layout. In our datasets, 40% of each sequence of 10 aircrafts is considered to have connections with other flights, in the way that flights 6, 7, 8, and 9 have connections with flights 1, 2, 3, and 4 and it is repeated for subsequent sequences of every 10 flights. However, our model is not size-dependent.

Figure 3 is the Gantt chart which illustrates the gate assignments obtained for different sets of aircraft.

As can be seen in Figure 3, the first arrival flight in the sequence of 20 aircraft is assigned to gate 4. Checking the aircraft type and gate type illustrates that both are the same size, small. This can be generalized to all flight sizes as none of the constraints are violated in our model. Considering connection flights associated with aircraft size, aircraft 6 has connection with aircraft 1, meanwhile, aircraft 16 has connection with 6. Among the available size-matched options which are all large gates including 3, 8, 9, and 10, flight 6 is assigned to gate 3. The walking distances between gate 4 and gates 3, 8, 9, and 10 are 6, 26, 13, and 38, so flight 6 is assigned to nearest possible gate. The second connection, aircraft 16, has the same size-matched options and it is assigned to gate 9. Following the same pattern for the second connection, the distances between gate 3 and the available options are 35, 13, and 38 for gates 8, 9, and 10 respectively, so gate 9 has the shortest distance with gate 3. All other gate assignments can be investigated in a similar way for different aircraft sequences being considered.

To investigate the ability of the proposed method in finding the non-dominated Pareto fronts, some other well-known multi-objective optimization algorithms, including Multi-objective Bat (MOBAT), Particle Swarm Optimization (MOPSO), Ant Colony Optimization (MOACO), and Artificial Bee Clooney (MOABC) are considered for comparison purposes.

Considering the amount of time spent, it is important to keep the non-dominated set as diverse as possible. To evaluate the distribution and spread of the non-dominated fronts generated by NSGA-II and four other meta-heuristic algorithms, three major metrics are considered as follows (*50*):

- Elapsed/computational time
- Diversity
- Spacing

The amount of time required to come up with the solution is measured by elapsed time. Diversity measures how well the true Pareto front is covered by the approximation set by summing up Euclidean distances of each solution from other solutions. Spacing measures how evenly the solutions obtained distribute themselves in the search space. This metric is obtained from distance standard deviation of each solution to the nearest solution in which the lower amount indicates that the obtained solutions are better organized. Moreover, the higher number of obtained Pareto frontiers indicates a better search and more feasible search space.

The comparison of the results is summarized in Table 1.

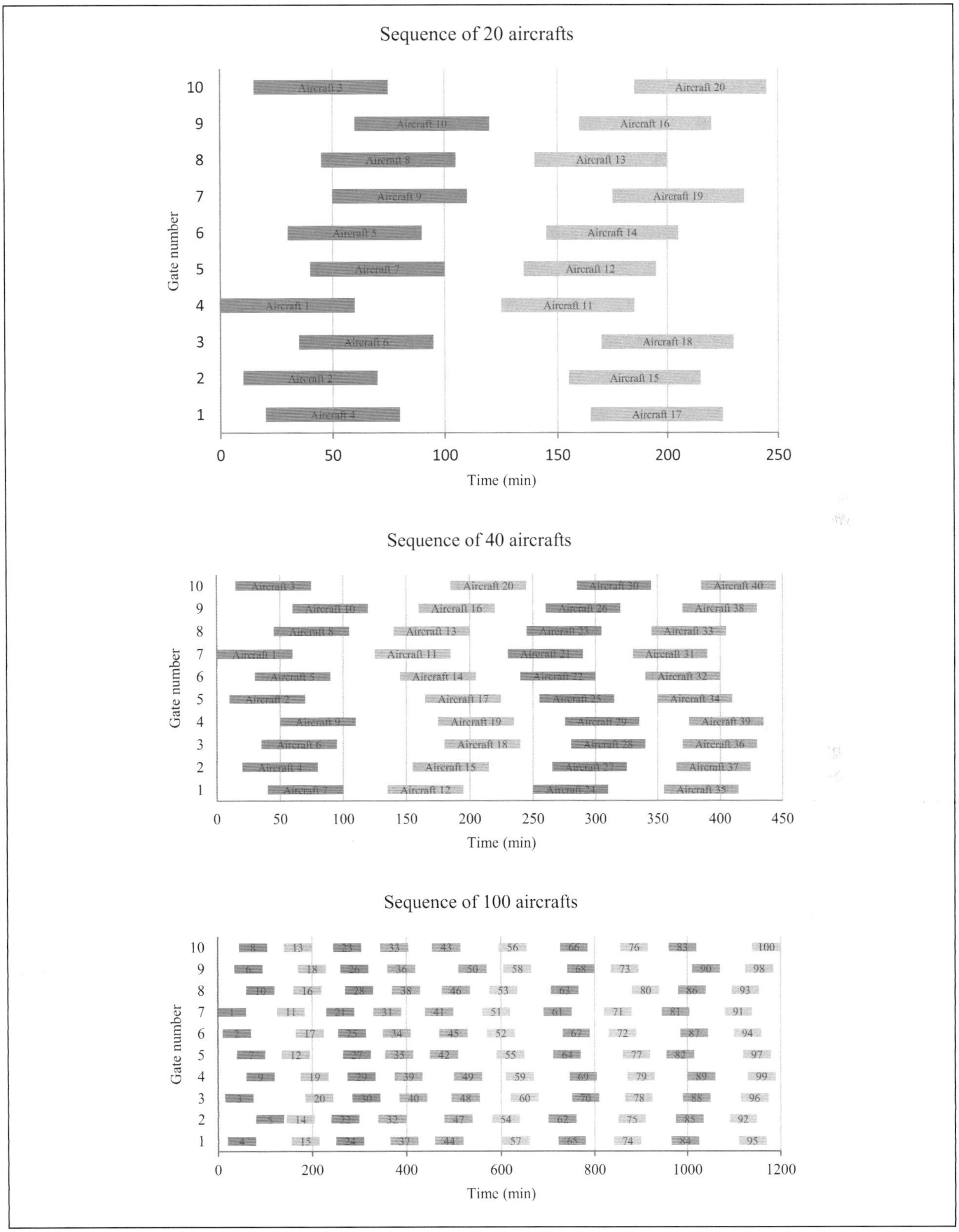

Figure 3. Gate assignment model result for different sequences of aircraft.

Table 1. Heuristics Computational Results

Algorithms	Number of flights	Number of iterations	Initial populations	Number of obtained solutions	FI members	Diversity	Spacing	Elapsed time (seconds)
Non-dominated sorting genetic algorithm (NSGA-II)	20	40	20	4	20	94868	1.33	2.61
Multi-objective BAT (MOBAT)				4	20	54772	1.33	2.86
Multi-objective particle swarm optimization (MOPSO)				3	20	54772	1	2.65
Multi-objective ant colony optimization (MOACO)				6	6	94868	1.6	517.17
Multi-objective artificial bee colony (MOABC)				2	2	54772	0	2.48
Non-dominated sorting genetic algorithm (NSGA-II)	40	60	40	7	40	144910	1.33	17.18
Multi-objective BAT (MOBAT)				7	40	54773	1.66	17.72
Multi-objective particle swarm optimization (MOPSO)				4	40	77460	1.33	17.69
Multi-objective ant colony optimization (MOACO)				11	11	284600	1.6	149.14
Multi-objective artificial bee colony (MOABC)				6	6	122470	1.2	16.07
Non-dominated sorting genetic algorithm (NSGA-II)	100	120	100	24	100	561250	1.65	356.23
Multi-objective BAT (MOBAT)				13	100	444970	1.04	349.7
Multi-objective particle swarm optimization (MOPSO)				19	21	268330	1.77	340.04
Multi-objective ant colony optimization (MOACO)				8	8	279280	0.95	908.29
Multi-objective artificial bee colony (MOABC)				6	6	225830	0.93	331.88

Results from Table 1 illustrate that NSGA-II gives acceptable computational time with good diversity and spacing to explore better solutions. All other approaches are lacking in certain areas. For instance, MOACO has the highest amount of computational time, while MOPSO gives low diversity and high spacing in large size problems. MOABC has a lower number of obtained results in all problem sizes. However, MOBAT has the potential to compete with NSGA-II in large sample size scenarios.

Conclusions

This paper has presented a modified NSGA-II to solve the flight-gate assignment problem in which the consideration is given to minimizing: total walking distance (representing passenger satisfaction); taxi conflicts (which demonstrate assignment robustness to keep the airport schedule well maintained); and airport/airline traditional costs. To the best of authors' knowledge and according to a well-elaborated literature review, most of the studies reviewed have overlooked one or some of the major concerns, applicable variables, and objectives in one model structure and therefore do not capture the true multi-criteria nature of the gate assignment problem. In addition, considering the NP-hard nature of the GAP problem, the authors believe that a meta-heuristic approach is a more appropriate candidate as it can solve the problem efficiently. The results of a recent study illustrate the computational inefficiency of a set of exact algorithms for practical implementation (*29*).

The proposed model was further tested under different simulated datasets that were created based on a sample airport layout. However, our model can simply be generalized to other terminal configurations. The model run results demonstrated good capabilities of NSGA-II in solving the model with no constraint violation. It means that the algorithm is capable of finding a series of acceptable feasible solutions in a practical amount of time. In the case of multi-objective optimization, providing decision makers with a list of feasible solutions is a substantial step in selecting the final optimal solution from the Pareto frontier, usually based on policy-related considerations.

Further research may include a real-life dataset for a specific hub airport layout to test the model efficiency. Other potential solution techniques such as hybrid meta-heuristics can also be investigated to get the higher accuracy solutions and validate the efficiency and effectiveness of the proposed algorithm. Moreover, common types of decision analysis methods can be applied to find an appropriate solution from the Pareto frontiers obtained, which supports its practical usability for future decision-making tools.

References

1. http://data.worldbank.org/indicator/IS.AIR.PSGR. Accessed July 5, 2017.
2. Lindh, A., T. Andersson, P. Värbrand, and D. Yuan. Intelligent Air Transportation—A Resource Management Perspective. *Proc., 14th World Congress on Intelligent Transport Systems*, Beijing, China, 2007.
3. Hu, X.-B., and E. Di Paolo. An Efficient Genetic Algorithm with Uniform Crossover for the Multi-objective Airport Gate Assignment Problem. *IEEE Congress on Evolutionary Computation, 2007, CEC 2007*, IEEE, 2007, pp. 55–62.
4. Dorndorf, U., A. Drexl, Y. Nikulin, and E. Pesch. Flight Gate Scheduling: State-of-the-art and Recent Developments. *Omega*, Vol. 35, No. 3, 2007, pp. 326–334.
5. Guépet, J., R. Acuna-Agost, O. Briant, and J.-P. Gayon. Exact and Heuristic Approaches to the Airport Stand Allocation Problem. *European Journal of Operational Research*, Vol. 246, No. 2, 2015, pp. 597–608.
6. Yamada, T., and R. Nakano. Job Shop Scheduling. *IEEE Control Engineering Series*, 1997, pp. 134–134.
7. de Man, S. *Flight to Gate Assignment: Solution Methods and Complexities at Planning Horizons*. Airport Capacity Planning, 2009.
8. Braaksma, J. Reducing Walking Distances at Existing Airports. *Airport Forum. No.* 7, 1977.
9. Babić, O., D. Teodorović, and V. Tošić. Aircraft Stand Assignment to Minimize Walking. *Journal of Transportation Engineering*, Vol. 110, No. 1, 1984, pp. 55–66.
10. Mangoubi, R., and D. F. Mathaisel. Optimizing Gate Assignments at Airport Terminals. *Transportation Science*, Vol. 19, No. 2, 1985, pp. 173–188.
11. Gosling, G. D. Design of an Expert System for Aircraft Gate Assignment. *Transportation Research Part A: General*, Vol. 24, No. 1, 1990, pp. 59–69.
12. Srihari, K., and R. Muthukrishnan. An Expert System Methodology for Aircraft-gate Assignment. *Computers & Industrial Engineering*, Vol. 21, No. 1–4, 1991, pp. 101–105.
13. Cheng, Y. A Knowledge-based Airport Gate Assignment System Integrated with Mathematical Programming. *Computers & Industrial Engineering*, Vol. 32, No. 4, 1997, pp. 837–852.
14. Bihr, R. A. A Conceptual Solution to the Aircraft Gate Assignment Problem Using 0, 1 Linear Programming. *Computers & Industrial Engineering*, Vol. 19, No. 1–4, 1990, pp. 280–284.
15. Bandara, S., and S. Wirasinghe. Walking distance minimization for airport terminal configurations. *Transportation Research Part A: Policy and Practice*. Vol. 26, No. 1, 1992, pp. 59–74.
16. Xu, J., and G. Bailey. The Airport Gate Assignment Problem: Mathematical Model and a Tabu Search Algorithm. *Proc., 34th Annual Hawaii International Conference on System Sciences, 2001, IEEE*, 2001. p. 10.
17. Ding, H., A. Lim, B. Rodrigues, and Y. Zhu. New Heuristics for Over-constrained Flight to Gate Assignments. *Journal of the Operational Research Society*, Vol. 55, No. 7, 2004, pp. 760–768.
18. Ding, H., A. Lim, B. Rodrigues, and Y. Zhu. The Over-constrained Airport Gate Assignment Problem. *Computers & Operations Research*, Vol. 32, No. 7, 2005, pp. 1867–1880.
19. Drexl, A., and Y. Nikulin. Multicriteria Airport Gate Assignment and Pareto Simulated Annealing. *IIE Transactions*, Vol. 40, No. 4, 2008, pp. 385–397.
20. Dorndorf, U., F. Jaehn, and E. Pesch. Modelling Robust Flight-gate Scheduling as a Clique Partitioning Problem. *Transportation Science*, Vol. 42, No. 3, 2008, pp. 292–301.
21. Prem Kumar, V., and M. Bierlaire. Multi-objective Airport Gate Assignment Problem in Planning and Operations. *Journal of Advanced Transportation*, Vol. 48, No. 7, 2014, pp. 902–926.
22. Yu, C., and H. Y. Lau. A Hybrid Algorithm for Solving Gate Assignment Problem with Robustness and Tow Considerations. In M. Bramer & M. Petridis (eds.), *Research and Development in Intelligent Systems XXXI*, Springer, Cham, 2014, pp. 269–279.
23. Marinelli, M., G. Palmisano, M. Dell'Orco, and M. Ottomanelli. Fusion of Two Metaheuristic Approaches to Solve the Flight Gate Assignment Problem. *Transportation Research Procedia*, Vol. 10, 2015, pp. 920–930.
24. Ghazouani, H., M. Hammami, and O. Korbaa. Solving Airport Gate Assignment Problem Using Genetic Algorithms Approach. *Proc., 4th International Conference on Advanced Logistics and Transport (ICALT), 2015, IEEE*, 2015. pp. 175–180.
25. Yu, C., and H. Lau. Airport Gate Reassignment Based on the Optimization of Transfer Passenger Connections. *Proc., 4th International Conference on Traffic and Logistic Engineering*, Vol. 3, No. 1, pp. 25–30, June 2015. doi: 10.12720/jtle.3.1.25-30.
26. DAŞ, G. S. Multi Objective Gate Assignment Models for Increasing Shopping Revenues at Airports. In *Proc., 4th International Conference on Traffic and Logistic Engineering, 2016*, Rotterdam, Netherlands, p. 60.
27. Liu, S., W.-H. Chen, and J. Liu. Robust Assignment of Airport Gates with Operational Safety Constraints. *International Journal of Automation and Computing*, Vol. 13, No. 1, 2016, pp. 31–41.
28. Neuman, U. M., and J. A. Atkin. Airport Gate Assignment Considering Ground Movement. *International Conference on Computational Logistics*, Springer, 2013. pp. 184–198.
29. Yu, C., D. Zhang, and H. Lau. MIP-based Heuristics for Solving Robust Gate Assignment Problems. *Computers & Industrial Engineering*, Vol. 93, 2016, pp. 171–191.
30. Kim, S. H., E. Feron, and J.-P. Clarke. Gate Assignment to Minimize Passenger Transit Time and Aircraft Taxi Time. *Journal of Guidance Control and Dynamics*, Vol. 36, No. 2, 2013, pp. 467–475.
31. Genç, H. M., O. K. Erol, İ. Eksin, M. F. Berber, and B. O. Güleryüz. A stochastic neighborhood search approach for airport gate assignment problem. *Expert Systems with Applications*. Vol. 39, No. 1, 2012, pp. 316–327.
32. Kaliszewski, I., J. Miroforidis, and J. Stańczak. The Airport Gate Assignment Problem–Multi-Objective Optimization Versus Evolutionary Multi-Objective. *Computer Science*, Vol. 18, No. 1, 2017, pp. 41–52.

33. Lü, R., X. Guan, X. Li, and I. Hwang. A Large-scale Flight Multi-objective Assignment Approach Based on Multi-island Parallel Evolution Algorithm with Cooperative Coevolutionary. *Science China Information Sciences*, Vol. 59, No. 7, 2016, p. 072201.
34. Aktel, A., B. Yagmahan, T. Özcan, M. M. Yenisey, and E. Sansarcı. The Comparison of the Metaheuristic Algorithms Performances on Airport Gate Assignment Problem. *Transportation Research Procedia*, Vol. 22, 2017, pp. 469–478.
35. Deng, W., H. Zhao, X. Yang, J. Xiong, M. Sun, and B. Li. Study on an Improved Adaptive PSO Algorithm for Solving Multi-objective Gate Assignment. *Applied Soft Computing*, Vol. 59, 2017, pp. 288–302.
36. Zhang, D., and D. Klabjan. Optimization for Gate Reassignment. *Transportation Research Part B: Methodological*, Vol. 95, 2017, pp. 260–284.
37. Yu, C., D. Zhang, and H. Y. Lau. An Adaptive Large Neighborhood Search Heuristic for Solving a Robust Gate Assignment Problem. *Expert Systems with Applications*, Vol. 84, 2017, pp. 143–154.
38. Marinelli, M., M. Dell'Orco, and D. Sassanelli. A Metaheuristic Approach to Solve the Flight Gate Assignment Problem. *Transportation Research Procedia*, Vol. 5, 2015, pp. 211–220.
39. Kim, S. H., E. Feron, J.-P. Clarke, A. Marzuoli, and D. Delahaye. Airport Gate Scheduling for Passengers, Aircraft, and Operations. *Journal of Air Transportation*, Vol. 25, No. 4, 2017, pp. 109–114.
40. Benlic, U., E. K. Burke, and J. R. Woodward. Breakout Local Search for the Multi-objective Gate Allocation Problem. *Computers & Operations Research*, Vol. 78, 2017, pp. 80–93.
41. Hu, X.-B., and E. Di Paolo. Genetic Algorithms for the Airport Gate Assignment: Linkage, Representation and Uniform Crossover. *Linkage in Evolutionary Computation*, 2008, pp. 361–387.
42. Mokhtarimousavi, S., H. Rahami, and A. Kaveh. Multi-Objective Mathematical Modeling of Aircraft Landing Problem on a Runway in Static Mode, Scheduling and Sequence Determination Using NSGA-II. *International Journal of Optimization in Civil Engineering*, Vol. 5, No. 1, 2015, pp. 21–36.
43. Hu, X.-B., and E. Di Paolo. Binary-representation-based Genetic Algorithm for Aircraft Arrival Sequencing and Scheduling. *IEEE Transactions on Intelligent Transportation Systems*, Vol. 9, No. 2, 2008, pp. 301–310.
44. Mokhtarimousavi, S., H. Rahami, M. Saffarzadeh, and S. Piri. Determination of the Aircraft Landing Sequence by Two Meta-heuristic Algorithms. *International Journal of Transportation Engineering*, Vol. 1, No. 4, 2014, pp. 271–284.
45. El Moudani, W., C. Alberto Nunes Cosenza, M. De Coligny, and F. Mora-Camino. A Bi-criterion Approach for the Airlines Crew Rostering Problem. *Evolutionary Multi-criterion Optimization*, Springer, 2001, pp. 486–500.
46. Souai, N., and J. Teghem. Genetic Algorithm Based Approach for the Integrated Airline Crew-pairing and Rostering Problem. *European Journal of Operational Research*. Vol. 199, No. 3, 2009, pp. 674–683.
47. Srinivas, N., and K. Deb. Muiltiobjective Optimization Using Nondominated Sorting in Genetic Algorithms. *Evolutionary Computation*. Vol. 2, No. 3, 1994, pp. 221–248.
48. Deb, K., A. Pratap, S. Agarwal, and T. Meyarivan. A fFast and Elitist Multiobjective Genetic Algorithm: NSGA-II. *IEEE Transactions on Evolutionary Computation*. Vol. 6, No. 2, 2002, pp. 182–197.
49. Li, H., K. Liu, and X. Li. *Computational Intelligence and Intelligent Systems*. Berlin/Heidelberg: Springer, 2010.
50. Cococcioni, M., P. Ducange, B. Lazzerini, and F. Marcelloni. A Pareto-based Multi-objective Evolutionary Approach to the Identification of Mamdani Fuzzy Systems. *Soft Computing-A Fusion of Foundations. Methodologies and Applications*, Vol. 11, No. 11, 2007, pp. 1013–1031.

The Standing Committee on Airfield and Airspace Capacity and Delay (AV060) peer-reviewed this paper (18-03302).

Article

TRR
JOURNAL OF THE TRANSPORTATION RESEARCH BOARD

Transportation Research Record
2018, Vol. 2672(23) 71–81

DOI: 10.1177/0361198118782272
journals.sagepub.com/home/trr
SAGE

Exploration of the Evolution of Airport Ground Delay Programs

Kexin (May) Ren[1], Amy M. Kim[1], and Kenneth Kuhn[2]

Abstract

This study introduces a novel method of merging disparate but complementary datasets and applying machine learning techniques to ground delay program (GDP) data. More specifically, it aims to characterize GDPs with respect to changing weather forecasts, GDP plan parameters, and operational performance. The analysis aims to gain insights into GDP usage patterns (implementation and revisions), with respect to these key dimensions. It also aims to gain insights into how GDP cancelations and revisions correlate with operational efficiency and predictability. The results could be used to help traffic managers and air carriers understand complex patterns in the evolution of GDPs, so that they might, for example, better anticipate or even plan a response to a change in weather conditions. The focus is on GDPs at Newark Liberty International Airport (EWR), from 2010 through 2014. A master dataset was generated by merging several datasets on GDPs, weather forecasts, and individual flight information. Several scenarios of GDP evolution were then identified by reducing the dimensionality of the master GDP dataset, then applying cluster analysis on the lower dimensional data. It was found that GDPs at EWR can be categorized into 10 types based on weather forecasts, realized weather, GDP scope, arrival rates, and duration. The characteristics of these 10 GDP clusters were further explored by examining the relationships between GDP scenarios and their performance. It was found that GDPs under stable, low-severity weather and with large scope may score higher on the efficiency metric than expected. When GDPs called in the same weather conditions have high program rates, medium durations, and narrow scopes, capacity utilization was higher than expected—less affected flights lead to fewer cancelations and more arrivals (albeit delayed), and therefore, higher capacity utilization. Results also suggest that program rates are set more conservatively than needed for some poor weather conditions that end earlier than expected. GDPs with fewer revisions were associated with a higher predictability score but lower efficiency score. These findings can provide greater insights and knowledge about GDPs for future planning purposes. More specifically, the findings could, for example, be used to support discussion around, or even future guidance regarding, how to set and adjust GDP program rates. In future work additional data could be utilized to provide a more comprehensive operational picture of GDPs, and a wider range of performance metrics could be considered. It is also recommended that the patterns of how GDPs evolve over their lifetimes be further explored using other machine learning techniques that may provide new and useful insights.

This work applies machine learning techniques to describe ground delay programs (GDPs) with respect to changing weather forecasts, realized weather, and GDP characteristics and performance. Our purpose is to gain insights into GDPs with respect to these key dimensions, by describing GDP performance in response to these (changing) variables. These insights could be used to start discussions with traffic managers and air carriers that allow all to gain a greater understanding of complex patterns in the evolution and performance of GDPs. Although there has been some work in evaluating GDP performance retrospectively (*1*), there has been little to no exploration into how GDPs evolve over the course of their lifetimes (typically, a day). This research characterizes GDPs with respect to weather, operational parameters, and performance, focusing on GDPs at Newark Liberty International Airport (EWR) from 2010 through 2014. We created a comprehensive master dataset of GDP initiatives, weather forecasts, and individual flight data, merged from several datasets obtained from various sources; identified GDP evolution scenarios through cluster analysis based on data visualization and the results of data dimensionality reduction; and explored the relationships between GDP scenarios and performance using statistical analysis.

A brief introduction to the literature is followed by a section describing the datasets used. Then we describe the machine learning techniques used to classify EWR GDPs into 10 types based on weather forecasts and GDP plan

[1]Department of Civil & Environmental Engineering, University of Alberta, Canada
[2]RAND Corporation, Santa Monica, CA

Corresponding Author:
Address correspondence to Amy M. Kim: amy.kim@ualberta.ca

parameters, performance metrics calculated for these 10 GDP types, and how the metrics' values compare with expectation.

Background

A GDP delays flights at their departure airports in order to control arrival demand at an airport where an imbalance of capacity and demand is anticipated (*2*, *3*). The capacity shortfalls that trigger GDPs are typically attributable to adverse weather conditions forecasted for the arrival airport (*4*). Planned airport capacity and GDP duration are determined by the Federal Aviation Administration (FAA)'s Air Traffic Control System Command Center (ATCSCC) based on predicted conditions at the affected airport. Considering the extensive use of GDPs and their significant operational impacts within the National Airspace System, they have been the subject of much attention in the literature. The majority of the existing literature focuses on improving GDP planning, accounting for airport capacity uncertainty caused by adverse weather conditions (*5*, *6*), with one (delay minimization) (*7*) or more performance objectives (*8*). Research efforts have focused on generating airport capacity profiles from weather forecasts, to aid GDP planning (*9–11*). Several researchers have looked at evaluating GDP performance retrospectively (*1*, 12, 13), whereas others have attempted to classify days at airports based on weather and GDP characteristics (*14–18*). Overall, the existing studies have provided much insight for GDP planning and prediction, and measuring performance, which have guided this work. Although this work most closely follows the last set of papers, it differs in that the analysis of GDPs is based on the (changing and realized) weather conditions, GDP plan parameters, and operational performance over the lifetime of each GDP.

The ATCSCC issues advisories modifying GDPs in response to changing weather and traffic conditions. Modifications are quite common; in this dataset (introduced in the next section) it was found that the average number of modifications per GDP was 1–2. This research attempts to characterize GDPs by these changing aspects, focusing on GDPs at EWR from 2010 through 2014. EWR is one of three major airports serving the New York metropolitan area, and one of the busiest airports in the US, serving over 35 million passengers in 2015 (*19*). EWR is also frequently subject to adverse weather conditions; the Traffic Management Initiative (TMI) advisory dataset used for this study indicates that 15% of all GDPs implemented in the US from 2010 to 2014 occurred at EWR. Operational problems at any of the three major New York airports have been demonstrated to have wide-reaching effects (*20*); Liu and Hansen (*1*) apply their GDP performance metrics at EWR as well, providing a useful point of comparison for this study of EWR GDPs.

Data

The data used in this study include GDP advisory data (from the FAA TMI Advisories database), weather forecast data [Terminal Aerodrome Forecast database (TAF)], flight data [Individual Flight (IF) dataset from the FAA's Aviation System Performance Metrics database], and observed weather data [Aviation Routine Weather Report (METAR)]. These four datasets were combined into a master dataset.

Data Sources

The TMI Advisories database contains ATCSCC advisories reporting TMI plans, modifications, and cancelations. Twenty-one variables were extracted from the original dataset, including advisory type, dates, times, causes, affected scopes, and program rates. Filtering by advisory category "GDP" and control element "EWR/ZNY" yielded 2,410 advisories (765 which were root advisories) from 2010 through 2014.

A TAF report is issued by the U.S. National Weather Service, and contains forecasted meteorological conditions at major U.S. airports. The forecast pertains to visibility, ceiling, winds, and other meteorological features of interest (*21*). TAFs are issued at least once every six hours and generally cover a 24- to 30-hour period following the forecast (*22*). Twenty-eight variables were extracted from the TAF dataset including TAF issue and forecast coverage times, and forecast visibility, ceiling, winds, and precipitation. After removing duplicate reports as well as those with illogical durations, the final dataset contained 96,829 forecasts from January 2010 through August 2014.

The FAA's Aviation System Performance Metrics includes a database of individual flights, which provides detailed information including various departure and arrival timing points (scheduled gate out, flight plan gate out, actual gate out, scheduled wheels off, etc.) and flight delays reported from Traffic Flow Management System (*23*), Out, Off, On, In (OOOI) and airline system quality performance records. Thirty-six variables were selected from this dataset, and information extracted for 879,507 flights inbound to EWR.

METARs report observational surface weather data and are generated and published hourly by the U.S. National Weather Service (*24*). Fifteen variables and 46,481 records were extracted from this dataset from January 2010 through August 2014.

For all airports with departing flights destined for EWR, geographic data [longitudes, latitudes, countries, and air route traffic control centers (ARTCCs)] and their great-circle distances from EWR were combined to create a dataset called Airport Information (AI). By doing so, the GDP parameter "departure scope"—usually defined as a radius from the GDP airport or a set of ARTCCs—can be redefined as GDP-affected departure airports.

The data included in the merged master dataset, with original sources, are listed in Table 1. Note that each data point drawn from the TMI dataset describes a single GDP; the TAF and METAR data sets describes a single weather report; the IF data set describes a flight; and the AI data set describes an airport.

Data Preparation

To prepare the datasets for merging, they were first filtered and cleaned, and time zones unified. Filtering was performed to include data from January 2010 through August 2014, TMI advisory category "GDP," and control element "EWR/ZNY." After filtering, illogical entries, such as TAFs or TMIs with abnormal (too long or negative) durations and duplicates were removed. Finally, all datasets were unified into local New York time.

Also calculated were several new variables from the original datasets, for the purposes of data merging and describing GDP features:

- TMI dataset: the authors added the planned advisory/initiative durations, number of revisions, and early cancelation time (if there should be one) based on the original GDP data. Then the actual GDP advisory end times were calculated; this is the advisory begin time of the subsequent revision advisory belonging to the same GDP (if there should be one). The actual initiative end times and advisory/initiative durations were thus generated. Also matched was the GDP departure scope to the affected departure airports.
- TAF and METAR: calculated a crosswind variable was calculated based on wind speed, wind angle, and runway direction. Precipitation was defined to consist of RA (rain), DZ (drizzle), SN (snow), SG (snow grains), GR (hail), GS (snow pellets), IC (ice crystals), and UP (unknown precipitation) (*25*).
- For each flight in the IF data, departure airport ARTCC and country, and flight distance to EWR was added.

The TAFs and METARs were matched to GDPs (in the TMI dataset) by time, and IFs matched to GDPs by both time and geography. The steps for matching the TAFs to GDPs include: (1) for each GDP advisory, select the TAFs issued before the GDP send time, and with a start time earlier than the GDP end time or an end time later than the GDP start time; (2) for each hour of the GDP, select the TAFs with a start time earlier than the last minute of the hour, and with an end time later than the first minute of the hour; (3) if, for a GDP hour, there are several TAF records, then choose the TAF with the latest issue time and match this to the GDP.

To match the METARs to GDPs, for each hour of a GDP, select the METARs issued during the hour. If there is more than one METAR for a GDP hour, select and use the most severe observed weather.

The IF data was matched to GDP data through the following steps: (1) pick out flights with base estimated time of arrival (scheduled gate-in time) falling between the GDP start and end times; (2) check whether the flights were GDP exempted; (3) attach the flights affected by a GDP to the GDP.

The additional variables generated from the merging of IFs and GDPs include initially scheduled arrivals during a GDP, arrivals affected by a GDP, ground delays, planned total delays, and actual total delays. Thus, a GDP advisory dataset was obtained with GDP advisories matched to weather forecasts and operational parameters. A dataset was then constructed where each row represents an hour when a GDP initiative was in place, and the GDP advisories data were reorganized into this time-based format. The final dataset contains 11,177 rows and 38 columns.

Data Description

From 2010 through 2014, 89% of EWR GDPs were initiated as a result of adverse weather, confirming that it was the dominant cause of GDPs from 2010 through 2014. Notable weather characteristics obtained from METAR data included the following. First, precipitation was the most common adverse weather condition from 2010 to 2014, followed by strong crosswinds (i.e., >15 knots) to parallel runways 4/22, and low ceiling/visibility (LCV) [causing instrument meteorological conditions (IMC)]. Second, weather conditions in December to May were generally worse than in other months. However, thunderstorms were more prevalent at EWR in the summer months, consistent with general knowledge about thunderstorms across the eastern states (*26*). Third, adverse weather was experienced more frequently in 2010–2011 than 2012–2014. Year 2010 experienced more strong crosswinds to runways 4/22 and precipitation, whereas precipitation and IMC were prevalent in 2011. These observations are consistent with reports from the National Oceanic and Atmospheric Administration (*27*). However, while adverse weather is the most common cited reason for GDP issuance, GDPs are typically caused by a combination of weather and heavy flight demands (*28*). Thus, GDP characteristics were explored using the METAR and individual flights datasets as well.

There are five observations to be made. Firstly, weather factors, especially crosswinds to runways 4/22 and LVC, were the most common causes of GDPs, consistent with previous findings (*29*, *30*). Although thunderstorms occurred with the lowest frequency of all adverse conditions, they caused a significant number of summer GDPs (June–August). Thunderstorms typically led to low GDP arrival rates and, therefore, significant arrival delays (*31*). Secondly, although the TMI data demonstrated that weather was the major cause of GDPs at EWR, it is known that GDPs would not be as prevalent if flight demands were lower. The advisories and IF data indicate that more GDPs were initiated in the spring

Table 1. Original Data Used in Master Dataset

Name	Source	Description
Year	TMI	Advisory send year
AdvisoryDate UTC	TMI	Advisory send date
AdvisoryNumber	TMI	Label of the advisory
SendDate.Time.UTC	TMI	Advisory send date and time (time zone = GMT)
AdvisoryCategory	TMI	TMI category; GDPs only used here
AdvisoryType	TMI	Advisory type, "GDP" or "GDPX" (GDP cancelation)
ControlElement	TMI	ARTCC which issued the advisory. Here, it should be "EWR/ZNY"
RootAdvisoryDate.UTC	TMI	Send date of this advisory's root advisory
RootAdvisoryNumber	TMI	Advisory Number of this advisory's root advisory
Derived.BgnDate.Time.UTC	TMI	The begin time of the GDP or GDPX advisory (time zone = GMT)
Derived.EndDate.Time.UTC	TMI	The end time of the GDP or GDPX advisory (time zone = GMT)
Is.RootAdvisory	TMI	Whether this advisory is a root advisory ("Yes" or "No")
Canadian.Dep.Arpts.Included	TMI	Affected Canadian departure airports included in the advisory
Dep.Scope	TMI	Affected departure scope: radius or a set of ARTCCs
GDP.Bgn.Date.Time.UTC	TMI	GDP begin time (time zone = GMT)
GDP.End.Date.Time.UTC	TMI	GDP end time (time zone = GMT)
GDPX.Bgn.Date.Time.UTC	TMI	GDP cancel begin time (time zone = GMT)
GDPX.End.Date.Time.UTC	TMI	GDP cancel end time (time zone = GMT)
Impacting.Condition	TMI	Causes of the advisory
Program.Rate	TMI	Hourly arrival capacity to GDP airport, for each hour
Exempt.Dep.Facilities	TMI	Airports exempt by the advisory
Issued date & time	TAF	TAF issue Year, Month, Day, Hour, Minute
From date & time	TAF	Forecast start Year, Month, Day, Hour, Minute
To date & time	TAF	Forecast end Year, Month, Day, Hour, Minute
Wind Angle	TAF	Forecasted wind angle (degrees)
Wind Speed	TAF	Forecasted wind angle (knots)
Visibility	TAF	Forecasted visibility (miles)
Ceiling	TAF	Forecasted ceiling (100 feet)
RA	TAF	Forecasted occurrence of rain (1 = yes, 0 = no)
DZ	TAF	Forecasted occurrence of drizzle (1 = yes, 0 = no)
SN	TAF	Forecasted occurrence of snow (1 = yes, 0 = no)
SG	TAF	Forecasted occurrence of snow grains (1 = yes, 0 = no)
GR	TAF	Forecasted occurrence of hail (1 = yes, 0 = no)
GS	TAF	Forecasted occurrence of snow pellets (1 = yes, 0 = no)
IC	TAF	Forecasted occurrence of ice crystals (1 = yes, 0 = no)
UP	TAF	Forecasted occurrence of unknown precipitation (1 = yes, 0 = no)
TS	TAF	Forecasted occurrence of thunderstorm (1 = yes, 0 = no)
start.time	METAR	Start date and time of the METAR observation
end.time	METAR	End date and time of the METAR observation
Wind.Angle	METAR	Observed wind angle (degrees)
Wind.Speed	METAR	Observed wind angle (knots)
Visibility	METAR	Observed visibility (miles)
Ceiling	METAR	Observed ceiling (100 feet)
RA	METAR	Observed occurrence of rain (1 = yes, 0 = no)
DZ	METAR	Observed occurrence of drizzle (1 = yes, 0 = no)
SN	METAR	Observed occurrence of snow (1 = yes, 0 = no)
SG	METAR	Observed occurrence of snow grains (1 = yes, 0 = no)
GR	METAR	Observed occurrence of hail (1 = yes, 0 = no)
GS	METAR	Observed occurrence of snow pellets (1 = yes, 0 = no)
IC	METAR	Observed occurrence of ice crystals (1 = yes, 0 = no)
UP	METAR	Observed occurrence of unknown precipitation (1 = yes, 0 = no)
TS	METAR	Observed occurrence of thunderstorm (1 = yes, 0 = no)
DEP_YYYYMM	IF	Scheduled Departure Year and Month (Local Date)

(continued)

Table 1. (continued)

Name	Source	Description
DEP_DAY	IF	Scheduled Departure Day (Local Day)
DEP_HOUR	IF	Scheduled Departure Hour (Local Hour)
DEP_QTR	IF	Scheduled Departure Quarter Hour (Local Qtr)
ARR_YYYYMM	IF	Scheduled Arrival Year and Month (Local Date)
ARR_DAY	IF	Scheduled Arrival Day (Local Day)
ARR_HOUR	IF	Scheduled Arrival Hour (Local Hour)
ARR_QTR	IF	Scheduled Arrival Quarter Hour (Local Qtr)
OFF_YYYYMM	IF	Actual Wheels Off Year and Month (ASQP/OOOI Off Local Date)
OFF_DAY	IF	Actual Wheels Off Day (ASQP/OOOI Off Local Day)
OFF_HOUR	IF	Actual Wheels Off Hour (ASQP/OOOI Off Local Hour)
OFF_QTR	IF	Actual Wheels Off Quarter Hour (ASQP/OOOI Off Local Qtr)
ON_YYYYMM	IF	Actual Wheels on Year and Month (ASQP/OOOI On Local Date)
ON_DAY	IF	Actual Wheels on Day (ASQP/OOOI On Local Day)
ON_HOUR	IF	Actual Wheels on Hour (ASQP/OOOI On Local Hour)
ON_QTR	IF	Actual Wheels on Quarter Hour (ASQP/OOOI On Local Qtr)
FAACARRIER	IF	Flight Carrier Code – ICAO
FLTNO	IF	Flight Number
Dep_LOCID	IF	Departure Location Identifier
Arr_LOCID	IF	Arrival Location Identifier
SchOutTm	IF	Scheduled Gate Departure Time (Local) HH:MM
FPDepTm	IF	Flight Plan Gate Departure Time HH:MM
ActOutTm	IF	Actual Gate Out Time HH:MM
SchOffTm	IF	Scheduled Wheels Off Time HH:MM
FPOffTm	IF	Flight Plan Wheels Off Time HH:MM
ActOffTm	IF	Actual Wheels Off Time HH:MM
DlaSchOff	IF	Airport Departure Delay Minutes (Based on Schedule)
DlaFPOff	IF	Airport Departure Delay Minutes (Based on Flight Plan)
DELAY_AIR	IF	Airborne Delay Minutes
EDCTOnTm	IF	Wheels on Time HH:MM (Filed on EDCT)
ActOnTm	IF	Actual Wheels on Time HH:MM
SchInTm	IF	Scheduled Gate-In HH:MM
FPInTm	IF	Flight Plan Gate-In HH:MM
ActInTm	IF	Actual Gate-In Time HH:MM
DlaSchArr	IF	Arrival Delay in Minutes (Compared with Scheduled)
DlaFPArr	IF	Arrival Delay in Minutes (Compared with Flight Plan)
Country	AI	The country in which the airport is located
City	AI	The city in which the airport is located
Latitude	AI	Airport latitude
Longitude	AI	Airport longitude
ARTCC	AI	ARTCC which the airport belongs to (for U.S. & Canadian airports only)
Distance	AI	Great-circle distance between the airport and EWR airport (in miles)

(March–May) months, and on weekdays as a result of heavier flight schedules. Thirdly, GDPs were typically sent in the late morning, initiated around noon, modified in the afternoon, and finished by late evening. Fourthly, most GDPs were initially planned for duration of 8–12 hours, are typically extended in a revision reaching planned durations of 10–13 hours, and actually run about 6–11 hours. Finally, each GDP had an average of 1.16 revisions, while 95% were canceled an average of 2 hours early. This seems to suggest that air traffic controllers were either conservative in their GDP planning, TAF forecasts are conservative, or both.

Methods and Results

GDP features were first extracted with the purpose of dimensionality reduction. GDP evolution scenarios were then identified using cluster analysis. Finally, correlations were examined between GDP types (as per 10 scenarios identified

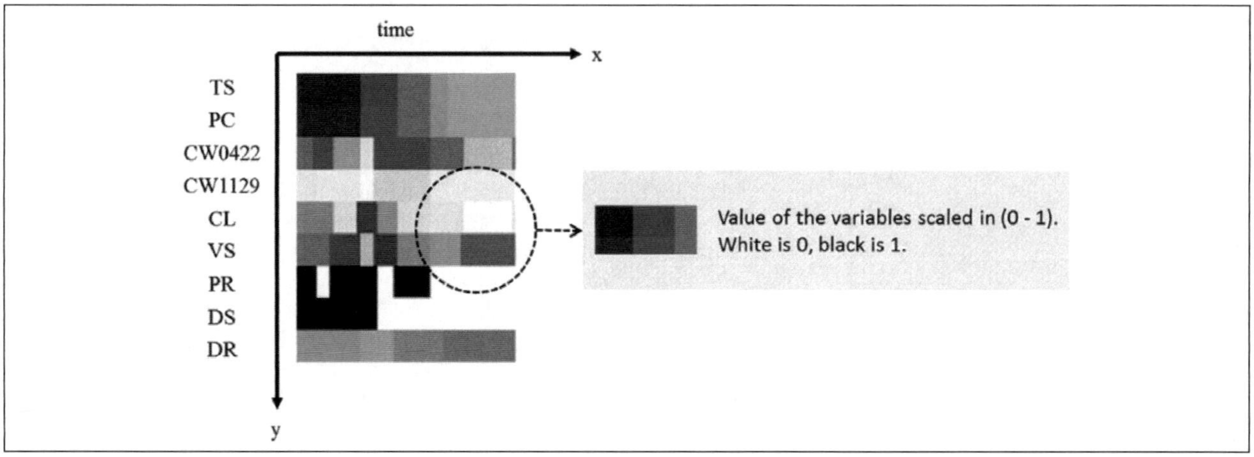

Figure 1. GDP grayscale image.

in the cluster analysis) and metrics calculated to describe GDP operational performance.

Data Feature Extraction

Nine important variables were identified in the merged dataset describing the GDPs by their forecasted weather conditions and corresponding GDP operational parameters. Six pertain to weather conditions: thunderstorm (TS), precipitation (PC), crosswind strength to runways 4/22 (CW0422), crosswind strength to runway 11/29 (CW1129), ceiling (CL), and visibility (VS). The remaining three variables pertain to GDP parameters, including: GDP program rate (PR, allowed flight arrival rate), departure scope (DS, represented by the number of GDP-affected flights), and planned initiative duration (DR). We represented GDPs using 2D grayscale images with these nine weather/GDP parameters represented on the y-axis (each parameter normalized to [0,1] and represented by the greyscale) and time (also normalized) on the x-axis.

Figure 1 shows an example of one observed GDP, as represented in this study. As the study aims to characterize GDPs not by their durations but by how the nine identified features evolve over their duration, the time period during which each GDP was active was divided into 65 equal-length intervals. For example, if a GDP was in effect from 9:00 am until 2:25 pm, the first time interval would cover 9:00–9:05 am and the 65th and final time period would cover 2:20–2:25 pm. The features described in the previous paragraph as observed during each time interval were then examined. In this way, each GDP is represented by a 9×65 matrix (585 cells), where rows represent features and columns represent time intervals. After removing GDPs with discontinuous weather forecasts, 495 GDPs remained in the dataset.

With the goal of clustering these 585-dimensional GDPs, a dimensionality reduction was performed on the GDP data, to identify the most important variables that describe GDPs at EWR. Dimensionality reduction is the process of reducing the number of variables describing some phenomenon, by selecting a subset of the original data (feature selection) or transforming the data to a lower dimensional space (feature extraction). The transformation can be linear or nonlinear. As linear methods can be restrictive, a technique that does not make a linearity assumption, called autoencoder, was used. An autoencoder is an artificial neural network which learns the features of inputs using a backpropagation algorithm (*32*). An autoencoder includes an input layer, one or more hidden layers, and an output layer. From input layer to hidden layer, the autoencoder learns representation for a dataset; from hidden layer to output layer (encoder), it is trained to optimize a loss function which measures how well the data are reconstructed based on the encoder representation. Autoencoders have been applied to reduce dimensionality for characterizing time-varying data in many studies; for example, Shin et al. applied autoencoders to automatically classify tissue types by the change in brightness of resonance images (*33*). By comparing the clustering results with different autoencoder forms, an autoencoder neural network was finally constructed with the following structural characteristics: one input layer with 585 neurons, three hidden layers with 300, 2, and 200 neurons in each successive layer, and one output layer with 585 neurons. The original 585-dimensional data was compressed to two in the second hidden layer. The use of autoencoder allowed for data dimensionality reduction (allowing for compact representation of the original dataset while minimizing information loss) to facilitate cluster analysis.

Cluster Analysis

To characterize evolving GDPs under changing weather forecasts, the study attempted to identify GDP evolution scenarios through cluster analysis based on the compressed

Table 2. GDP Cluster Characteristics

Characteristic	Values
Forecasted adverse weather	Crosswinds (CW) >9.4 knots
	Precipitation (PC) accounting for >30% of GDP duration
	Thunderstorms (TS) accounting for >30% of GDP duration
	Low visibility/ceiling (LVC): <3 miles, <1000 feet
Weather severity	Less severe: only strong crosswinds (>15 knots), low ceiling (<1000 feet) or low visibility (<4 miles) forecasted (*35*)
	Severe: precipitation plus strong crosswinds, low ceiling or low visibility (< 4 miles) forecasted
	Very severe: thunderstorms forecasted
Weather stability across time	Stable: no weather variables expected to change significantly over time
	Medium: 1 weather variable expected to change significantly over time
	Unstable: ≥2 weather variables expected to change significantly over time
GDP program rate	Low/Medium: hourly program rate ≤35 arrivals/hour
	High: hourly program rate >35 arrivals/hour
GDP departure scope	Narrow: number of affected flights <100
	Medium: number of affected flights between 100–130
	Wide: number of affected flights >130
GDP planned duration	Short: planned duration <9 hours
	Medium: planned duration 9–11 hours
	Long: planned duration >11 hours

Table 3. Cluster Descriptions

Cluster	Weather types	Weather severity	Weather stability	GDP Type	# Obs
1	CW	Less severe	Stable	High, Wide, Medium	110
2	LVC, CW	Less severe	Stable	High, Narrow, Short	39
3	CW	Less severe	Stable	High, Medium, Short	151
4	PC, CW	Severe	Unstable	Low, Wide, Long	23
5	LVC, PC	Severe	Unstable	Medium, Wide, Long	46
6	PC, LVC	Severe	Unstable	Medium, Wide, Long	34
7	PC, LVC	Severe	Medium	Low, Wide, Long	36
8	PC, LVC	Severe	Medium	Low, Medium, Medium	37
9	TS, PC, LVC	Very severe	Unstable	Medium, Medium, Short	26
10	TS, PC, LVC	Very severe	Medium	Low, Narrow, Short	10

2-dimensional data. Three classes of clustering methods (k-means, partitioning around medoids (PAM), and hierarchical clustering) were applied, and k-estimation methods (average silhouette and gap statistic) were used to determine the optimal number of clusters. By comparing their results, k was judged to be a good candidate. However, values of k between 8 and 12 were further explored, by comparing the similarity of images within the same group as well as the differences between images in different groups. It was found that for $k = 8$ or 9, some clusters appeared to hold very different images and thus were candidates for further division into more groups; with $k = 11$ or 12, several different clusters appeared quite similar and candidates for combining into a single cluster. Finally, with the PAM clustering method and $k = 10$, the grayscale images were such that GDPs within a group were quite similar whereas those in different groups were more distinguished.

To understand the general features of the clusters, the average of each of the nine variables for GDPs of a cluster was calculated. Also calculated was the average of the variance of each variable to report the dispersion of the variables in each cluster. The clusters were characterized by forecasted weather, weather severity, weather stability across time, and GDP parameters of program rate, departure scope and duration, which are further described in Table 2. The characteristics of the 10 clusters are shown in Table 3. For more detailed results, refer to Ren (*34*).

It was found that the clusters could be categorized into three groups based on forecasted weather conditions: (1) less severe and stable weather, with LVC or strong crosswinds (CW) as the main adverse weather condition; (2) severe and unstable weather, with precipitation (PC) as the main adverse weather with CW or LVC occurring together; and (3) very severe and unstable weather, with thunderstorms (TS) as the

Table 4. CFA Results and Cluster Performance

Cluster	Weather forecast	GDP parameters	CFA results 1	2	3	4	5	Mean values 1	2	3	4	5
1	Less severe, stable	High, wide, med	High	-	-	≥2	-	1.03	0.55	0.50	2.20	1.15
2	Less severe, stable	High, narrow, short	-	High	-	-	0	1.02	0.74	0.34	1.88	0.31
3	Less severe, stable	High, med, short	High	-	-	-	0	1.05	0.64	0.44	1.94	0.64
4	Severe, unstable	Low, wide, long	-	-	-	-	≥2	1.00	0.46	0.54	1.48	1.70
5	Severe, unstable	Low, wide, long	-	-	-	-	≥2	0.99	0.43	0.48	1.79	1.66
6	Severe, unstable	Low, wide, long	Low	-	High	-	-	0.97	0.51	0.54	1.35	1.09
7	Severe, medium	Low, wide, long	Low	-	-	-	≥2	0.95	0.58	0.52	1.92	1.50
8	Severe, medium	Low, med, med	-	Low	-	-	≥2	0.93	0.41	0.45	1.65	1.57
9	Very severe, unstable	Low, med, short	-	-	-	-	-	0.99	0.46	0.53	1.60	1.38
10	Very severe, unstable	Low, narrow, short	-	-	-	-	0~1	0.91	0.62	0.43	2.08	0.30

Note: 1: Efficiency (planned/actual arrivals; unitless); 2: Capacity utilization (ratio; unitless); 3: Predictability (ratio; unitless); 4: Early CNX time (hrs); 5: Revisions (no.). - = occurred as expected.

main adverse weather, with PC and LVC occurring together. The first category, which includes clusters 1–3, contains the most GDPs. GDPs in clusters 1–3 were all planned with high program rates, and short to medium durations. Also, GDPs in clusters 1 and 3 had medium to wide departure scopes whereas those in cluster 2 (smallest membership) had narrow scopes. The second category, consisting of clusters 4–8, was the second most frequently occurring group. All the GDPs in this category had medium to low program rates, medium to wide departure scopes, and medium to long planned durations. The third category, which includes clusters 9 and 10, occurred with the lowest frequency. GDPs in this category had medium to low program rates, medium to narrow departure scopes, and short planned durations. The clustering results were used to assess expected performance as described next.

GDP Performance Assessment

The GDPs in each of the 10 clusters were evaluated using the efficiency, capacity utilization, and predictability metrics proposed by Liu and Hansen (*1*). Early cancelation time and number of revisions were also explored. Then, a series of configural frequency analysis (CFA) tests were conducted to assess the relationships between GDP clusters and the expected outcome of each performance metric. CFA is a widely used, parameter-free multivariate data analysis method, which can be applied to any set of data regardless of its statistical distribution. It identifies values of metrics that occur statistically more, equal to, or less than expected under the assumption that there is no relationship between (for example) GDP clusters and values of a performance metric. Table 4 contains the results of CFA applied to the clustering results. Columns 1–3 contain cluster number, weather conditions, and GDP operational parameters (shown as rate, scope, duration). Columns 4–6 show the CFA results (comparisons with expected metric performance). A "High" ("Low") score means that, for the given metric, the observed value of the metric is higher (lower) than its expected value and that this result is statistically significant. To obtain these results, the expected frequencies were calculated using a first-order CFA model, significance of the difference between observed and expected frequencies was tested using the Z-test, and statistically significant configurations (at a 90% confidence level) were identified. The mean metrics values calculated for each cluster are also provided and compared against those found by Liu and Hansen (*1*). However, their results are for 2011 only; this study's metrics, when calculated for 2011, are similar. The cells highlighted in gray are of particular interest and, therefore, discussed below.

Recall that CFA tests whether a configuration occurred statistically significantly higher than expected. For example, the "efficiency" metric was divided into three bins of equal size—high, medium, and low efficiency. Table 4 indicates that for cluster 6, the number of observations in the "low efficiency" bin was statistically significantly higher than expected; thus, cluster 6's "efficiency" is marked "Low." The cells marked with "-" indicate that the results are as expected. Table 4 contains a rich set of results to discuss and synthesize; however, owing to limited space, two sets of results of particular note are discussed here.

The first observations pertain to the results for clusters 1–3, highlighted in light gray (and bordered in dark gray). The weather forecast that occurs with the GDPs in these clusters is less severe and stable, such that initiation of GDPs in this group may be attributed more to high demands rather

Table 5. CFA Results Summary (Key Observations)

Clusters	Weather forecasts	GDP features	Possible reason
1–3	less severe stable	Affecting more flights More efficient than expected	Despite a wide scope, stable weather conditions led to more stable GDPs.
		Affecting fewer flights Higher capacity utilization than expected	Smaller number of affected flights led to fewer cancelations and more arrivals.
5–8	severe unstable	Affecting more flights Less efficient than expected	Unstable weather conditions and a wide scope led to more volatile and rapidly changing GDP, and further (airborne) delays.
		Affecting fewer flights Lower capacity utilization than expected	Program rates are set more conservatively than actually needed for some poor weather conditions that end earlier than expected; GDP canceled early as well.

than severe weather. When those GDPs have high program rates, short-medium durations, and medium-wide scopes (cluster 1 and 3), the efficiency metric is significantly higher than expected (as per the CFA results). Comparing with cluster 2 (high, short, narrow GDPs), this suggests that GDPs with larger scope (i.e., larger geographic scope and therefore, more affected flights) may be more efficient (ratio of GDP-induced departure over arrival delay) than would be expected. This could be attributed to the fact that, despite a wide scope, stable weather conditions lead to more stable GDPs. When these GDPs have high program rates, medium durations, and narrow scopes (cluster 2), capacity utilization is significantly higher than expected (based on CFA results). Comparing with clusters 1 and 3, this result could be attributable to these high program rate GDPs with narrower scopes involving less flights, leading to fewer cancelations and more arrivals (albeit delayed), and therefore, higher capacity utilization.

The second set of observations pertains to the results for clusters 6–8, highlighted in darker gray. The GDPs of clusters 6–8 are distinguished by weather forecasted to be severe and unstable (i.e., rapidly changing). When a GDP with low program rate, wide departure scope, and long duration (clusters 6 and 7) occurs, the efficiency metric values are found to be lower than expected. When compared with cluster 8 GDPs (low, medium, medium), this result may be attributed to unstable weather conditions and a wider scope leading to a more volatile and rapidly changing GDP, which will lead to further delays in the air, and therefore, a lower efficiency score. When a GDP with low program rate, medium departure scope, and medium duration (cluster 8) occurs, capacity utilization is lower than expected. With longer duration the capacity utilization is as expected. This seems to suggest that program rates are set more conservatively than actually required for some poor weather conditions that end earlier than expected, with early GDP cancelation as well. These two sets of findings are summarized in Table 5.

Different revision decisions may involve a trade-off between predictability and efficiency. Clusters 6–8 have similar forecasted weather (severe and unstable with PC and LVC). By comparing these clusters, a trade-off was found to exist between high (2 or more) and low number of modifications; fewer revisions were associated with higher predictability but lower efficiency.

These results suggest the joint impact of GDP plans and weather forecasts on GDP efficiency; when weather is predicted to be less severe, a wide GDP departure scope would lead to higher-than-expected efficiency, whereas when weather is predicted to be severe and unstable over time, it would lead to lower-than-expected efficiency. It may be interpreted that, under less severe and stable forecasted weather conditions, GDPs with wider departure scope would lead to higher efficiency because they can absorb the airborne delays almost completely on ground by delaying numerous flights at their departure airport instead of en route; under long-term severe and unstable weather, fewer of flights' airborne delays may be transferred to the ground, as a result of the uncertainties induced by the long-term unstable conditions.

Practical Application of This Work

These results include clusters that show typical GDP types and weather patterns observed at EWR, and could be used to help traffic managers save time when planning future GDPs. A recommendation engine could highlight a typical GDP or modifications to a GDP based on the observed or forecasted weather. These results could also be used by airlines, for example to generate a set of scenarios representing plausible combinations of GDPs and weather patterns. The airlines could plan against these scenarios and develop operational strategies. The results also include details about the performances of different types of GDPs. These results could be used to start data-driven discussions with traffic managers and policy makers, which could lead to more consistent, predictable, and/or efficient GDPs.

Concluding Remarks

This research explored the characteristics of GDPs and weather conditions as realized during the lifetimes of the GDPs. In particular, modifications made to GDPs were considered, and the focus was not restricted to GDPs as planned initially. Also examined were the correlations between GDP characteristics and performance. Based on TMI advisory, weather forecast, and flight data at EWR from 2010 through 2014, machine learning techniques were applied to better observe the characteristics of GDPs as they evolved over a day at EWR. A master dataset was first developed through the merging of weather forecasts, realized weather, TMI advisories, and individual flights information datasets. Second, the GDP evolution data were visualized in order to support data processing process and clustering results. Third, autoencoder was used to reduce 585 dimensions of GDP evolution into two. Fourth, GDP evolution scenarios were identified through cluster analysis based on the compressed 2-dimensional data. Finally, correlations were assessed between the identified GDP clusters and GDP performances, using CFA.

The data confirmed that, as expected, various indications of inclement weather were determined to be the most frequent causes of GDPs. After dimensionality reduction, GDPs were clustered into 10 scenarios according to weather type, severity, and stability over time, in addition to GDP duration, scope, and program rate. The results of the CFA suggest that GDPs under stable, low-severity weather and with large scope (i.e., more affected flights) may score higher on the efficiency metric than expected. This could be attributed to the fact that stable weather conditions lead to more stable GDPs. When these GDPs have high program rates, medium durations, and narrow scopes, capacity utilization is higher than expected—less affected flights lead to fewer cancelations and more arrivals (albeit delayed), and therefore, higher capacity utilization. Results also suggest that program rates are set more conservatively than needed for some poor weather conditions that end earlier than expected, with GDP being canceled early as well. GDPs with fewer revisions were associated with a higher predictability score but lower efficiency score.

The results of this work could be used to raise awareness of typical and unusual patterns in how GDPs are revised in response to changing weather conditions. The methodology could be applied to study other forms of air traffic flow management, to study how, for example, FAA playbook routes and reroute initiatives are used. For future work, it is recommended that additional data be utilized to provide a more comprehensive operational picture of GDPs, and that a wider range of performance metrics be considered in the CFA analysis. In addition, it is also recommended that the patterns of how GDPs evolve over their lifetimes, with respect to several key variables identified using statistical analysis and dimensionality reductions, be further explored using other novel machine learning techniques that may provide new and useful insights.

Acknowledgments

The authors would like to acknowledge financial support for this work from new faculty start up funds at the University of Alberta.

Author Contributions

The authors confirm contribution to the paper as follows: study conception and design: Amy Kim, Kenneth Kuhn; data collection: Kenneth Kuhn, Amy Kim, Kexin Ren; analysis and interpretation of results: Amy Kim, Kexin Ren, Kenneth Kuhn; draft manuscript preparation: Amy Kim, Kexin Ren, Kenneth Kuhn.

All authors reviewed the results and approved the final version of the manuscript.

References

1. Liu, Y., and M. Hansen. Evaluation of the Performance of Ground Delay Programs. *Transportation Research Record: Journal of the Transportation Research Board*, 2013. 2400: 54–64.
2. Ball, M. O., and G. Lulli. Ground Delay Programs: Optimizing Over the Included Flight Set Based on Distance. *Air Traffic Control Quarterly*, Vol. 12, No. 1, 2004, pp. 1–25.
3. Federal Aviation Administration. *Traffic Flow Management in the National Airspace System*. Federal Aviation Administration, Washington, D.C., 2009.
4. Federal Aviation Administration. *Ground Delay Program*. http://www.fly.faa.gov/Products/AIS_ORIGINAL/shortmessage.html. Accessed August 2, 2017.
5. Richetta, O., and A. R. Odoni. Solving Optimally the Static Ground-Holding Policy Problem in Air Traffic Control. *Transportation Science*, Vol. 27, No. 3, 1993, pp. 228–238.
6. Mukherjee, A., and M. Hansen. A Dynamic Stochastic Model for the Single Airport Ground Holding Problem. *Transportation Science*, Vol. 41, No. 4, 2007, pp. 444–456.
7. Inniss, T. R., and M. O. Ball. Estimating One-Parameter Airport Arrival Capacity Distributions for Air Traffic Flow Management. *Air Traffic Control Quarterly*, Vol. 12, 2004, pp. 223–252.
8. Liu, J., K. Li, M. Yin, X. Zhu, and K. Han. Optimizing Key Parameters of Ground Delay Program with Uncertain Airport Capacity. *Journal of Advanced Transportation*, Vol. 2017, 2017, pp. 1-9.
9. Buxi, G., and M. Hansen. Generating Probabilistic Capacity Profiles from Weather Forecast: A Design-of-Experiment Approach. *In USA/Europe Air Traffic Management Research & Development Seminar*, Berlin, Germany, 2011.
10. Liu, P. B., M. Hansen, and A. Mukherjee. Scenario-Based Air Traffic Flow Management: From Theory to Practice. *Transportation Research Part B: Methodological*, Vol. 42, No. 7, 2008, pp. 685–702.
11. Richetta, O., and A. R. Odoni. Dynamic Solution to the Ground-Holding Problem in Air Traffic Control. *Transportation Research Part A: Policy and Practice*, Vol. 28, No. 3, 1994, pp. 167–185.
12. Hoffman, R. L., and M. O. Ball. Measuring Ground Delay Program Effectiveness Using the Rate Control Index. *The Journal of Air Traffic Control*, Vol. 42, No. 2, 2000, pp. 19–23.
13. RAND Corporation. *Performance Metric Ranking of Air Traffic Flow Management Plans*. 2016. Deliverable for NASA Project NNH14ZEA001N-CTD1.
14. Hoffman, B., J. Krozel, S. Penny, A. Roy, and K. Roth. A Cluster Analysis to Classify Days in the National Airspace

System. *AIAA, In Guidance, Navigation, and Control Conference and Exhibit*, Austin, Tex., 2003.
15. Grabbe, S., B. Sridhar, and A. Mukherjee. Clustering Days with Similar Airport Weather Conditions. *AIAA, In Aviation Technology, Integration, and Operations Conference*, Atlanta, Ga., 2014.
16. Mukherjee, A., S. Grabbe, and B. Sridhar. Classification of Days Using Weather Impacted Traffic in the National Airspace System. *AIAA, In Aviation Technology, Integration, and Operations Conference*, Los Angeles, Calif., 2013.
17. Mukherjee, A., S. Grabbe, and B. Sridhar. Predicting Ground Delay Program at an Airport Based on Meteorological Conditions. *AIAA, In Aviation Technology, Integration, and Operations Conference*, Atlanta, Ga., 2014.
18. Kuhn, K. D. A Methodology for Identifying Similar Days in Air Traffic Flow Management Initiative Planning. *Transportation Research Part C: Emerging Technologies*, Vol. 69, 2016, pp. 1–15.
19. The Port Authority of New York and New Jersey. *Airport Traffic Report*. The Port Authority of New York and New Jersey, 2015.
20. Hansen, M., and Y. Zhang. Operational Consequences of Alternative Airport Demand Management Policies: Case of LaGuardia Airport, New York. *Transportation Research Record: Journal of the Transportation Research Board*, 2005. 1915: 95–104.
21. National Weather Service. *National Weather Service Instruction*. Publication 10–813. National Weather Service, 2016.
22. Federal Aviation Administration, and National Weather Service. *Aviation Weather Services. Advisory Circular*. Publication AC 00-45G. Federal Aviation Administration and National Weather Service, 2010.
23. Federal Aviation Administration. *Traffic Flow Management System (TFMS) - ASPMHelp*. http://aspmhelp.faa.gov/index.php/Traffic_Flow_Management_System_(TFMS). Accessed August 2, 2017.
24. UQAM Atmosphere Sciences Group. *Aviation Routine Weather Report (METAR)*. http://meteocentre.com/doc/metar.html. Accessed August 2, 2017.
25. National Weather Service. *AWC - TAF Decoder*. http://www.aviationweather.gov/static/help/taf-decode.php, Accessed June 5, 2018.
26. Kim, A., and M. Hansen. Deconstructing Delay: A Non-Parametric Approach to Analyzing Delay Changes in Single Server Queuing Systems. *Transportation Research Part B: Methodological*, Vol. 58, 2013, pp. 119–133.
27. GlobalChange.gov. *National Climate Assessment*. http://nca2014.globalchange.gov/node/1961, 2014, Accessed June 5, 2018.
28. Jonkeren, O., P. Rietveld, and J. van Ommeren. Climate Change and Inland Waterway Transport: Welfare Effects of Low Water Levels on the River Rhine. *Journal of Transport Economics and Policy*, Vol. 41, No. 3, 2007, pp. 387–411.
29. Wang, Y., and D. Kulkarni. Modeling Weather Impact on Ground Delay Programs. *Presented at SAE 2011 Aero Tech Congress and Exhibition*, Toulouse, 2011.
30. Grabbe, S., B. Sridhar, and A. Mukherjee. Similar Days in the NAS: An Airport Perspective. In *AIAA Aviation Technology, Integration, and Operations*, AIAA, Los Angeles, 2013, pp. 1–14.
31. Allan, S. S., J. A. Beesley, J. E. Evans, and S. G. Gaddy. Analysis of Delay Causality at Newark International Airport. *In 4th USA/Europe Air Traffic Management R&D Seminar*, Santa Fe, N.Mex., 2001.
32. Hinton, G. E., and R. R. Salakhutdinov. Reducing the Dimensionality of Data with Neural Networks. *Science*, Vol. 313, No. 5786, 2006, pp. 504–507.
33. Shin, H.-C., M. Orton, D. J. Collins, S. Doran, and M. O. Leach. Autoencoder in Time-Series Analysis for Unsupervised Tissues Characterisation in a Large Unlabelled Medical Image Dataset. *Proc., 20th International Conference on, Machine Learning and Applications and Workshops (ICMLA)*, IEEE, Honolulu, Hawaii, 2011, pp. 259–264.
34. Ren, K. *Exploration of the Evolution of Airport Ground Delay Programs*. University of Alberta, Edmonton, 2017.
35. Federal Aviation Administration. *Newark Liberty International Airport Capacity Profile*. 2014. Federal Aviation Administration.

The Standing Committee on Airfield and Airspace Capacity and Delay (AV060) peer-reviewed this paper (18-06610).

Article

TRR
JOURNAL OF THE TRANSPORTATION RESEARCH BOARD

Transportation Research Record
2018, Vol. 2672(23) 82–94
© National Academy of Sciences:
Transportation Research Board 2018
Article reuse guidelines:
sagepub.com/journals-permissions
DOI: 10.1177/0361198118791624
journals.sagepub.com/home/trr
SAGE

Design and Construction of the World's First Full-Scale Electrically Conductive Concrete Heated Airport Pavement System at a U.S. Airport

Hesham Abdualla[1], Halil Ceylan[2], Sunghwan Kim[3], Mani Mina[4], Kristen S. Cetin[5], Peter C. Taylor[6], Kasthurirangan Gopalakrishnan[5], Bora Cetin[5], Shuo Yang[5], and Akash Vidyadharan[7]

Abstract

Airport agencies spend millions of dollars to remove ice and snow from airport pavement surfaces to achieve accessible, safe, and sustainable operations during the winter. Electrically conductive concrete (ECON) based heated pavement system (HPS) has gained attention as a promising alternative technology for preventing snow and ice accumulation by maintaining pavement surface temperatures above the freezing point. The objective of this study was to demonstrate the world's first full-scale ECON-based HPS at a U.S. airport. Two ECON slabs were designed and constructed in the General Aviation (GA) apron at the Des Moines International Airport (DSM), Iowa in 2016. Systematic design components were identified, and construction procedures were developed and implemented for ECON-based HPS. Using collected sensor data, the performance of the constructed and remotely-operated ECON slabs was evaluated under real weather conditions at DSM in the 2016–2017 winter season. The results demonstrate that ECON-based HPS have promising deicing and anti-icing capacities, promising to provide uniform heat distribution and prevent snow and ice accumulations on the entire area of application under various winter weather conditions, including extreme cold weather (i.e., arctic blasts).

Airport agencies spend millions of dollars per year to remove ice and snow from airport pavement surfaces to achieve accessible, safe, and sustainable operations during the winter season. The presence of ice or snow on such surfaces has the potential to cause flight delays impacting travel throughout the U.S. and worldwide, and may contribute to airplane incidents and accidents (*1*, *2*). Using snow removal equipment such as brooms and plows to remove snow/ice from congested areas is a challenging task and could also lead to accidents (*3*). The use of sand/chemical mixtures has the potential to cause foreign object damage to aircraft engines, cause corrosion to the overall airplane structure, and lead to environmental issues including possible contamination of groundwater and nearby bodies of water (*4*).

In recent years, new alternative approaches are receiving increased attention for mitigating problems associated with the presence of ice/snow on the paved areas of airfields and roadways, including engineering surfaces through application of nanotechnology for repelling water/ice (*5*) and heated pavement systems (HPS) (*6*). Developed technologies for HPS include hydronic heating (*7*), resistive cables embedded inside concrete structures (*8*), and electrically conductive concrete (*9–11*).

Electrically conductive concrete (ECON)-based HPS have gained attention as a promising alternative technology for preventing snow and ice accumulation by maintaining pavement surface temperatures above the

[1]Department of Civil, Construction and Environmental Engineering (CCEE), Iowa State University, Ames, IA
[2]CCEE, Program for Sustainable Pavement Engineering and Research (PROSPER), Iowa State University, Ames, IA
[3]CCEE, Institute for Transportation, Iowa State University, Ames, IA
[4]Department of Electrical and Computer Engineering, Iowa State University, Ames, IA
[5]CCEE, Iowa State University, Ames, IA
[6]National Concrete Pavement Technology Center, Iowa State University, Ames, IA
[7]InfraDrone LLC., Des Moines, IA

Corresponding Author:
Address correspondence to Hesham Abdualla: abdualla@iastate.edu

freezing point. The advantages of such methods include environmental friendliness, extended life and durability of concrete pavements, and, most importantly, melting ice and snow in a short period of time (*9*). ECON, by virtue of its lower electrical resistivity compared with that of conventional Portland cement concrete (PCC), behaves like an electrical conductor. As electric current flows through the conductive material network inside ECON, its resistance results in conversion of electrical energy into thermal energy (heat).

A previous study on ECON HPS implementation for transportation systems was focused only on bridge deck winter maintenance (*12*). However, the use of steel fibers as the conductive material for ECON HPS in that study cannot be used for airport pavement applications since steel fiber has potential to corrode, creating foreign objective debris and damaging aircraft tires. In addition, in the previous study, the electrode configuration and installation approach isolated each pair of electrodes, requiring supplemental construction procedures, which result in additional cost and time penalties. Recently, an ECON HPS system design using a newly developed ECON mix design recipe for airport pavement application (*9*, *13*) has been developed and evaluated by an Iowa State University (ISU) research team, using a prototype ECON heated slab. Even though there has been no guidance with respect to ECON HPS construction practices reported at actual airports, it is imperative to investigate full-scale implementation of HPS through field demonstrations in real airport environments. The outcome of such research will result in a thorough and practical understanding of heated concrete pavement system operations at actual airport sites.

Objective and Scope

The objective of this study was to demonstrate field implementation of ECON HPS in real airport environments. Two ECON test slabs were constructed in 2016 in the northern General Aviation (GA) apron of Des Moines International Airport (DSM) located in Iowa. To construct these slabs, the best design approach under the given airport conditions was first identified; then, accordingly, systematic construction details (electrode installation, slab instrumentation, power supply, controlling unit, etc.) were planned and executed. Following construction, the slabs were constantly monitored using an automated data acquisition system. The gathered data were analyzed, and the performance of the ECON slabs was thoroughly evaluated for a range of winter weather conditions characteristic of cold climate zones (ASHRAE Climate Zone 5, 6 and 7) (*14*), including different snow precipitation rates, ice formation conditions, and arctic-blast weather. The rest of the paper is organized in sections as follows: full-scale ECON HPS demonstration overview, system design, description of materials, instrumentation plan and installation methods, construction and instrumentation, DSM ECON heated pavement performance evaluation, and conclusions and recommendations.

Full-Scale ECON HPS Demonstration Overview

Construction Site Descriptions

The location for the ECON HPS field demonstration was in the north GA apron reconstruction area at the DSM in Iowa (see Figure 1). The selected location was close to the main power supply located in Building 69, approximately 15 ft from the ECON slabs. As part of a GA apron reconstruction project (construction project contractor: Foth Infrastructure and Environment, LCC, located in Johnston, Iowa), two slabs were utilized for the ECON HPS implementation and demonstration conducted by ISU researchers. Each slab was 15 ft long × 12.5 ft wide × 7.5 in. thick.

Key Components of Full-Scale ECON HPS Demonstration

The overall components of the ECON HPS for large-scale construction are illustrated in Figure 2. These include ECON as a conductive paving material (heating element), electrodes, temperature sensors, power supply, control unit, and PVC conduits and junction boxes. ECON for melting ice and snow on the surface can be placed in a thin concrete layer on top of a thicker PCC layer in the HPS structure to save on construction costs while providing adequate pavement structural capacity. The construction of the ECON layer should be isolated from embedded pavement light and cabling (*15*). For activation and deactivation of the ECON system under certain conditions, temperature sensors installed in the ECON layer are used to sense predetermined setpoint temperatures which should be defined for turning the system "on" and "off" (*14*, *15*). The performance of an ECON system can be monitored using sensors and a data acquisition system to estimate energy density and energy consumption.

System Design

ECON HPS system design is a procedure for determining HPS geometric features (slab dimension, layer thickness, etc.) and electrode configuration consistent with given airport environments (i.e., winter weather conditions,

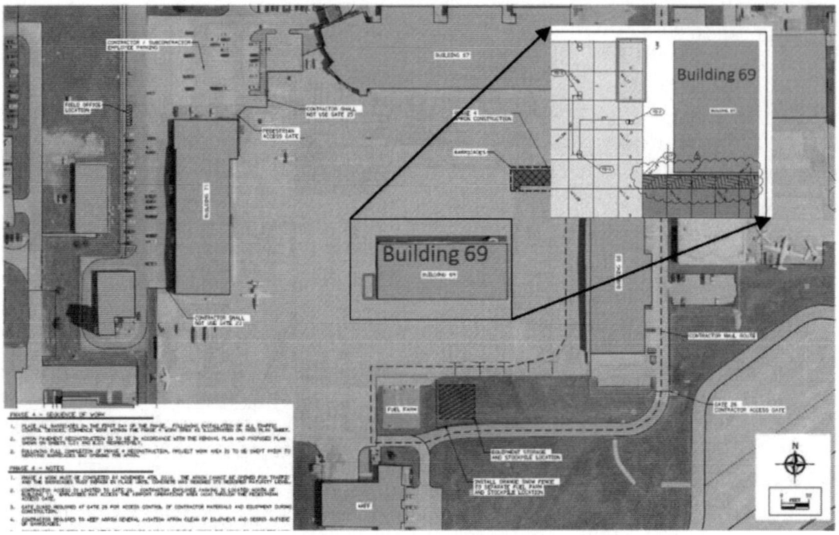

Figure 1. ECON HPS construction location at DSM.

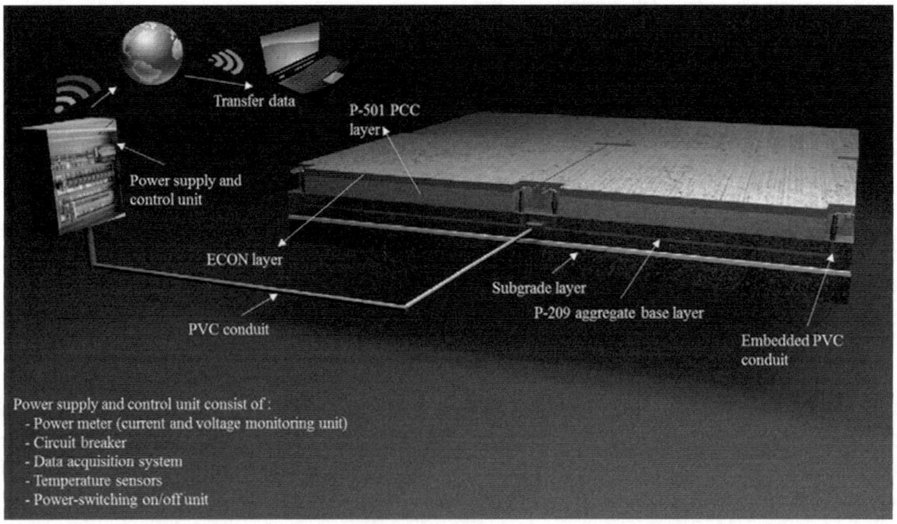

Figure 2. 3-D visualization schematic of ECON HPS.

estimated power density, electric power supply capacities available) to ensure adequate heating and structural performance during service life.

Design of ECON HPS Structures

To provide a clear understanding of ECON HPS construction, Figure 3 illustrates the overall 3-D design plan visualization and the structural layer. Each slab is 15 ft long × 12.5 ft wide × 7.5 in. thick and each has six electrodes. The designed ECON HPS structure consists of a two-layer 7.5-in-thick combined layer comprised of a 3.5 in. ECON top layer and a 4 in. P-501 PCC bottom layer, an 8 in. thick P-209 aggregate base course, and subgrade (Figure 3*a*). The purpose of only a thin ECON layer was to lower overall construction costs and to heat the surface as much as required to melt snow and ice without having to also heat additional concrete beyond the

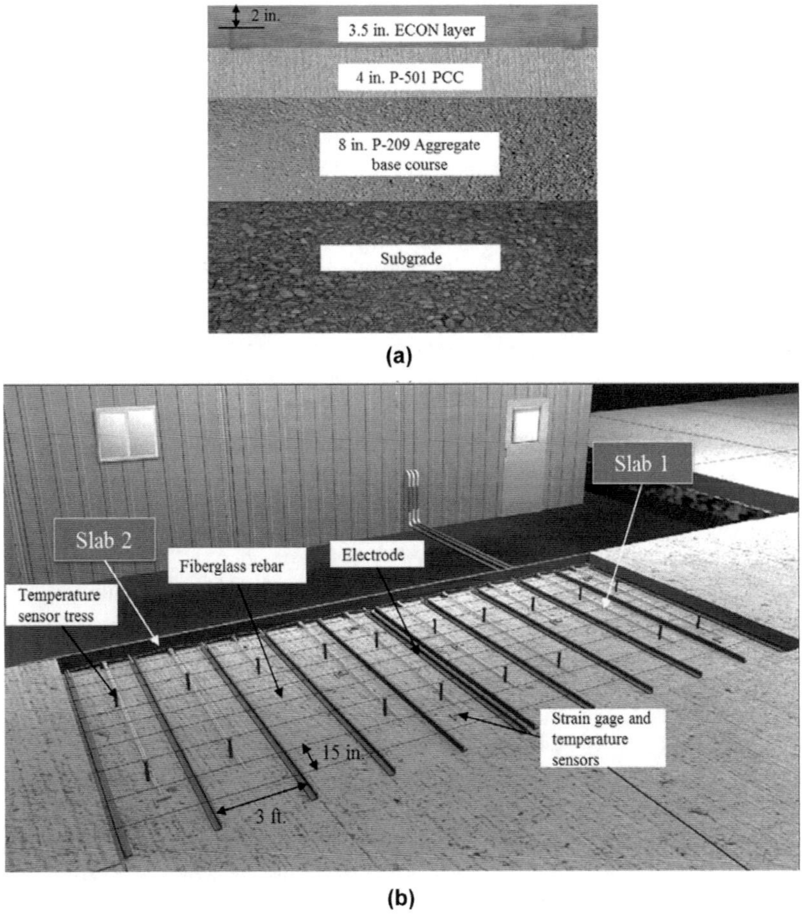

Figure 3. ECON HPS design plan visualization: (*a*) ECON HPS layer structure, and (*b*) 3-D installation plan for ECON HPS.

ECON layer. The concept of implementing two layers of concrete placed one over another is not new and has been successfully used as a sustainable construction technique for placement of concrete such as with concrete overlay or two-lift concrete paving (*16*).

To provide electric power to the ECON layer, six angle-shaped perforated electrodes (1.5 in. long × 1.5 in. wide × 1/8 in. thick) per slab were embedded 3 ft apart in the ECON layer (Figure 3*b*). Sensor systems were designed for measuring temperature and strain changes in the HPS during operation. An electric power sensor system was also designed to monitor and measure power, voltage, and current to evaluate the power density and efficiency of the ECON HPS.

Design of Electrode Configuration

The electrode spacing of 3 ft was determined based on the ECON material properties, ECON thickness, slab size, and the required amount of electric power. The electrode spacing was designed using a design flow (*9*) that met the design criteria, including the amount of snow/ice to be melted and the associated power density requirements. The power density requirement (design load) can either be experimentally obtained or estimated using the steady-state energy balance equation for required pavement heat output that considers expected rate of snowfall, air temperature, relative humidity, wind speed, and dimensions as well as pavement material characteristics (*14*, *15*). A finite-element (FE) method can also be used as a tool for estimation of the required power density and the snow/ice melting time for ECON HPS (*17*).

Description of Materials

ECON/P-501 PCC Mix Design

The mix design of ECON was developed and tested by researchers at ISU through a sequence of more than 40 trial mixes in the laboratory to attain a balance between

Table 1. ECON Mix Components and Proportions for Field Implementation

Components		Type	Content
Basic	Coarse aggregate	3/4 in. concrete stone	1,001.0 lb/yd^3
	Intermediate aggregate	3/8 in. chips	499.0 lb/yd^3
	Fine aggregate	Concrete sand	1,134.0 lb/yd^3
	Cement	Holcim type I/II	800.0 lb/yd^3
	Water	Tap water	337.0 lb/yd^3
Admixtures	Methylcellulose	Fiber dispersive agent	1.6 lb/yd^3
	DCI admixture	30% calcium nitrite solution, corrosion inhibitor and conductivity improving agent	42.0 lb/yd^3
	Carbon fiber 0.23 in.	Synthetic carbon fiber	1.0 (% Vol.)

Note: Aggregate contents presented are in saturated surface dry condition (SSD).

workability, mechanical properties, and electrical conductivity (*13*). The final ECON mix design for field implementation is described in Table 1. The mix properties and materials were checked and tested for conformance to FAA advisory circular P-501 (*18*). The 28-day electrical resistivity value of ECON mix was about 307 Ω-in. at 73°F ambient room temperature. The 28-day compressive strength and flexural strength of ECON mix were approximately 5,319 and 1,111 psi, respectively. The mix contained 1.0% carbon fibers by volume of the total concrete mix; 70% of carbon fiber was 0.23 in. length and 30% was 0.12 in. length and methylcellulose was used as an agent to evenly disperse the carbon fiber particles and thereby improve electrical conductivity. Derex corrosion inhibitor (DCI) is a liquid admixture meeting the requirement of ASTM C1582 that can both enhance the conductivity of the ECON mixture and mitigate the corrosive action of chlorides on reinforcing steel. The mix design for the P-501 PCC (bottom layer) was provided by the construction project contractor (i.e., Foth Infrastructure and Environment, LLC) in charge of the design of the reconstruction of the north GA apron at the DSM airport.

Electrodes

Although electrodes made from metallic materials are capable of enhancing electric current flow into an ECON layer since their conductivity is higher than that of ECON (*9*), the electrode material should have high resistance to corrosion so that electrodes will not degrade or crack, reducing efficiency and effectiveness. Angle-shaped perforated stainless steel 316L had 13 gaps (3/8 in. diameter) to allow ECON particles to interlock with the electrodes and to provide superior corrosion resistance (*9*, *19*). The electrodes should be sufficiently bonded with the ECON to provide sufficient heat performance. A nylon rod was selected to anchor the electrodes both because of its resistance to corrosion and its insulating properties that will prevent current leakage into the ground. The nylon-threaded rod and nut size was 4 in. × 3/8 in. diameter.

Sensors and Data Acquisition System

A set of sensors, both wired and wireless, was used to monitor temperature, moisture, strain, voltage, and electric current changes in HPS during operations. Before being selected as field instruments for the ECON slabs in DSM, these sensors had been evaluated using results from previous field studies to fulfill the needs of reliable long-term monitoring (*20*, *21*). The wired sensors installed in DSM ECON HPS included Arduino-based temperature sensors, Geokon strain gages, Sensirion relative humidity sensors, and Decagon electrical conductivity sensors. The data acquisition system combined a Campbell Scientific CR6 data logger with a Campbell Scientific AM16/32 multiplexer.

A wireless-based sensing system can provide advantages of being less time-consuming and less labor-intensive than a wired-sensor system (*20*, *21*). In this study, Monnit wireless sensors were used, including industrial and coin cell temperature, humidity, voltage, and current sensors. Each Monnit wireless sensor consisted of a sensing unit (usually a probe), a cable, a transmission head, and an antenna, with measurements first taken by the probe, then sent to the sensor head that would process and then transmit the data wirelessly through the equipped external antenna. A gateway was also used to capture data emitted from the collection of wireless sensors and upload them to pre-subscribed web-based software for real-time monitoring.

Remote Monitoring and Control System

The ECON HPS in DSM was equipped with surveillance cameras to monitor the ECON slabs, a control system to automatically turn the system on/off based on a setpoint temperature, and a remote access system for collecting sensor measurements and manually operating the ECON HPS using an off-site smart phone or computer.

Two on-site surveillance HD cameras were installed to provide reliable weather information from the slab site, to monitor the ECON slabs, and to capture the melting process activity during snow events.

Arduino board microcontrollers were used to control the ECON HPS through the universal serial bus (USB) communication interfaces provided by the on-site laptop by sending a signal in a code based on a set temperature. The code, developed using Arduino 1.6.13 software, based on C programming, was used to control the measuring interval and to convert data readings into the desired format and units through the serial USB connection to the laptop.

The remote access system used Splashtop software to provide access to all data collected through the airport Wi-Fi from an off-site system from which it could be controlled and monitored as needed. The operation of the ECON system could be chosen as either automated or manual using a remote access system through a smart phone or computer.

Other Materials

Fiberglass rebar installed in the ECON HPS (see Figure 3*b*) is highly corrosion resistant and coated with coarse quartz sand to provide bond adhesion to concrete. Fiberglass rebar is also electrically and thermally non-conductive and has a high tensile stress. ¼ in. diameter fiberglass rebar rods were continuously placed perpendicular on top of the electrodes and secured using nylon cable ties to prevent potential cracking due to environmental load effects. PVC conduit was used to house and secure wires for all sensor and electrode systems.

Instrumentation Plan and Installation Methods

Following material selection for the ECON HPS construction, the instrumentation plan and installation methods for electrodes, sensors, data acquisition system, and power supply system integrated with a remote-monitoring control system were developed based on extensive design and field construction experience on the part of the research team. Figure 4 depicts the installation methods and procedures used for electrodes and sensors, and Figure 5 shows the power supply panel integrated with the remote-monitoring control system.

Electrodes Instrumentation Plan and Installation Methods

A nylon rod 4 in. long and with a 3/8 in. diameter was used to anchor and fix the electrodes to prevent movement during the placement of the ECON layer, as shown in Figure 4*a*. The electrodes can be anchored to the existing pavement using an electric drill to drill holes to fix the electrodes to (Figure 4*b*). After fixing the electrode, electrical wire should be connected to the electrodes using gauge-ring wire connectors to provide power to the ECON layer for generation of heat (Figure 4*c*).

Sensors Instrumentation Plan and Installation Methods

Sensor locations were selected based on comprehensive plans, ensuring that critical locations in terms of temperature, strain, moisture, humidity, and electrical conductivity measurements were covered. Sensors were installed after electrode placement by first marking their locations on a temperature sensor tree, then mounting the temperature sensors. A ¼ in. stainless steel rod was used for the temperature sensor tree; it was covered with plastic tubing to eliminate its effect on electric current during ECON operation. Figure 4*b* shows the installation of the temperature sensor tree with a clip used to anchor the wires of the temperature sensors.

Strain gages were fixed using plastic chairs and steel plates at the surface of the PCC layer. These plastic chairs held strain gages during ECON placement. An electric drill was used to anchor the strain gages that were then placed in a perpendicular orientation to capture the strain measurements in different directions (Figure 4*e*).

After the installation of sensor and electrode systems, the wires for sensor and electrode systems were sorted and placed in PVC conduits connected to a junction box to prevent the wires from potential damage during construction. To prevent interruption or noise in the sensor data, the electrical wires of the electrodes were placed into separate conduits to prevent interference between sensor signals (Figure 4*f*).

Integration of Power Supply System and Remote Monitoring Control System

A single-phase power panel (120/240 V, 200 A power supply) was available in Building 69 near the ECON slabs. It was integrated with the electric sensors, data logger system, and laptop computer to support remote control and operation of the ECON slabs (see Figure 5). The power supply had six contactors connected to the electrodes through electrical wires, with each contactor connected to two electrodes. The contactors can receive remote

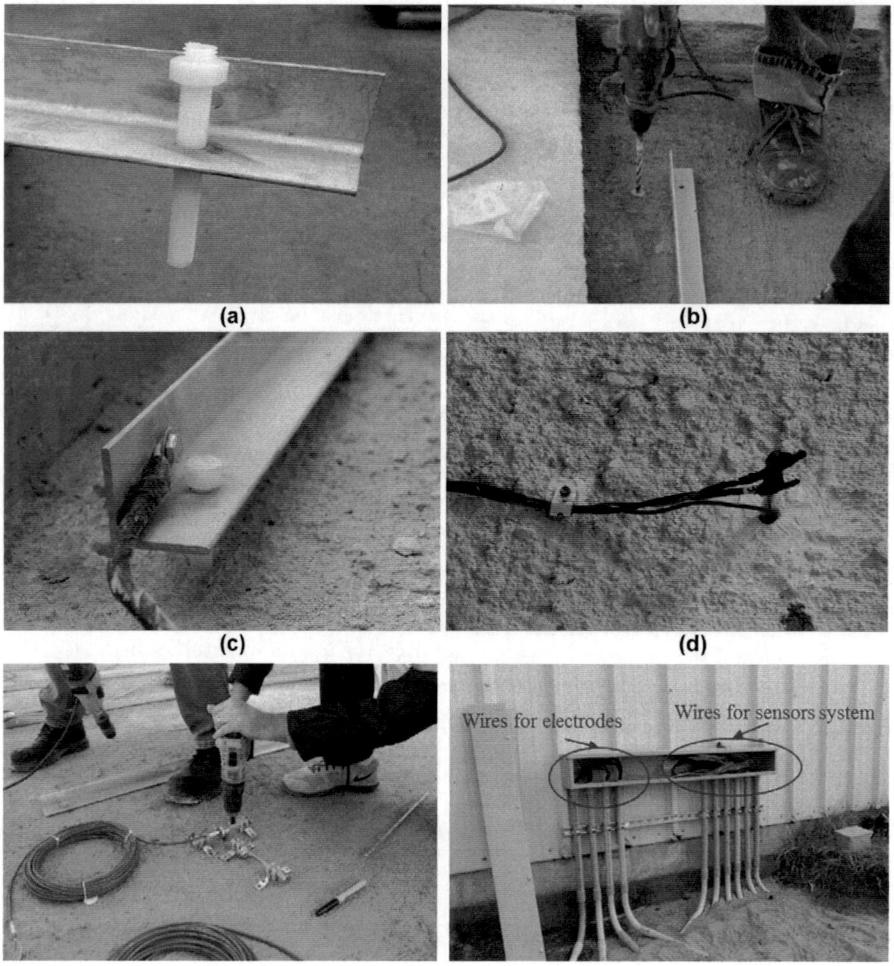

Figure 4. Electrode and sensor installations: (*a*) electrode with anchor, (*b*) electrodes installation, (*c*) electrode with electrical wire connection, (*d*) temperature sensor tree installation, (e) strain gages installation, and (*f*) PVC conduit and junction box.

signals to turn the system on and off based on temperature measurements from Arduino-based temperature sensors. Current and voltage sensors were installed to monitor and measure power density and consumption for the two ECON slabs, and a 60 A circuit breaker was used to control the current in each ECON slab.

Construction and Instrumentation

Figure 6 illustrates the construction process of the ECON HPS at DSM. Fixed-form paving was used with a form set on one side of the graded base for the entire 7.5 in. thickness. A 4 in.-thick layer (bottom layer) of P-501 PCC was placed on a prepared 8 in.-thick layer of P-209 aggregate based course (Figure 6*a*). Hand screed and hand-operated vibrators were used to strike the bottom layer off at the desired elevation and consolidate it. The surface of the bottom layer was screeded using brooms to create a rough surface to enhance the bonding with the ECON layer (Figure 6*b*). Six angel-shaped perforated electrodes per slab were placed on top of the PCC layer (Figure 6*c*), then sensors were placed in different locations (Figure 6*d*) and fiberglass rebar rods were continuously placed perpendicular on the electrodes (Figure 6*e*). Prior to placing the ECON layer, the surface of the bottom layer and the electrodes were thoroughly cleaned using an air-blower, brooms, and a pressure washer (Figure 6*f*).

A vibrating screed (Figure 6*g*), normally used for thin concrete pavements, was chosen to pave the two ECON layers, since the area of the ECON slabs was small. A

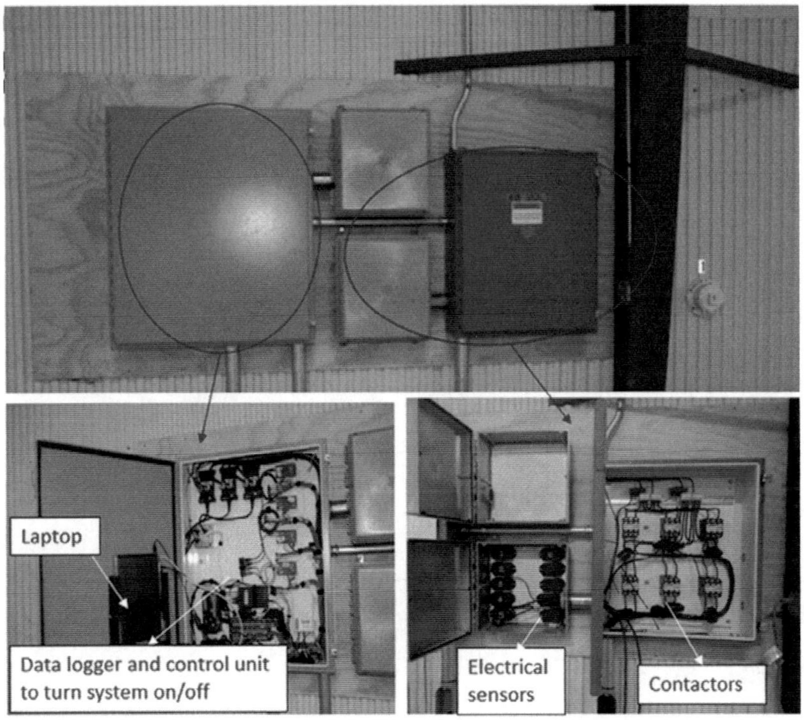

Figure 5. Power supply panel integrated with remote monitoring control system.

3.5 in.-thick ECON layer was paved on top of a 4 in.-thick P-501 PCC layer on November 03, 2016 (Figure 6*h*). The vibrating screed was attached with a vibrating motor to help consolidate and smooth the ECON layer. Caution was taken during the placement of the ECON layer to prevent any damage to the electrode and sensor systems. The ECON electric power supply system was tested to check wiring connections to the electrodes before placing the ECON layer.

A white-pigmented concrete compound was applied to the surface as a curing compound to avoid plastic and drying shrinkage and thermal stress cracking. A saw-cutting device was used to construct joints for control of cracking and to allow expansion and contraction due to temperature and moisture changes. The joints were cleaned and sealed using backer rod material and joint sealant. The bottom layer (4 in. PCC layer) was saw-cut at the joint so that both layers behave as monolithic pavement structures.

DSM ECON Heated Pavement Performance Evaluation

To evaluate the performance of the constructed ECON slabs in DSM under winter weather conditions, the ISU team operated the ECON HPS and collected data from December 2016 to February 2017.

Figure 7 illustrates the excellent ECON HPS performance results. The performance of the ECON HPS was very successful and capable of preventing snow accumulation on the ECON surface even while the apron all around the test slabs was covered with 1.2 in. of snow in December 10, 2016 (Figure 7*a* and *b*). Figure 7*c* shows the performance of the traditional snow removal method, *viz*., blows and brooms vehicles to remove snow, in comparison with the ECON HPS, that provides a safe environment and a reliable system, while mitigating the adverse effects of using traditional snow removal methods. To provide evidence of the temperature uniformity of the ECON surface, an infrared (IR) thermographic camera was used to capture the heat distribution during the operation (Figure 7*d*), and this IR heat map demonstrates that the ECON slab could generate sufficient heat for preventing snow accumulation over the entire slab surface.

Table 2 summarizes the constructed ECON slab operation performance results under various winter weather conditions during 2016–2017, ranging from light to severe snow storms and including arctic blast weather conditions that continued for two days (December 18 to 19 in 2016) (*22*). Two scenarios—anti-icing and deicing

Figure 6. ECON HPS construction procedures at DSM: (*a*) P-501 PCC placement, (*b*) screed PCC surface, (*c*) electrode installation, (*d*) sensors installation, (*e*) fiberglass and nylon cable tie installation, (*f*) ECON surface cleaning, (*g*) vibrating screed, and (*h*) ECON placement.

(a) (b) (c) (d)

Figure 7. ECON HPS performance results: (*a*) a 1.2 in. thickness of snow accumulation at DSM in December 10, 2016, (*b*) ECON performance results in December 10, 2016, (*c*) ECON performance vs. traditional method, and (*d*) infrared thermographic image of the ECON slabs.

Table 2. Summary of ECON Performance

Date	Condition	Average air temp. (°F)	Wind speed (mph)	Total snow thickness (in.)	Average power density (W/ft^2)	Operation time minutes (h)	Energy consumption (kWh/ft^2)
Dec. 10, 2016	Anti-icing	25	8	1.2	38	420 (7)	0.27
Dec. 18–19, 2016	Deicing	−11	16	1	39	1,286 (21.4)	0.84
Jan. 25, 2017	Deicing	14	14	0.5	38	90 (1.5)	0.06
Feb. 08, 2017	Deicing	14	13	1.5	35	210 (3.5)	0.12
Feb. 24, 2017	Deicing	22	17	0.8	37	150 (2.5)	0.09
Mar. 13, 2017	Deicing	21	13	1.3	33	120 (2)	0.07

Note: The anti-icing case is defined as preventing ice or snow accumulation on paved surface by preheating the ECON surface prior to snowfall or ice and the deicing case is defined as melting ice or snow following ice or snow accumulation on a paved surface.

cases—were executed to examine the ECON HPS operation under different weather conditions. The anti-icing case refers to preheating the ECON slabs before snowfall, while the deicing case refers to heating the ECON slabs during/or after the snowfall. The wireless electric sensors installed measure electric current changes during ECON HPS operations while applying a 210 V voltage. The average electric current measurements during ECON HPS operations were used to estimate the average power density and the energy consumption values listed in Table 2.

The ECON performance results of December 10, 2016 demonstrate how ECON HPS works under anti-icing conditions. The ECON HPS was preheated to prevent ice and snow accumulation on ECON slabs before snowfall, then turned on at 5:40 a.m. (i.e., about 3 h before the snowfall event) based on a weather forecast reporting that snowfall was expected to begin at 7:40 a.m. The actual snowfall occurred from 9:00 a.m. to 12:40 p.m. (i.e., about 4 h snowfall event), thus the ECON HPS operational time was from 5:40 a.m. to 12:40 p.m. (i.e., about a 7 h operation), sufficient to prevent snow accumulation on the ECON surface. The energy consumption for this operation was 0.22 kWh/ft^2, while the average air temperature, relative humidity, and wind speed were 23°F, 74%, and 8 mph respectively.

The ECON performance results of December 18 to 19, 2016 demonstrate how ECON HPS worked under arctic blast weather conditions. The ECON HPS was turned on after snow and ice had accumulated on ECON slabs, and outdoor ambient temperatures were very low (i.e., an average of −11°F). The melting process took longer in comparison with other events (Table 2), although eventually snow and ice were melted. The energy consumption for this operation was 0.59 kWh/ft^2.

As seen for other dates of operation, under deicing conditions existing from January 25, 2017 to March 13, 2017, the ECON HPS appeared to be operating effectively under these weather conditions. The lowest energy consumption was 0.05 kWh/ft^2 to melt 0.5 in. of snow accumulation, while the highest energy consumption was 0.10 kWh/ft^2 to melt 1.5 in. of snow accumulation on the ECON slabs. The energy consumption varied due to the impact of the snowfall time, air temperature, operation time, and wind speed.

Conclusions and Recommendations

The goal of this study was to demonstrate system design, construction, and performance of an ECON HPS, the first such facility built at a U.S. airport. A full-scale implementation of two ECON slabs was designed and constructed at DSM, Iowa in 2016. Sensor instrumentations and a control unit were incorporated into the design of the ECON HPS to monitor the performance and support remote operation of the system. The ECON's power density, energy consumption, and performance during deicing and anti-icing were summarized and evaluated based on actual data obtained from the set of sensors implemented. The following summarize the results discussed in this work:

- Two ECON test slabs were constructed in the northern GA apron of DSM, Iowa in 2016 as the first full-scale ECON heated airport pavement system in a US airport as well as in the world. The constructed ECON HPS demonstrated deicing and ant-icing capacities for providing uniform heat distribution and preventing snow and ice accumulations on the entire area of application under various winter weather conditions, including extremely cold weather (i.e., arctic blast).
- The energy consumption of the ECON HPS for a deicing application ranged from 0.05 kWh/ft^2 to 0.10 kWh/ft^2 required to melt 0.5 to 1.5 in. of snow/ice accumulations under various winter weather conditions, except the arctic blast event that required 0.59 kWh/ft^2 to melt 1 in. of snow/ice accumulations. The energy consumption of the ECON HPS for the anti-icing application was estimated at 0.22 kWh/ft^2 for preventing snow and ice accumulations for 7 h.
- The ECON HPS in DSM was constructed using a two-lift approach that placed a thin ECON layer (top layer) on a P-501 PCC layer (bottom layer). This approach can be effective not only to save in construction costs instead of constructing a single thick ECON layer, but also to facilitate the electrode installation and electrical wire connections.
- The surface of the P-501 PCC layer (bottom layer) was screeded to create a rough surface to provide satisfactory bonding between the two layers.
- Perforated stainless steel was used for the electrodes because of its high corrosion resistance, and so that electrodes will not be reduced in efficiency; the use of stainless steel also eliminated cracking potential that would be possible with steel or galvanized steel.
- Electrodes should be placed at least 1 to 2 in. below the ECON surface so as not to interfere with a paver's vibrator, pan, and augur when slipform paving is used during large-scale construction.

Acknowledgments

This paper was prepared from a study conducted at Iowa State University under the Federal Aviation Administration (FAA) Air Transportation Center of Excellence Cooperative Agreement 12-C-GA-ISU for the Partnership to Enhance General Aviation Safety, Accessibility and Sustainability (PEGASAS). The authors would like to thank the current project Technical Monitor, Mr. Benjamin J. Mahaffay, and the former project Technical Monitors, Mr. Jeffrey S. Gagnon (interim), Mr. Donald Barbagallo, and Dr. Charles A. Ishee, for their invaluable guidance on this study. The authors also thank Gary L. Mitchell at the American Concrete Pavement Association, and Mr. Gordon Smith and Mr. Dan King in ICPA for valuable discussions and comments on concrete pavement construction. The authors also would like to thank the PEGASAS Industry Advisory Board members for their valuable support. The authors also would like to thank Mr. Bryan Belt, Mr. Mark Duffy, Mr. William Konkol at the Des Moines International Airport, and Mr. Adam Wilhelm and Mr. Andrew Gettler, Foth Infrastructure and Environmental, LLC, and Mr. Dan Hutton of Kingston Services, LLC, for their full support during construction. Special thanks are expressed to Zoltek, the Candlemakers Store (TCS), and WR Grace & Co for providing carbon fiber, methyl cellulose, and corrosion inhibitor admixture, respectively. The authors would like to express their sincere gratitude to Mr. Robert F. Steffes, ISU CCEE PCC Lab Manager, for his significant assistance with the lab and field investigations, and other ISU graduate and undergraduate student research assistants.

Author Contributions

The authors confirm contribution to the paper as follows: study conception and design: Hesham Abdauual, Halil Ceyaln, Sunghwan Kim, Kasthurirangan Gopalakrishnan, Kristen S. Cetin, Peter C. Taylor, Bora Cetin; data collection: Hesham Abdualla, Shuo Yang, Halil Ceyaln, Sunghwan Kim, Akash Vidyadharan; analysis and interpretation of results: Hesham Abdauual, Halil Ceyaln, Sunghwan Kim, Kasthurirangan Gopalakrishnan; draft manuscript preparation: Hesham Abdauual, Halil Ceyaln, Sunghwan Kim, Kasthurirangan Gopalakrishnan, Peter C. Taylor. All authors reviewed the results and approved the final version of the manuscript.

References

1. McCartney, S. The Case for Heated Runways. *Wall Street Journal*, February 2014. http://www.wsj.com/news/article_email/SB10001424052702304914204579392883809689994-lMyQjAxMTA0MDIwMDEyNDAyWj. Accessed July 2017.
2. *Airport Winter Safety and Operation*. Advisory Circular 150/5200-30C. Federal Aviation Administration, U.S. Department of Transportation, 2008.
3. Hatch, C., S. Hartley, O. Roberto, and A. Alcala. *National Transportation Safety Board Aviation Accident Factual Report*. National Transportation Safety Board, December 2015. https://app.ntsb.gov/pdfgenerator/ReportGeneratorFile.ashx?EventID=20151224X81125&AKey=1&RType=Factual&IType=LA. Accessed July 2017.
4. Xi, Y., and P. J. Olsgard. *Effect of De-Icing Agents (Magnesium Chloride and Sodium Chloride) on Corrosion of Truck Components*. Final Report for Report No. CDOT-DTD-R-2000-10. Colorado Department of Transportation, Denver, CO, 2000.
5. Arabzadeh, A., H. Ceylan, S. Kim, K. Gopalakrishnan, and A. Sassani. Superhydrophobic Coatings on Asphalt Concrete Surfaces. *Transportation Research Record: Journal of the Transportation Research Board*, 2016. 2551: 10–17.
6. Ceylan, H., K. Gopalakrishnan, S. Kim, and W. Cord. Heated Transportation Infrastructure Systems: Existing and Emerging Technologies. *Proc., 12th International Symposium on Concrete Roads*, Prague, Czech Republic, September 2014. http://lib.dr.iastate.edu/ccee_conf/23.
7. Hoppe, E. J. Evaluation of Virginia's First Heated Bridge. *Transportation Research Record: Journal of the Transportation Research Board*, 2001. 1741: 119–206.
8. Joerger, M. D., and F. C. Martinez. *Electrical Heating of I-84 in Land Canyon, Oregon*. Report No. FHWA-OR-RD06-17. Oregon Department of Transportation, Salem, 2006.
9. Abdualla, H., H. Ceylan, S. Kim, K. Gopalakrishnan, P. C. Taylor, and Y. Turkan. System Requirements for Electrically Conductive Concrete Heated Pavements. *Transportation Research Record: Journal of the Transportation Research Board*, 2016. 2569: 70–79.
10. Gopalakrishnan, K., H. Ceylan, S. Kim, S. Yang, and H. Abdualla. Electrically Conductive Mortar Characterization for Self-Heating Airfield Concrete Pavement Mix Design. *International Journal of Pavement Research and Technology*, Vol. 8, No. 5, 2015, pp. 315–324.
11. Gomis, J., O. Galao, V. Gomis, E. Zornoza, and P. Garcés. Self-Heating and Deicing Conductive Cement. Experimental Study and Modeling. *Construction and Building Materials*, Vol. 75, 2015, pp. 442–449. https://doi.org/10.1016/j.conbuildmat.2014.11.042.
12. Tuan, C. Y. *Implementation of Conductive Concrete for Deicing (Roca Bridge)*. Final Report for Project No. SPR-PL-1(04) P565. University of Nebraska-Lincoln, NE, 2008.
13. Sassani, A., H. Ceylan, S. Kim, K. Gopalakrishnan, A. Arabzadeh, and P. C. Taylor. Influence of Mix Design Variables on Engineering Properties of Carbon Fiber-Modified Electrically Conductive Concrete. *Construction and Building Materials*, Vol. 152, 2017, pp. 168–181.
14. ASHRAE. *ASHRAE Handbook - HVAC Applications American Society of Heating, Chapter 51 – Snow Melting and Freeze Protection*. American Society of Heating, Refrigeration and Air-Conditioning Engineer, Inc., Atlanta, GA, 2015, pp. 51.1– 51.20.
15. *Airside Use of Heated Pavement Systems*. Advisory Circular 150/5370-17. Federal Aviation Administration, U.S. Department of Transportation, 2011.
16. Gillen, S. L., A. S. Brand, J. R. Roesler, and W. R. Vavrik. Sustainable Long-Lift Composite Concrete Pavement for Illinois Tollway. *Proc., International Conference on Long-Lift Concrete Pavements*, Seattle, Washington, September 2012, pp. 18–21.
17. Abdualla, H., K. Gopalakrishnan, H. Ceylan, S. Kim, M. Mina, P. C. Taylor, and K. S. Cetin. Development of a Finite Element Model for Electrically Conductive Concrete Heated Pavements. Presented at 96th Annual Meeting of the Transportation Research Board, Washington, D.C., 2017.
18. *Standards for Specifying Construction of Airports*. Advisory Circular 150/5370-10G. Federal Aviation Administration, U.S. Department of Transportation, 2014.
19. Hilbert, L. R., D. Bagge-Ravn, J. Kold, and L. Gram. Influence of Surface Roughness of Stainless Steel on Microbial Adhesion and Corrosion Resistance. *International Biodeterioration and Biodegradation*, Vol. 52, 2003, pp. 175–185. https://doi.org/10.1016/S0964-8305(03)00104-5.
20. Ceylan, H., K. Gopalakrishnan, S. Kim, P. C. Taylor, M. Prokudin, and A. F. Buss. Highway Infrastructure Health Monitoring Using Micro-electromechanical Sensors and Systems (MEMS). *Journal of Civil Engineering and Management*, Vol. 19, 2013, pp. S188–S201. http://dx.doi.org/10.3846/13923730.2013.801894.

21. Yang, S., H. Ceylan, K. Gopalakrishnan, and S. Kim. Smart Airport Pavement Instrumentation and Health Monitoring. Presented at 2014 FAA Worldwide Airport Technology Transfer Conference, Galloway, NJ, 2014.
22. The Weather Company. Bitter Cold Arctic Air Sets Dozens of Record Lows in the Midwest and Plains. *Weather Underground*, December 19, 2016. https://www.wunderground.com/news/back-to-back-arctic-cold-blasts-midwest-east-december-2016. Accessed July 2017.

The Standing Committee on Aircraft/Airport Compatibility (AV070) peer-reviewed this paper (18-00578).

Although the FAA has sponsored this project, it neither endorses nor rejects the findings of this research. The presentation of this information is in the interest of invoking comments by the technical community with respect to the results and conclusions of the research.

TRR
JOURNAL OF THE TRANSPORTATION RESEARCH BOARD

Article

Transportation Research Record
2018, Vol. 2672(23) 95–105

DOI: 10.1177/0361198118796066
journals.sagepub.com/home/trr
SAGE

Development of Smartphone Accelerometer-Based Airfield Friction Assessment Tools

Andrew B. Ward[1] and Craig A. Rutland[2]

Abstract

Airfield managers have identified a need for contingency/emergency friction assessment tools that are both cost-effective and commonly found. Presently, airfield managers and pavement inspection teams prefer to measure friction at high speed with continuous friction measuring equipment (CFME) like the Findley Irvine GripTester. This device and the labor required to operate it are cost prohibitive. This paper explores an alternative, more economical measure of a surface's frictional characteristics, that is, smartphone accelerometers. Accelerometers were standard friction assessment tools before CFMEs became available. Accelerometer-based friction testers measure the peak deceleration of a vehicle during braking. This acceleration, specifically the vehicle's deceleration, is proportional to the surface's friction coefficient. This paper covers research conducted to verify that smartphone accelerometers have sufficient accuracy and precision to operate as deceleration-based friction assessment tools. Findings reported here show that smartphone accelerometers, given sufficient conservatism, can function as deceleration-based friction assessment tools. Along with existing models, new regressions are presented that show correlation between smartphone deceleration measurements and high-speed GripTester measurements. The present work found that improved correlation was limited by common problems associated with poor reproducibility in braking tests.

Friction surveys are a necessary part of airfield management and pavement inspection in general. The frictional properties of a surface determine vehicle skid and rolling resistance. Friction surveys can also give a measure of a surface's relative safety in cases of surface contamination such as rubber, snow, or oil. A widely used device by both military and civilian airfield managers to conduct friction surveys is the Findley Irvine Mark 2 GripTester (GT) operating at 40 and 60 mph. The GT is in the class of continuous friction measuring equipment (CFME). It is typically run behind a pickup truck or van. Onboard the driving vehicle is a water system that sends a 1-mm water-film depth to the rear testing tire of the GT, as per ASTM E2340. High-speed GT friction surveys require not only an expensive friction tester but also multiple technicians to operate it. There are few options in emergency situations to make a measure of a surface's friction response when such equipment and technician support are unavailable. One such alternative is a slower walk-behind CFME like the Findley Irvine micro GripTester (mGT) or the ASFT T2Go. These friction testers operate under the same fixed-slip principle as the full-size GT but at much lower speeds, that is, 1 to 3 mph. Though these walk-behind devices operate at lower, more convenient speeds, they are still expensive, require trained technician support, are not widely available, and may not readily correlate to the faster, preferred, GT.

A second alternative relies on the principle that vehicle deceleration on a surface is related to the frictional properties of that surface. Decelerometers are devices that measure the deceleration of a vehicle while braking. These devices rely upon an inertial sensor, accelerometer, to measure deceleration. The accelerometers are usually packaged alongside other inertial sensors into a single inertial measurement unit (IMU) chip. It is these small IMUs that are found inside modern smartphones. Testing of these smartphone IMUs is necessary to determine suitability for friction testing. Throughout this

[1]Airfields and Pavements Branch, Geotechnical and Structures Laboratory, U.S. Army Engineer Research and Development Center, Vicksburg, MS
[2]Civil Engineering Branch, Engineering Division, Air Force Civil Engineer Center, FL

Corresponding Author:
Address correspondence to Andrew B. Ward: andrew.b.ward@usace.army.mil

article, the terms accelerometer and decelerometer are considered synonymous.

Objectives and Scope

The primary objective of this work was to evaluate the use of modern smartphones as deceleration-based friction assessment tools. Android and Apple-based apps were developed and various smartphones were tested for accuracy and repeatability of deceleration measurements. Braking tests were performed on concrete surfaces in dry pavement conditions. Smartphone accelerometers were tested against ASTM-accepted deceleration equipment, that is, Bowmonk Airfield Friction Meter 2 Mark 3 (Bowmonk). Correlation was attempted between the smartphone accelerometers and the more popular GT. Methods for smartphone accelerometer friction testing and satisfactory correlations with accepted friction testing equipment are presented here. Satisfactory correlations were those with a Pearson's r-value of 0.8 or greater.

Literature Review

Pavement Friction

Frictional characteristics are of considerable interest to both aeronautic and ground-based agencies. Of particular interest in this article is the interaction of rubber tires with a pavement's surface. A measurement of friction from such an interface relies on both the properties of the rubber tire and the surface. The rubber tire has a much lower elastic modulus (Young's modulus) than the pavement surface and therefore undergoes most of the deformation during the interaction, whereas the size and amount of aggregate particles in the pavement dictate the maximum friction available. A measurement of friction depends on a myriad of factors including but not limited to temperature of surface and tire, molecular makeup of surface (texture) and tire, driving force of tire, and surface contamination. Although these factors are many and difficult to control, early work (*1–3*) shows that much of this tire–surface interaction can be described empirically as the sum of two distinct forces, that is, hysteresis and adhesion. The hysteresis force, F_h, is a result of the bulk deformation of the rubber tire about the aggregate particles present in the pavement's surface and the corresponding loss of kinetic energy as thermal energy. More generally, F_h takes into account the internal friction of the rubber tire during deformation and allows some of the tire's energy to be dissipated as heat. The adhesion force, F_a, results from the microscale bonding of the tire to the pavement surface (*4*). It should be noted here that friction is rarely presented as a measure of force. It is often presented as a dimensionless coefficient written as μ. This friction coefficient is defined as the quotient of the total frictional force between some interface and the total vertical load on the interface (Equation 1):

$$\mu = \frac{F_f}{F_N} \tag{1}$$

where F_N is the total vertical load, or normal force, acting on the tire and F_f is the total friction force acting on the tire. This friction coefficient is broken into two sub categories: static and kinetic. The coefficient of static friction describes the frictional force between two objects that are stationary with respect to each other, for example, a tire rolling without slipping over pavement. The coefficient of kinetic friction describes the frictional force between two objects that are moving with respect to each other, for example, a tire skidding across pavement.

A key contributor to friction response is the texture of the surface. Table 1 shows the wavelength thresholds for pavement texture as defined by the World Road Association (PIARC) (*5*).

The adhesion force is largely dependent on surface microtexture. The mean profile depth (MPD) and wavelength of a surface microtexture determine the magnitude of the adhesion force. The hysteresis force is largely dependent on the larger textures, that is, macrotexture and above. The size of a pavement's aggregate particles (>0.5 mm) dictate the magnitude of the bulk deformation of the rubber tire and therefore the magnitude of the hysteresis force. Surface texture becomes particularly important when considering wet pavement conditions. During such conditions, macrotexture governs how efficiently water is drained from the surface. At low speeds in which the tire is allowed ample time to expel the majority of water from the tire–pavement contact area, the microtexture contribution to friction grows. Higher speeds shorten the time allowed for the tire to expel water from the tire–pavement contact area reducing the microtexture contribution and, in extreme cases, eliminates it altogether during wet skidding or hydroplaning (*6*).

Friction Testing Equipment

Texture measurements are typically difficult and time consuming to obtain. It is often preferred to measure the friction coefficient directly or analogously. There are a number of full-scale friction testers available that operate at traffic speeds. These friction testers can be broken

Table 1. Pavement Texture Thresholds

Texture	Wavelength (mm)
Microtexture	0.0–0.5
Macrotexture	0.5–50
Megatexture	50–500
Roughness	>500

down into five types: locked-wheel, fixed-slip, variable-slip, side-force, and decelerometer. Of particular importance to the present work are the fixed-slip and decelerometer types.

The braking slip ratio is defined as the ratio of the angular velocity of the tire under application of torque and the angular velocity of the tire un-braked. Fixed-slip friction testers measure friction by slipping a test tire at a fixed-slip ratio. The slip ratio is set to 10% to 20%. The GripTester is set to 15% slip. These friction testers have one or more driving tires in the front and one smooth test tire in the rear. It is the smooth test tire that is slipped by chain. A water system is held on board the driving vehicle and delivers water to the pavement just in front of the test tire (*6*). For the GripTester, the driving vehicle typically runs at two speeds, 40 and 60 mph.

Decelerometers were popularized in the mid to late 20th century as a measure of pavement friction and aircraft stopping performance. Decelerometers measure the deceleration of a vehicle while braking and report the value as a percentage of the acceleration of gravity, $\%g$. These devices were used to report the runway condition reading (RCR) of a section of pavement. The RCR value is determined by fitting a decelerometer to an aircraft or ground vehicle and measuring the deceleration as the vehicle enters the locked-wheel state. The peak deceleration value was then reported as the RCR in units of ft/s^2. Correlation of standard RCR measurements with runway surface conditions (RSC) are as follows; $12 < \text{RCR} < 23$ has an RSC of "Dry", $5 < \text{RCR} < 12$ has an RSC of "Wet", and $\text{RCR} < 5$ has an RSC of "Icy".

Decelerometer measurements were known to have poor correlation with actual aircraft stopping performance in wet and flooded pavement conditions (*7*, *8*). The RCR was known, however, to report conservative values for aircraft stopping performance on snow- and ice-covered runways (*8*). As such, the FAA does not approve decelerometer measurements for condition surveys during wet pavement conditions, but does approve them during winter conditions (*9*).

Models of Friction

Due to friction being a function of speed, some theoretical considerations are necessary in order to correlate friction measurements made at different speeds. One of the most commonly used models of friction on pavement surfaces is the Penn State Model (*10*):

$$\mu = \mu_o e^{-\frac{PNG}{100}S} \tag{2}$$

where

μ is the friction coefficient
μ_o is the zero-speed friction coefficient
S is the slip speed

PNG is the percent normalized gradient as defined in Equation 3:

$$PNG = -\frac{100}{\mu}\frac{d\mu}{dS} \tag{3}$$

A more compact form of Equation 2 can be found in Equation 4 (*6*):

$$\mu = \mu_o e^{-S/S_p} \tag{4}$$

where S_p is the speed constant. Equation 4 was later used as the basis for a more standardized model of friction and texture called the International Friction Index (IFI) model seen in equations 5, 6, and 7:

$$S_p = a + bTX \tag{5}$$

$$FR(60\ kph) = FR(S)e^{-(S-60)/S_p} \tag{6}$$

$$F(60\ kph) = A + B\ FR(60\ kph) \tag{7}$$

where a and b are calibration constants defined by what texture measurement device is used. The standard source of this macrotexture measurement is the Nippo Sangyo Circular Track/Texture Meter (CTM) that yields $a = 14.2$ and $b = 89.7$ (ASTM E1960). TX is the texture measurement provided as MPD. MPD is defined as the deviation of the profile peak from the profile mean. As per ASTM E1845, the profile is split into 100-mm segments, any slope is removed by linear regression, and the 100-mm segments are further divided into halves. The mean of the 100-mm segment is removed from the average of the two half-peak heights and reported as the MPD in units of mm. $FR(60\ kph)$ is the friction coefficient adjusted to a slip speed of 60 kph, $FR(S)$ is the friction coefficient at slip speed, S. $F(60\ kph)$ is the predicted friction coefficient at a slip speed of 60 kph incorporating a texture measurement. Calibration constants A and B are given by ASTM E1960 in the case that a CTM is used for the macrotexture measurement and the Nippo Sangyo Dynamic Friction Tester (DFT) is used for the friction measurement at 20 kph: $A = 0.081$ and $B = 0.732$. Should another friction tester be used, A and B are found by fitting a linear regression to the standard CTM/DFT parameters. The values of S_p and $F(60\ kph)$ are reported as the IFI for a surface. Due to its inclusion of macrotexture, the IFI model is useful for correlating friction testers operating at different speeds.

Hardware/Software

Accelerometers

There are three types of accelerometers: mechanical, solid state, and microelectromechanical systems (MEMS)-based. Mechanical accelerometers operate under the

Table 2. Smartphone Inertial Sensors

Manufacturer	Model	Accelerometer	Magnetometer	Gyroscope
Samsung	Galaxy S7	STM LSM6DS3	AKM AK09911	STM LSM6DS3
LG	V20	Bosch LGE Accel.	Bosch LGE Magnet.	Bosch LGE Gyro.
Huawei	Nexus 6P	Bosch BMI160	Bosch BMM150	Bosch BMI160
HTC	10	hTC Corp Accel. Sensor	hTC Corp. Magnet. Sensor	hTC Corp. Gyro. Sensor
Apple	iPhone 7	Invensense 773C	Magnetometer Sensor	Invensense 773C

Note: All sensors listed are manufactured using MEMS technology. Each smartphone came with a magnetometer and an IMU containing at least one accelerometer and gyroscope per 3D axis.

principle of a mass on a spring. When this mass is acted on by an outside acceleration, it moves from its reference position. The displacement of the mass in three-dimensional (3D) space is proportional to the acceleration that acted on it. Solid state accelerometers operate under similar principles to mechanical accelerometers except, instead of measuring the displacement of a mass, they measure acoustic waves/vibrations, electrical response, and so forth, caused by outside acceleration. Lastly, MEMS-based accelerometers operate on the same principles as mechanical and solid state accelerometers at the micro scale (*11*). MEMS manufacturing techniques allow for mechanical and solid state accelerometers and other inertial sensors to be fabricated into small chips at the micro scale. Although MEMS-based inertial sensors are convenient and cost-effective, they exhibit a number of unique error characteristics. These error characteristics are explained in detail by Woodman (*11*). Of note are an offset bias in the sensor's output signal that is constant with time for acceleration measurements, a white noise perturbation in the output signal, and bias fluctuations in the output signal caused by temperature changes.

Android Development

Hardware. A market survey was conducted, and five of the most widely used smartphones were purchased. Among them were four Android devices: Samsung Galaxy S7, LG V20, Huawei Nexus 6P, and HTC 10. Wide variability in smartphone chipset was not considered a problem as processor/memory speed and memory capacity for each smartphone was found to be sufficient for rapid deceleration measurements. What was in question for each smartphone was IMU choice. Information is sparse regarding specific smartphone IMU hardware. Hardware information collected from the phones themselves is presented in Table 2.

The accelerometers on the IMU chip are in principle similar to full-size mechanical or solid state accelerometers. Of interest to this project are two sensors on board each smartphone (accelerometer and magnetometer). The accelerometer is used to measure acceleration in 3D. The magnetometer is used to measure

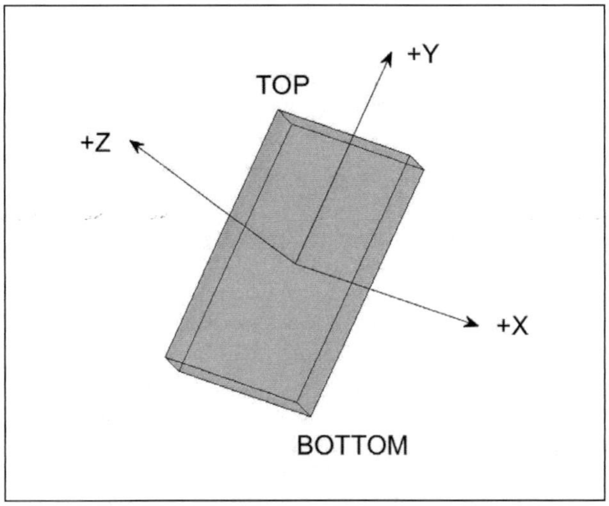

Figure 1. Three-dimensional phone orientation.

absolute rotation in 3D. On the Android platform, acceleration is presented in standard SI units of meters per second squared, $\frac{m}{s^2}$. The absolute rotation is presented as a quaternion as seen in Equation 8 (see phone orientation in Figure 1). In Figure 1, "TOP" represents the top left of the phone, "BOTTOM" represents the bottom right, " + X" represents the positive x-axis, " + Y" the positive y-axis, and " + Z" the positive z-axis:

$$\vec{q} = cos\left(\frac{\theta}{2}\right) + x\ sin\left(\frac{\theta}{2}\right)\hat{i} + y\ sin\left(\frac{\theta}{2}\right)\hat{j} + z\ sin\left(\frac{\theta}{2}\right)\hat{k} = q_r + q_x\hat{i} + q_y\hat{j} + q_z\hat{k} \quad (8)$$

where θ is the angle of rotation about $\vec{q}$. The quaternion, $\vec{q}$, can be easily converted into the three Euler angles (*12*) as seen in Equation 9:

$$\begin{bmatrix} \varphi \\ \theta \\ \psi \end{bmatrix} = \begin{bmatrix} arctan\left(\frac{2(q_r q_z + q_x q_y)}{1-2(q_y^2 + q_z^2)}\right) \\ arcsin(2(q_r q_y - q_x q_z)) \\ arctan\left(\frac{2(q_r q_x + q_y q_z)}{1-2(q_x^2 + q_y^2)}\right) \end{bmatrix} \quad (9)$$

where
φ is yaw (normal)
θ is pitch (transverse)
ψ is roll (longitudinal)

Software. Software capable of pulling data from the various sensors onboard the IMU was written for the Android platform in the form of an app. Android software was developed in Android's proprietary integrated development environment (IDE) Android Studio using the Java programming language. There were no software limitations to the IMU sensor sampling rate. A maximum sampling rate of approximately 200 Hz was achieved on all Android devices. The developed Android software had a minimum targeted software development kit (SDK) version of 15 (Android 4) and a targeted SDK version of 24 (Android 7). Android devices were kept to the most up-to-date Android version allowed by the manufacturer.

iOS Development

Hardware. One Apple smartphone was purchased as part of a market survey to find the most widely used smartphones. The Apple device along with its respective inertial sensors is listed in Table 2. Processor/memory speed and memory capacity was found to be sufficient for rapid deceleration measurements. Much like the Android devices, the Apple device came with two sensors of interest (accelerometer and magnetometer). On the Apple platform, acceleration is measured in units of acceleration of gravity, g. Absolute rotation is presented in Euler angles of yaw, pitch, and roll. Phone orientation is shown in Figure 1.

Software. Much like the Android development, software was developed for the Apple platform, in the form of an app, in order to pull acceleration and rotation data from the onboard inertial sensors. Apple software was written for the iOS operating system in Apple's proprietary IDE Xcode 8 using the Swift programming language. The Apple motion sensor manager, CoreMotion, limits by software the maximum IMU sampling frequency to around 100 Hz. It is unknown whether or not hardware limitations contribute to the maximum sampling frequency cutoff. The iOS deployment target was iOS 10.2 with a minimum deployment target of iOS 8.0. The Apple device was kept to the most up-to-date software pushed by the manufacturer.

Bowmonk Decelerometer

Hardware. For the purpose of validating smartphone accelerometer measurements, two Bowmonk Airfield Friction Meter 2 Mark 3 (Bowmonk) decelerometers were tested. The Bowmonk is considered the standard for deceleration-based measurements on airfields. Unlike smartphone accelerometers, the Bowmonk is designed for pavement deceleration measurements. The Bowmonk employs a solid state accelerometer operating at a sample rate of approximately 400 Hz. The Bowmonk claims a measurement range of 0.0 to 1.4 g (*13*).

Equipment Summary

A summary of friction testing equipment relevant to this study are in Table 3.

Field Tests

Test Site

Field tests took place at the Silver Flag (SF) test runway at Tyndall Air Force Base, Florida. The SF test runway contains six unique Portland Cement Concrete pavement sections, each 200-ft long by 8-ft wide. The test track is split into left and right lanes with three test sections each. Each test section has a unique texture/friction characteristic. See Table 4 for macrotexture and friction measurements from Elatextur and GT devices, respectively. For the purposes of this article, the SF test sections will be numbered from 1 to 6 representing the sections L1 to L3, R1 to R3.

Table 3. Friction Testing Equipment

Manufacturer	Model	Measurement type	Measurement hardware
Findley Irvine	Mk2 GripTester	Fixed-slip CFME	Load cell instrumented smooth test tire
Bowmonk	AFM2 Mk3	Decelerometer	Solid state accelerometer
Samsung	Galaxy S7	Decelerometer	MEMS-based accelerometer
LG	V20	Decelerometer	MEMS-based accelerometer
Huawei	Nexus 6P	Decelerometer	MEMS-based accelerometer
HTC	10	Decelerometer	MEMS-based accelerometer
Apple	iPhone 7	Decelerometer	MEMS-based accelerometer

Table 4. Aggregated Friction Data

Lane	Elatextur (MPD)	GT 60 mph	GT 25 mph	Bowmonk (%g)	iPhone 7 (%g)	Galaxy S7 (%g)	V20 (%g)	HTC 10 (%g)	Nexus 6P (%g)
1 (L1)	0.576	0.53	0.75	93.44	106.70	97.52	104.19	42.25	113.19
2 (L2)	1.245	0.73	0.76	93.31	100.23	97.59	104.44	44.69	110.13
3 (L3)	0.580	0.59	0.78	96.78	99.39	89.94	101.31	40.32	105.59
4 (R1)	0.346	0.31	0.57	80.54	94.91	90.44	94.33	41.11	97.51
5 (R2)	0.440	0.50	0.76	96.52	101.09	101.06	100.82	40.05	109.23
6 (R3)	0.112	0.13	0.14	70.98	80.89	76.80	84.52	41.06	82.37

Note: Peak deceleration measurements were taken from the average of at least 15 braking tests per lane. Measurements were subjected to a median absolute deviation (MAD) outlier analysis. Any results more than two absolute deviations from the median with respect to the MAD were rejected.

CFME Tests

A set of CFME data was made available for the SF test track (*14*). Of importance to the present work were GT results at multiple speeds. GT tests were conducted at five speeds (1.6, 3, 25, 40, and 60 mph). CFME tests were conducted on the SF test track in one direction multiple times on a smooth test tire leading to the aggregated CFME results in Table 4.

Decelerometer Tests

Vehicle braking tests were performed in order to test smartphone and Bowmonk decelerometers. Braking tests consisted of reaching a speed of 20 mph in a test vehicle and then braking hard to a complete stop. The brake-to-stop braking action is important as early results using a brake stab braking action were inconsistent. Bowmonk decelerometers were secured in the front passenger floorboard with Velcro. Smartphones were secured with Velcro in an acrylic mount held by the seatbelt in the front passenger seat with the direction of travel being the + Y direction (see Figure 1). Braking tests were performed in a full-size car, 2016 Chevy Malibu LT, with ribbed tires. Tire pressure was kept to the manufacturer recommended value of 36 psi. The test car weighed ~3,400 lb. Braking tests were also conducted in a compact size car, 2016 Nissan Versa, weighing ~2,400 lb. The compact size car braking test results were not included in this study as they indicated that the smaller vehicle could not realize all friction levels at the SF test track.

Results

Relevant numeric CFME and decelerometer data are in Table 4. It should be noted here that peak deceleration measurements may be sensitive to sensor noise at high sample rates. In such a situation where the sample rate is too high, practitioners should measure average peak deceleration to avoid the influence of sensor noise.

Reference Data

The Bowmonk Airfield Friction Meter 2 Mark 3 decelerometer is considered the standard for airfield deceleration-based friction measurement tools and is used, in this study, as a reference for deceleration-based tools to compare against the smartphone decelerometers. The GT is considered a high standard of continuous friction measurement for airfields and is used, in this study, for correlation with smartphone decelerometers.

Though the present article only used the GT for correlation at speeds of 60 and 25 mph, it is useful to observe the GT output at various speeds in order to understand friction's relationship with speed. Friction coefficient results by lane for the GT at five speeds (1.6, 3, 25, 40, and 60 mph) taken at the SF test track are illustrated in Figure 2*a*. Data in Figure 2*a* indicate that the friction coefficient has an inverse relationship with speed. Friction coefficient results by speed for the GT on all six test sections are shown in Figure 2*b*. Data in Figure 2*b* more clearly show the relationship between friction coefficient and speed. It is clear, from Figure 2*b*, that friction coefficient measurements made at around 20 mph differ from those made at 60 mph. It is likely, then, that braking tests performed at 20 mph correlate well with CFME tests at 25 mph. It is also likely that braking tests at 20 mph will not readily correlate with CFME tests at 60 mph and will require further consideration.

It should be noted that the two non-linear friction-speed curves in Figure 2*b* belong to the two extreme texture test sections: 2 (L2) at 1.245 MPD and 6 (R3) at 0.112 MPD. The friction response on test section 6 (R3) can be viewed as follows: lower speeds increase the microtexture contribution of the surface thereby increasing the friction coefficient, higher speeds diminish the microtexture contribution and increase the macrotexture contribution, of which the surface has very little, thereby decreasing the friction coefficient to a minimum value. The friction response in test section 2 (L2) is more nuanced. The apparent non-linear increase in friction at

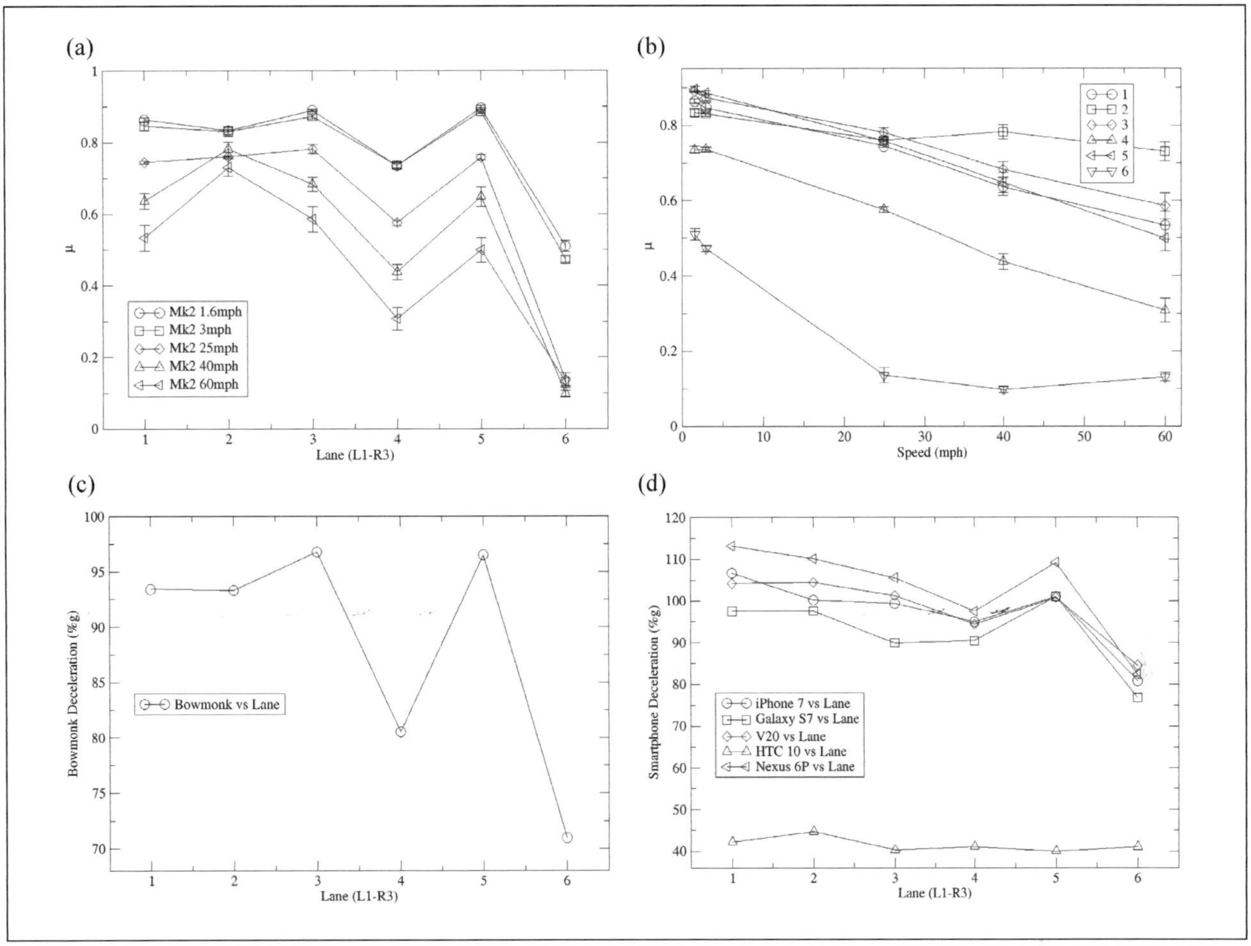

Figure 2. CFME and deceleration data by lane: (*a*) GT friction coefficient vs lane; (*b*) GT friction coefficient vs speed; (c) Bowmonk deceleration vs lane; (*d*) smartphone deceleration vs lane.

40 mph could be explained by the hindrance of high macrotexture at low speeds. At low speeds, the expected increase in the microtexture contribution could be impeded by the tire's inability to adhere to the surface due to large aggregate particles (macrotexture). This in conjunction with polished aggregate particles could lead to a lower than expected friction response at lower speeds. Bowmonk decelerometer results from Table 4 are illustrated in Figure 2*c*. Results in Figure 2*c* indicate that the friction response measured by the Bowmonk devices during 20 mph braking tests closely resembles the friction response measured by the GT device during 25 mph CFME tests.

Smartphone Data

Smartphone deceleration measurements by lane are in Figure 2*d*. Data shown in Figure 2*d* indicate that the HTC 10 smartphone malfunctioned in some way and was therefore excluded from the later analysis. Numerous updates were pushed to the HTC 10 device with no change in accelerometer output.

Given the relationship between friction coefficient and speed, correlation should be attainable between the 20 mph braking tests performed with decelerometers and the 25 mph CFME tests performed with the GT. It was predicted that the deceleration-based tools would be linearly proportional to each other and to the GT operating at 25 mph. It was found, however, that the Bowmonk devices had a strong exponential relationship with the GT 25 mph and the smartphone decelerometers. Figure 3 shows the exponential relationship between the Bowmonk and GT 25 mph devices, and the inset in Figure 3 shows an example of the relationship between the Bowmonk device and the smartphone decelerometers, in this case the iPhone 7.

Similar results to the Figure 3 inset were found between the Bowmonk device and the Galaxy S7, V20,

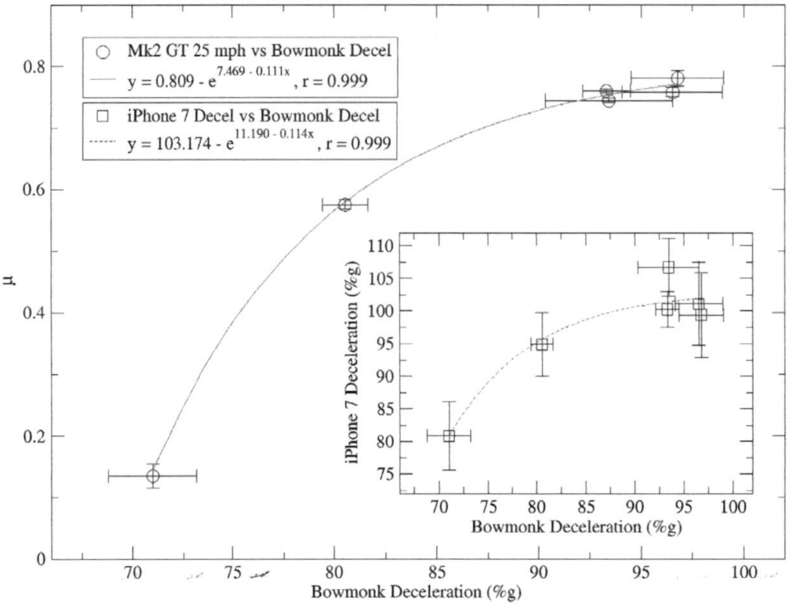

Figure 3. GT 25 mph vs Bowmonk exponential regression.

and Nexus 6P smartphones. Regression results in Figure 3 indicate that the solid state accelerometer in the Bowmonk device exponentially relates to the GT at 25 mph; however, linear regression results in Figure 4 indicate that the MEMS-based accelerometers in the smartphones linearly correlate to the GT operating at 25 mph. Linear correlation between the GT 25 mph and smartphone devices can be found in Figure 4.

Linear regressions in Figure 4 lead to the following relationships between the smartphone devices and the GT 25 mph:

$$\mu_{GT\ 25} = -\ 1.990 + 0.027a_{iPhone7} \quad (10)$$

$$\mu_{GT\ 25} = -\ 1.750 + 0.026a_{GalaxyS7} \quad (11)$$

$$\mu_{GT\ 25} = -\ 2.483 + 0.032a_{V20} \quad (12)$$

$$\mu_{GT\ 25} = -\ 1.540 + 0.021a_{Nexus6P} \quad (13)$$

where $\mu_{GT\ 25}$ is the predicted GT 25-mph friction coefficient, $a_{iPhone7}$ is the iPhone 7 deceleration from 20 mph, and so on for the Galaxy S7, V20, and Nexus 6P.

Smartphone correlation with higher speed CFMEs requires a texture measurement. The previously mentioned IFI model incorporates a macrotexture contribution into a friction measurement. Figure 5*a* details the IFI calculation for the four functioning smartphones: iPhone 7, Galaxy S7, V20, and Nexus 6P. Figure 5*b* shows smartphone GT 60-mph predictions on top of measured GT 60-mph data.

Linear regression results from Figure 5 lead to the following relationships in equations 14 to 17:

$$\mu_{GT\ 60} = -\ 0.097 + 1.888F60_{iPhone7} \quad (14)$$

$$\mu_{GT\ 60} = -\ 0.098 + 1.901F60_{GalaxyS7} \quad (15)$$

$$\mu_{GT\ 60} = -\ 0.099 + 1.907F60_{V20} \quad (16)$$

$$\mu_{GT\ 60} = -\ 0.099 + 1.919F60_{Nexus6P} \quad (17)$$

where μ_{GT60} is the predicted GT 60-mph friction coefficient, $F60_{iPhone7}$ is the IFI converted (see Equation 7) iPhone 7 measurement, likewise for the remaining smartphones. Equations 14 to 17 yield the following respective mean absolute error (MAE) values: 0.043, 0.039, 0.031, and 0.034. Low MAE values, along with high Pearson's *r*-values, indicate strong linear correlation between the predicted 60-mph GT calculations and actual 60-mph GT measurements. It should be noted that unlike the smartphone correlations to 25-mph GT measurements, in Figure 4, the IFI converted smartphone data can realize both low and high friction portions of the SF test sections indicating the strong relationship between friction and macrotexture.

Discussion of Results

Validation of smartphone decelerometers as friction assessment tools was sought through two means: correlation between existing deceleration-based equipment and

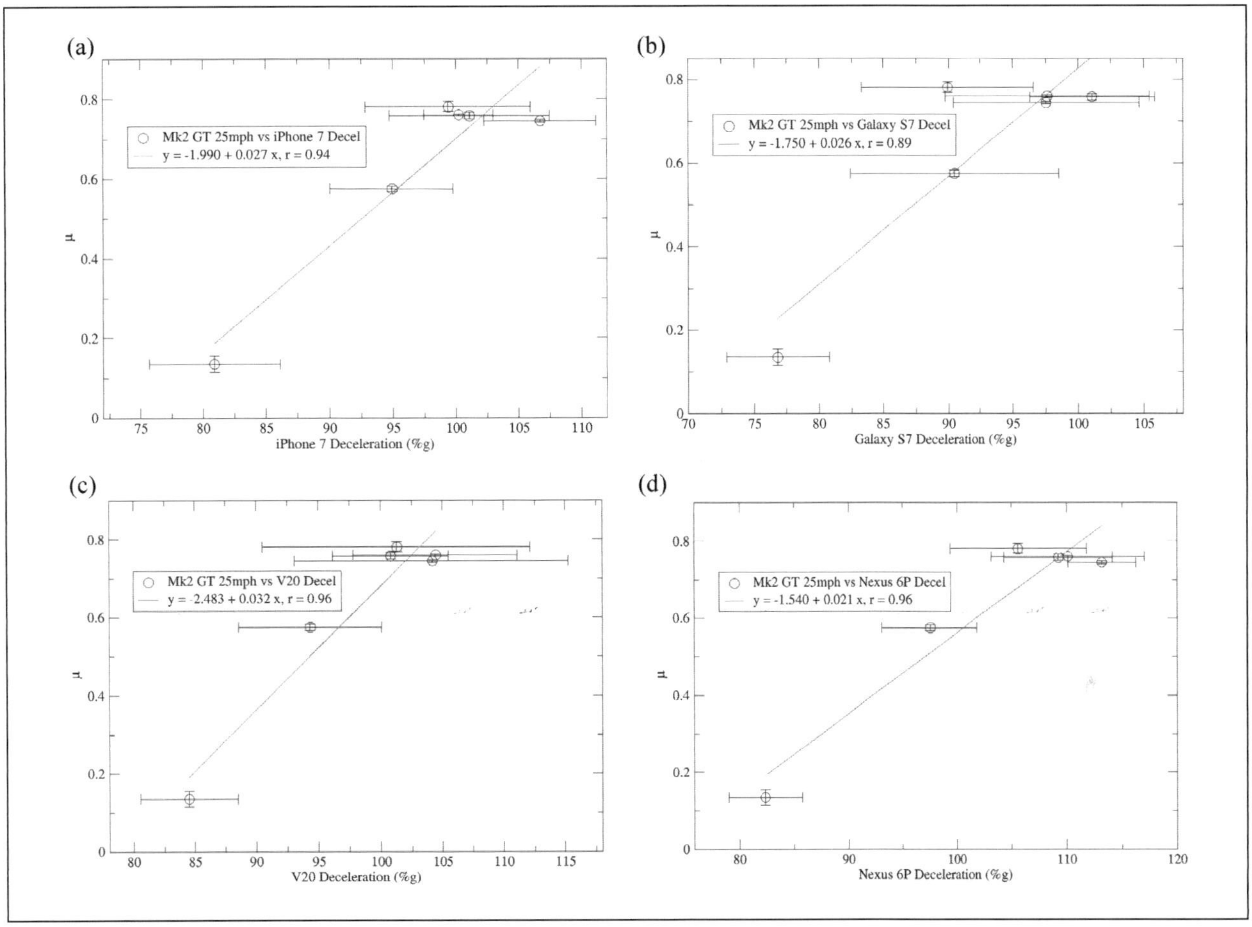

Figure 4. Smartphone/GT 25 mph linear regressions: (*a*) iPhone 7 vs GT 25 mph; (*b*) Galaxy S7 vs GT 25 mph; (*c*) V20 vs GT 25 mph; (*d*) Nexus 6P vs GT 25 mph.

correlation between existing CFME equipment. It was found that the Bowmonk decelerometers were exponentially related to the smartphone decelerometers and to the 25-mph GT (Figure 3). Unlike the Bowmonk decelerometers, the smartphone decelerometers were found to be linearly related to the 25-mph GT (equations 10 to 13, and Figure 4). Given a texture measurement, further correlation was sought and found between the smartphone decelerometers and the 60-mph GT (equations 14 to 17, and Figure 5).

Figure 4 indicates that the smartphone deceleration-based technique is, by itself, unable to differentiate between friction values greater than μ_{GT25}~0.57; however, this technique can easily distinguish $0.14 < \mu_{GT25} < 0.57$. This technique is useful, then, for contingency friction measurement on low friction/high contamination surfaces. Given a macrotexture measurement, the presented technique can accurately realize all friction levels of the SF test track in comparison with the GT at 60 mph, as seen in Figure 5.

It should be noted that the reproducibility of the presented results is considered poor due to a number of factors commonly plaguing deceleration-based measurements. These factors include but are not limited to driver braking action, vehicle weight/weight distribution, tire tread/temperature/pressure, IMU type, smartphone/IMU temperature, position of IMU relative to center of mass, and stability of IMU during braking action. As such, the presented deceleration-based measurements are only considered valid for standard size cars of weight ~3,400 lb and for a brake-to-stop braking action on ribbed tires.

Conclusions

This paper sought to develop smartphone accelerometer-based friction assessment tools and sufficiently correlate those tools to accepted deceleration and CFME equipment. This paper has successfully detailed software and hardware development of a friction assessment technique

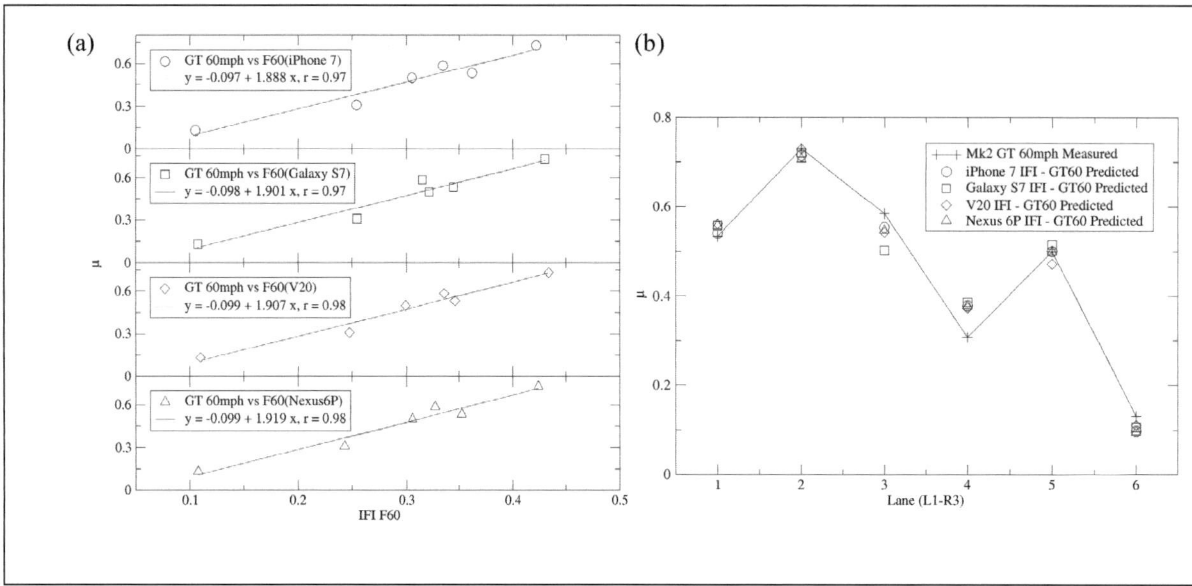

Figure 5. Smartphone IFI model implementation: (*a*) GT 60 mph vs smartphone IFI F60; (*b*) smartphone predictions of GT 60 mph.

for use with both Android and iOS smartphone devices. Using said deceleration-based friction measurement technique, a number of conclusions were reached and are as follows:

- Exponential correlation was found between the smartphone and Bowmonk decelerometers (Figure 3).
- Linear correlation was found between the smartphone decelerometers and the 25-mph GT equipment (equations 10 to 13, and Figure 4).
- Linear correlation results between the smartphone decelerometers and the 25-mph GT device are only useful for low friction measurements, $\mu_{GT25} < 0.57$.
- Linear correlation was found between the smartphone decelerometers and the 60-mph GT equipment by incorporating a texture measurement using the IFI model (equations 14 to 17, and Figure 5).
- Linear correlation results between the smartphone decelerometers and the 60-mph GT device were found to be useful for the entire friction range at the SF test track, $0.13 < \mu_{GT60} < 0.73$.
- Regardless of similar technology, Bowmonk and smartphone decelerometers respond differently to pavement friction characteristics.
- Reproducibility concerns, common with deceleration-based techniques, restrict the above considerations to standard size cars of weight ~3,400 lb on ribbed tires with a brake-to-stop braking action. The development of a calibration procedure to relate field conditions to the reported models could reduce the limitations of poor reproducibility.

Acknowledgments

The authors thank Aaron Pullen and Brian Cotter of Applied Research Associates (ARA) for their CFME data and Jeb Tingle of the U.S. Army ERDC for project management support. Permission to publish was granted by the Director, Geotechnical and Structures Laboratory, U.S. Army Engineer Research and Development Center.

Author Contributions

The authors confirm contribution to the paper as follows: study conception and design: ABW, CAR; data collection: ABW; analysis and interpretation of results: ABW, CAR; draft manuscript preparation: ABW. All authors reviewed the results and approved the final version of the manuscript.

References

1. Tabor, D. The Importance of Hysteresis Losses in the Friction of Lubricated Rubber. *Proc. 1st International Skid Prevention Conference*, Virginia Council of Highway Investigation and Research, Part 1, University of Virginia, Charlottesville, VA, 1958, pp. 211–218.
2. Grosch, K. A. The Relation between the Friction and Visco-elastic Properties of Rubber. *Proceedings of the Royal Society of London Series A*, Vol. 274, No. 1356, 1963, pp. 21–39.
3. Kummer, H. W., and W. E. Meyer. *Measurement of Skid Resistance*. ASTM Special Technical Publication No. 326 - Symposium on Skid Resistance, New York, NY, 1962, pp. 3–28.

4. Kummer, H. W., and W. E. Meyer. *NCHRP Report 37: Tentative Skid-resistance Requirements for Main Rural Highways.* HRB, National Research Council, Washington, D.C., 1967.
5. PIARC World Road Association. Report of the Committee on Surface Characteristics. *Proc., XVIII World Road Congress*, Brussels, Belgium, 1987.
6. Henry, J. J. *NCHRP Synthesis of Highway Practice 291: Evaluation of Pavement Friction Characteristics.* TRB, National Research Council, Washington, D.C., 2000.
7. NASA. *Runway and Highway Pavement Grooving and Traction Studies.* NASA-SP-5073. NASA, Washington, D.C., 1969.
8. Yager, T. J., W. P. Phillips, W. B. Horne, and H. C. Sparks. *A Comparison of Aircraft and Ground Vehicle Stopping Performance on Dry, Wet, Flooded, Slush-, Snow-, and Ice-covered Runways.* NASA Technical Note D-6098, Washington, D.C., 1970.
9. FAA. *Measurement, Construction, and Maintenance of Skid-Resistant Airport Pavement Surfaces.* Advisory Circular (AC) 150/5320-12C. FAA, Washington, D.C., 2004.
10. Leu, M. C., and J. J. Henry. Prediction of Skid Resistance as a Function of Speed from Pavement Texture Measurements. *Transportation Research Record: Journal of the Transportation Research Board*, 1978. 666: 7–13.
11. Woodman, O. J. *An Introduction to Inertial Navigation.* UCAM-CL-TR-696. University of Cambridge, Computer Laboratory, Cambridge, MA, 2007.
12. Blanco, J. L. *A Tutorial on SE(3) Transformation Parameterizations and On-Manifold Optimization.* Technical Report #012010. University of Málaga, Málaga, Spain, 2010.
13. *Bowmonk AFM2 Datasheet.* www.bowmonk.com/uploads/files/2016/06/afm2.pdf. Accessed July 17, 2017.
14. Pullen, A., C. Ishee, and B. Cotter. Evaluation of Portable Push-Mode Continuous Friction Measurement Devices for Airfield Use. Air Force Civil Engineer Center, Tyndall Air Force Base, FL, Manuscript in Preparation.

The Standing Committee on Aircraft/Airport Compatibility (AV070) peer-reviewed this paper (18-01229).

Article

TRR
JOURNAL OF THE TRANSPORTATION RESEARCH BOARD

Transportation Research Record
2018, Vol. 2672(23) 106–116

DOI: 10.1177/0361198118790134
journals.sagepub.com/home/trr
SAGE

Evaluation of Midpoint Runway Marking for General Aviation Airports

H. Gene Hawkins, Jr.[1] and Brian G. Dillman[2]

Abstract

The aviation industry continues to experience landing overruns at general aviation airports due in part to pilots touching down beyond the midpoint of the available runway. In some landing overruns, pilots were unaware that the midpoint was reached and nonetheless continued the landing instead of implementing a go-around/aborted landing procedure. Incident reports confirmed that forced landings were coupled with frequent overruns. Over the years airport operators and pilots have occasionally reported that knowing the location of the midpoint of runways would help them to minimize such accidents. There is a greater tendency for landing overruns at airports that primarily serve general aviation operations with runway lengths less than 4,200 ft in length and being utilized by student pilots and those pilots that primarily fly on the weekends only. In order to identify potential markings that could assist in mitigating this risk, the research team evaluated the design for and potential benefits of a marking placed at or near the midpoint of a general aviation runway that would inform pilots of the potential for an overrun/overshoot on a landing. The research team evaluated several preliminary prototype midpoint runway markings using stakeholder input and an online survey of pilots. The best performing marking was further evaluated in an aviation training device study and in field evaluations at six airports. The study results found that the numeric ½ located to the left of the runway centerline provided pilots with useful information on the remaining length of runway.

Landing a general aviation (GA) aircraft is a complex maneuver that begins well before the touchdown point. Ideally, per federal regulation, a pilot should account for all flight and environmental conditions prior to the approach and execute a safe landing. Unfortunately, runway overruns continue to be a concern for all aviation segments, not just the GA community. Statistics indicate a greater number of runway overruns/excursions on GA runways, visual and non-precision runways of less than 4,200 ft in length, and airports with weekend and student pilots. Although larger airports with longer runways have aiming point markings, they are not present on many GA runways.

In some GA landing overruns/excursions, pilots were unaware that the midpoint of the runway was reached and nonetheless "forced" the landing instead of implementing the go-around procedure. Over the years, airport operators and pilots have occasionally reported to the Federal Aviation Administration's (FAA) Airport Engineering Division that knowing the location of the midpoint of runways would help them to minimize such accidents. A unique runway marking intended to reduce the occurrence of runway overruns has the potential to provide a cost-effective means of addressing this serious aviation safety issue. The goal of this research was to evaluate the effectiveness of prototype runway midpoint markings with the thought that such a marking would help pilots on visual flight rules (VFR) approaches make better and safer landing decisions. The research was conducted through the Partnership to Enhance General Aviation Safety, Accessibility and Sustainability (PEGASAS) university research program. The major activities of the research effort included reviewing background information, developing prototype markings, evaluating the prototype marking through a survey of pilots, evaluating the preferred midpoint marking in a simulator environment, and performing a field evaluation of the recommended midpoint marking.

Background and Prototype Marking Development

At the start of the project, the research team gathered information from a variety of sources to help identify

[1]Zachry Department of Civil Engineering, Texas A&M University, College Station, TX
[2]Department of Aviation Technology, Purdue University, West Lafayette, IN

Corresponding Author:
Address correspondence to H. Gene Hawkins: gene.hawkins@tamu.edu

current practices and to guide in the development of markings that were worthy of evaluation as part of the research.

Background

The authors were not able to identify any uses of midpoint markings in the United States. They were able to identify a midpoint marking at commercial airports in Japan including Ryugasaki Airfield and Tokyo's Haneda International Airport, although they could not identify documentation supporting the use of these markings. The marking used at the Japanese airports was three closely-spaced transverse lines across the entire runway at the midpoint. Some form of midpoint markings may also be used in other countries. The research team also reviewed GA landing overrun incident reports provided by the FAA. One of the challenges identified in this review and subsequent discussions with stakeholders is that overrun incidents are not typically reported unless they result in damage to the plane, damage to the airport, or an emergency response from airport/responder personnel. In the next activity, researchers contacted various stakeholders and discussed the issue of GA landing overruns, potential corrective actions, the potential issues associated with adding a midpoint marking, and possible designs for such a marking.

Prototype Midpoint Markings Development

After taking into account the factors above and their impacts on the design of the proposed runway midpoint marking, the team identified three prototype midpoint markings. One of the issues related to a midpoint marking at GA airports that the researchers tried to address in at least one design is the placement and recognition of the marking when there are displaced thresholds and the midpoints of each approach on the runway are separated from one another. Full-scale mockups were created and photographed on an abandoned runway, as shown in Figure 1. The three markings included:

- Triangles: This marking (Figure 1*a*) consists of two triangles located at the midpoint and placed on the edges of the runway surface with vertices pointed toward the runway centerline.
- Circular dot: This marking (Figure 1*b*) is a single filled circle positioned laterally in the center of the landing area and located at the runway midpoint. The marking appears as a circle when viewed from overhead, but as an oval when viewed from runway level.
- Numeric ½: This marking (Figure 1*c*) is the fraction one-half located at the midpoint of the runway length. As illustrated in later figures in this report, the marking is placed on the left side of the centerline. An inverted numeric marking would face pilots landing from the other direction. The fraction provides a directional context that the other prototype markings did not provide.

Prototype Midpoint Marking Evaluation: Survey of Pilots

After developing the prototype midpoint markings, the research team developed a survey to determine which of the prototype markings exhibited the highest level of

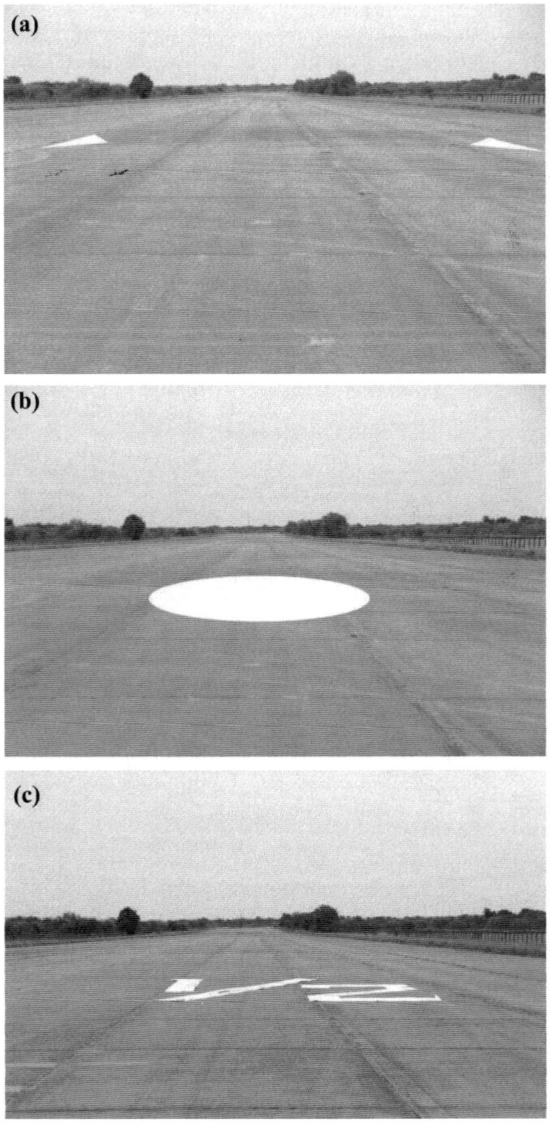

Figure 1. Prototype midpoint markings: (*a*) triangles; (*b*) circular dot; (*c*) numeric ½.

Table 1. Marking Interpretation Categories

No response given	Helicopter landing
No meaning	Emergency landing
Threshold	Directional indication
Lateral runway characteristics	Drop target
Midpoint	Parachute target
Hazard	Arresting gear
Aiming point/touchdown zone	Runway maintenance
Distance marker	Runway designation
Ground traffic marking	Weight restriction
Inactive runway	Non-instrument approach

understanding. The nine-question survey included multiple-choice, level of agreement with a statement, and open response questions. All nine questions were administered to each respondent. All questions provided the opportunity for open comments. The first four survey questions addressed demographic characteristics: plane weight, plane horsepower, runway length at home airport, and number of hours flown. The next three questions asked the respondent to provide an interpretation of the meaning of each of the three prototype runway midpoint markings. Each of the prototype marking questions asked for an open-ended response to one of the photographs from Figure 1. The order of these three questions was randomized for each respondent. The final two questions asked the respondent to indicate the level of agreement with regard to general statements about the landing maneuver.

The survey notice was distributed by email to several different groups (airport contacts, PEGASAS site directors, FAA staff, and selected professional contacts) with a request to share with other individuals and flying organizations, including the Aircraft Owners and Pilots Association. The authors do not know the ultimate number of individuals who received the survey notice as it was distributed by forwarded emails and administered online; 593 responses were received by the end of the survey window. Responses were received from the United States and 15 other countries, with U.S. responses from 47 states and Washington, D.C. Nineteen of these states had 10 or more respondents.

The heart of the survey was the three questions that asked the meaning of each of the prototype midpoint markings. Because these were open-ended questions, there was significant variation in the exact wording of individual responses so the research team developed 20 general categories (see Table 1) that could be used to classify the open-ended responses for analysis. The results based on these response categories are summarized in Table 2.

These results indicate that the marking interpreted most frequently as representing the midpoint of the

Table 2. Summary of Runway Marking Interpretations

	Total responses			Percent of responses		
Response category	Marking 1 (triangles)	Marking 2 (dot)	Marking 3 (1/2)	Marking 1 (triangles)	Marking 2 (dot)	Marking 3 (1/2)
No response given	63	66	64	**11%**	**11%**	**11%**
No meaning	139	155	122	**23%**	**26%**	**21%**
Threshold	62	2	0	**10%**	0%	0%
Lateral runway characteristics	64	2	3	**11%**	0%	1%
Runway midpoint	31	9	346	5%	2%	**58%**
Hazard	6	7	0	1%	1%	0%
Aiming point/touchdown zone	184	243	3	**31%**	**41%**	1%
Distance marker	18	7	21	3%	1%	4%
Ground traffic marking	8	3	7	1%	1%	1%
Inactive runway	7	6	10	1%	1%	2%
Helicopter landing	1	75	0	0%	**13%**	0%
Emergency landing	0	1	0	0%	0%	0%
Directional indication	9	4	1	2%	1%	0%
Drop target	0	4	0	0%	1%	0%
Parachute target	0	6	0	0%	1%	0%
Arresting gear	0	2	0	0%	0%	0%
Runway maintenance	0	1	0	0%	0%	0%
Runway designation	0	0	15	0%	0%	3%
Weight restriction	0	0	1	0%	0%	0%
Non-instrument approach	1	0	0	0%	0%	0%
Total	593	593	593	100%	100%	100%

Note: Bold values indicate response percentages of 10% or more.

Table 3. Distribution of Responses to Landing Maneuver Survey Questions

Statement	Strongly disagree	Disagree	Neutral	Agree	Strongly agree	No response
Most landings can be completed with appropriate corrective actions (aborted landings can usually be avoided).	3%	16%	17%	49%	7%	8%
A marking that indicates the midpoint of the runway would be extremely helpful to me.	3%	14%	26%	37%	13%	8%

runway was the numeric ½ at 58%, whereas the triangles marking and the circular dot marking were most frequently interpreted to indicate a different function, being a runway aiming point or touchdown zone (31% and 41%, respectively) on the runway surface. Importantly, the results also indicated that the triangles and circular dot markings had other interpretations that could create confusion, whereas the numeric ½ had no other recurrent interpretation. The triangles marking was viewed as a lateral runway characteristic (edge) by 11% of the respondents whereas the circular dot marking was viewed as a helicopter landing spot by 13% of the respondents. Each of the three markings had about an equal percentage of responses (11%) indicating that the marking did not convey any meaning to the respondents.

The final analysis performed on the online survey was to determine the distribution of responses to the last two questions in the survey. These questions asked respondents to indicate their level of agreement with two statements. The results are indicated in Table 3. From the table it was observed that both statements had the highest percent of responses in the "Agree" response.

The purpose of the survey was to provide a large sample size perspective on pilots' understanding of the three prototype runway midpoint marking patterns without prior education or training on the intended meaning of the markings. The following represents the significant findings of the survey:

- The numeric ½ marking was the pattern with the highest understanding as a marking that indicates the midpoint of a runway (58%). The other two marking patterns had low understanding as a midpoint marking (5% or less).
- The triangles and circular dot markings were often misunderstood to represent a touchdown or landing point (31% and 41%) whereas the numeric ½ was interpreted in this manner by only 1% of respondents.
- The only other category that received at least a 10% response rate for any of the marking patterns was the triangles marking, which was interpreted as a threshold (10%) or lateral runway edge characteristic (11%), whereas the circular dot was interpreted as a heliport landing location (spot) (11%).
- For all three marking patterns, about 11% of respondents did not give a response.
- For all three marking patterns, 21% to 26% of respondents indicated the patterns had no meaning.
- The survey results indicate that the numeric ½ marking has the highest potential for a runway midpoint marking of the three evaluated.
- Of the three prototype midpoint markings, the numeric ½ allows the greatest application for runways with displace thresholds. That is, the shorter available landing distance associated with the displaced threshold (direction of operation) could be clearly indicated to pilots.

Prototype Midpoint Marking Evaluation: Simulation Evaluation

After identifying the numeric ½ as the best understood marking, the evaluation proceeded to a simulator environment to assess pilot behavior in response to that marking. A mobile aviation training device was used to present various landing scenarios with and without the prototype numeric ½ midpoint marking to evaluate pilot landing performance. The simulation evaluation attempted to address three research questions:

1. Is there a significant difference between pilot estimation of runway distance remaining when the runway midpoint marking was visible compared with when it was not visible?
2. Is there a significant difference in go-around decisions when the runway midpoint marking was visible and when it was not visible?
3. What were pilot perceptions of the runway midpoint marking during landings?

Methodology Development

The researchers selected a Frasca Level 5 Advanced Aviation Training Device (AATD) for conducting the

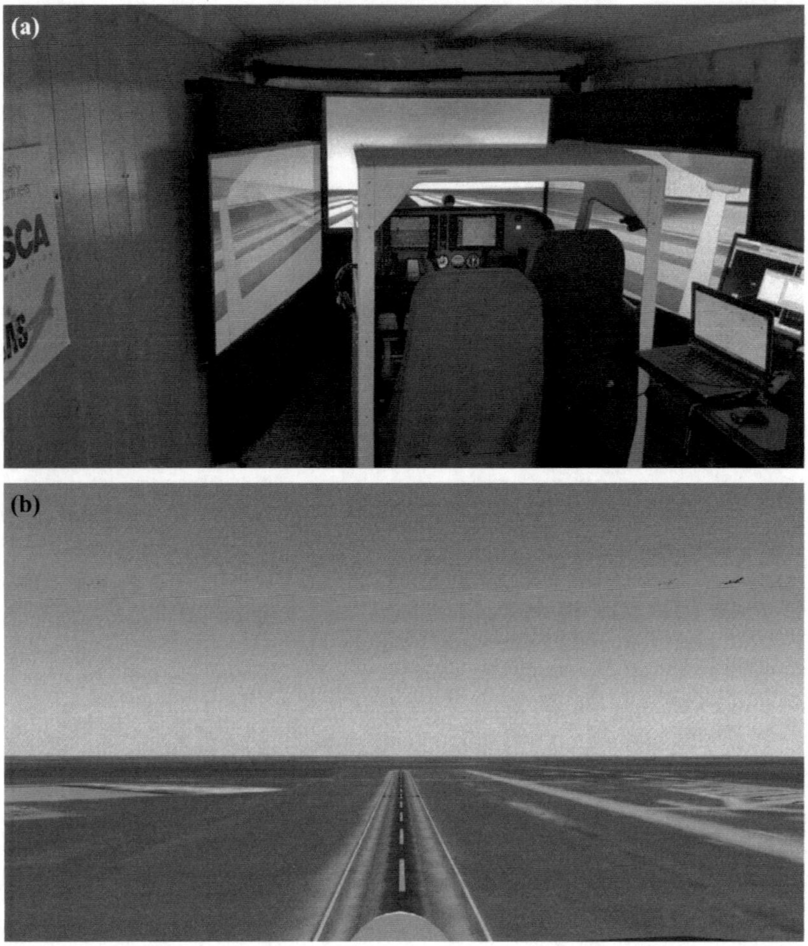

Figure 2. Advanced Aviation Training Device: (*a*) simulator setup; (*b*) simulation example (aircraft located 1,000 ft from midpoint marking).

evaluation. The AATD was housed in a trailer and traveled to each evaluation location. The AATD was a single-engine small aircraft that had a glass cockpit avionics suite with three visual screens at the front and sides, audio, TruVision terrain, and aerial mapping. Figure 2 illustrates the simulator used in the evaluation and a simulator graphic on approach.

The simulation evaluation consisted of two interactions for each subject. In the first, participants were asked to hand-fly the training device throughout the approach to landing. In the second, participants monitored prerecorded approaches to landings and pressed a button that stopped the simulation if he/she believed that a go-around was necessary at that point.

For the hand-flown approach interaction, participants completed three approaches. The first was a practice approach to landing (no data recorded), and the second and third were recorded for data analysis. For all of the approaches, the subject was placed in a plane that was 100 ft above the ideal approach on a three-degree glideslope 1.5 mi from the runway threshold at an above-ground level of 700 ft. After the practice approach was completed, the subject started the next approach with a 30 knot tailwind on final approach. The purpose was to create the illusion that the subject was high and fast on final approach and could land at or past the midpoint of the runway. The third approach was the same as the second, except for the presence of the midpoint marking. The presence of the midpoint marking flipped between the second and third approaches with each subject. The runway used in the simulation was 4,000-ft long. All other markings were removed from the runway to represent a basic VFR runway. The distance remaining signs on both ends of the runway were also removed to prevent those marking from affecting the decision-making process of the pilot participants. Each participant was briefed with a script prior to beginning their first approach.

For the monitored approaches, the researchers created three simulations of approaches landing on different areas of the runway. One approach landed before the midpoint marking and stayed on the runway, another approach landed on the midpoint marking and stopped within the last 100 ft of the runway, and the last approach landed after the midpoint marking and went off the end of the runway. The subject watched the approaches and if he/she decided to go around, he/she pressed a button on the yoke and it froze the playback. The researchers then asked questions about the pilot's reasoning behind the decision. These monitored approaches were presented in a random order for each subject with regards to the type of approach and also, whether or not the midpoint marking was displayed. Figure 2*b* represents the pilot's view on approach during both the hand-flown and monitored approaches with the midpoint marking displayed.

Simulation Administration

Participants were recruited from university students that had participated in two National Intercollegiate Flying Association Regional Flight Team competitions. All subjects had a pilot certificate. Pre-experiment briefing instructed each participant to fly like they would in real life and make decisions as if they were in a real airplane.

During the hand-flown approaches the flight sequence was recorded in the flight training device for distance measurements following data collection, along with critical data such as altitude, airspeed, groundspeed, and time. At the completion of each approach, whether hand-flown or monitored, a series of questions were asked to determine the impact that the midpoint marking might have made on the decision-making process for that landing. The participants were also asked how much runway length was remaining after they either landed or conducted the go-around, as well as how much runway is remaining upon coming to a complete stop on the runway if the participant completed the landing cycle for later comparison to the actual distance. Additional questions were asked of the participants as to why they decided to conduct a go-around if that decision was made. Seventy-five participants flew a total of 370 approaches (145 hand-flown and 225 monitored approaches) at the competitions.

Data Analysis and Results

After conducting the simulations, the researchers used the data to evaluate the three basic research questions.

Results: Estimate of Remaining Runway. The first question asked whether there was a significant difference between pilot estimation of runway distance remaining when the runway midpoint marking was visible compared with when it was not visible. For the 370 simulations, the researchers calculated the difference between the pilots' estimate of the remaining distance of runway at touchdown or go-around (provided by the test subject) and the actual remaining runway distance. The mean values for these are shown in the "all simulations" rows of Table 4. As indicated in the table, the difference between the estimated differences with and without the midpoint marking was about 53 ft. A paired t-test comparison of the differences in means indicated that the difference in means was not statistically significant (t-test statistic $=$ 0.676, $p = 0.500$).

There were 151 aborted approaches (go-arounds) conducted prior to even reaching the runway. Participants in these cases aborted the landing for reasons other than the impact of a midpoint marking. At locations prior to reaching the runway, the midpoint marking is not visible and therefore could not have had any bearing on the decision-making process of the pilot. The "at/beyond threshold" rows of Table 4 indicate the differences in the remaining runway distance for the 219 scenarios where the plane reached the runway threshold (2,000 ft from

Table 4. Remaining Runway Distance for All Approaches

Subset of data	Midpoint marking	*N*	Mean distance remaining (ft)*	Difference between midpoint marking present and not present (ft)
All simulations	Present	182	734.48	53.05
(*N* = 370)	Not present	188	787.53	
At/beyond threshold	Present	109	439.12	74.75
(*N* = 219)	Not present	110	513.87	
1,000 ft before threshold	Present	75	341.25	155.48
(*N* = 149)	Not present	74	496.73	

*Absolute value of difference between the actual remaining distance and the pilot estimate of remaining distance.

the marking). The difference between the scenarios with and without the midpoint marking was about 75 ft, which also was not statistically significant (t-test statistic = 1.282, p = 0.201).

Because the runway length was 4,000 ft, it was difficult for the pilots to read the numeric ½ midpoint marking at the runway threshold when the pilot was 2,000 ft from the marking. To provide a sense of the performance of the midpoint marking when it was becoming discernable to a pilot, the "1,000 ft from threshold" rows of Table 4 indicate the differences in the remaining runway distance for the 149 scenarios when the plane was 1,000 ft from the midpoint marking. The difference in remaining distance estimates was about 155 ft at this point. This is the largest differences of those differences evaluated and is statistically significant for α = 0.05 (t-test statistic = 2.474, p = 0.014). As seen from these other results, the differences between the estimated and actual remaining runway distance became more accurate as the pilot neared the marking.

Results: Impact of Marking on Go-Around Decision. The most important question of interest was whether the numeric ½ midpoint marking had an effect upon a pilot making a go-around decision late in the approach (essentially when over the runway). To address the question of whether there was a significant difference in go-around decisions when the runway midpoint marking was visible and when it was not visible, the points at which a decision to perform a go-around was initiated were plotted with respect to the location of the runway midpoint marking. During the hand-flown approaches there was also an evaluation of whether the pilot would allow the airplane to touch down after the midpoint of the runway with or without the runway midpoint marking being present.

Of the scenarios presented, only one showed any type of pattern in regard to this question. During the monitored approaches there was one approach where the plane landed at the approximate location of the runway midpoint. In instances where the runway midpoint marking was present there were no go-around decisions made after the midpoint location, but there were three instances where the pilot allowed the aircraft to touch down and then decided there was not enough runway to come to a safe stop and decided to abort the landing.

Decisions to allow the aircraft to touch down during hand-flown approaches were also of interest when the runway midpoint marking was present and not present. There were four pilots that allowed the airplane to touch down after the midpoint location when the midpoint marking was not present and four pilots that allowed the airplane to touch down after the midpoint location when the runway midpoint marking was present. There was a larger dispersion of landings without the runway midpoint marking. The small number of data points made it difficult for the researchers to address this research question with any authority.

Results: Pilots' Perceptions of Midpoint Marking during Landings. To address the question of pilot perceptions of the runway midpoint marking during approaches, if any, at distances where the marking was visible, additional information was collected regarding the participant's perception and general interaction with the runway midpoint marking on the various approaches. This information was collected by researchers through questions after the completion of the approach sequence in both the hand-flown and monitored approaches. The primary question of, "Why was the go-around initiated?" was followed up with asking whether the participant had any final comments. Additionally, researchers recorded any unsolicited comments that were made while the participant was flying the aircraft that related to the runway midpoint marking or any other aspects that could affect decision making.

Participants' comments regarding the runway midpoint marking generally fell into two categories. The pilots either made comments regarding the midpoint of the runway and its impact on their decision making or the pilots were confused as to what the marking represented and decided to abort the landing due to something being on the runway.

Summary

Based upon the results obtained from the simulation portion of the project, there appears to be value in the use of a numeric ½ midpoint marking to improve landing safety. Several participants mentioned the marking in their decision-making process during the approaches both during the hand-flown and the monitored approaches. Aborted landings occurred after the midpoint of the runway without the presence of a runway midpoint marking as well as more touchdowns occurring after the midpoint of the runway without the runway midpoint marking being displayed on the runway. Finally, there was statistical significance in the accuracy of the runway remaining estimates when a runway midpoint marking was present at distances were it was visible.

Preferred Midpoint Marking Evaluation: Field Evaluation

In the final evaluation activity of the project, the research team installed the numeric ½ marking at six airfields as field experiments to assess the value of the marking in a real-world application. The evaluation was a self-

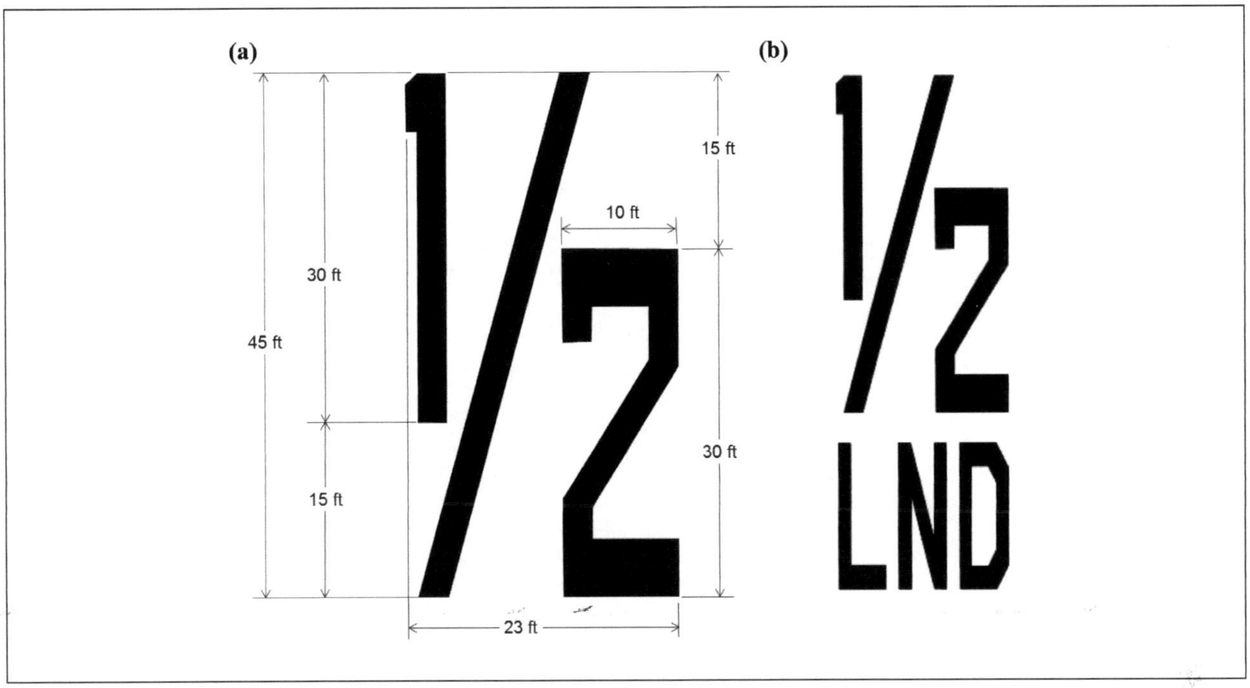

Figure 3. Layout for numeric ½ midpoint marking: (*a*) midpoint marking layout; (*b*) midpoint marking for thresholds.

administered pilot assessment of the effectiveness of the marking at assisting in landing maneuvers.

Airport Selection and Marking Installation

The first step in the field experimental effort was to identify airports that had runways of appropriate length and that met other criteria. The researchers ultimately identified six airports that were willing to cooperate in the evaluation. Four of the six had a single runway and the other two each had two runways. All runway lengths were between 3,074 ft and 4,051 ft and runway widths varied between 60 ft and 75 ft. One of the runways had displaced thresholds at each end (193 ft and 1,509 ft).

The runway midpoint markings installed at each airport were based on the design shown in Figure 3*a*. The overall size of the numeric ½ marking was 45-ft tall and 23-ft wide. The marking was installed on the left side of the runway at the midpoint of the runway based on any displaced threshold (if present). Where there was a displaced threshold, LND (20-ft tall) as shown in Figure 3*b*, was added below the midpoint marking to indicate that the midpoint marking applies to the runway midpoint for a landing. Figure 4 presents several photographs of midpoint markings installed for the field evaluation. If there was no displaced threshold, the midpoint is the same for a takeoff or landing maneuver and the midpoint markings for each direction were adjacent to each other but facing in opposite directions as shown in Figure 4, *b* and *c*. Figure 4*d* illustrates an installation with displaced thresholds in both directions.

Field Evaluation Results

Once the markings were installed, the field evaluation entered the data collection phase. At each airport, the researchers distributed a response form for pilots to evaluate the effectiveness and value of the numeric ½ midpoint marking. Pilots could also complete the response form online. A total of 98 responses were received. The response form asked for general demographic information and then asked the following questions:

- When you landed on the runway today, did you notice the ½ midpoint marking on the runway?
- If you did, was this the first time you had noticed the ½ midpoint marking when landing at this airport?
- Did the ½ midpoint marking have any impact on your landing today? (select response)
    - It helped me make a more informed landing maneuver.
    - It confused me.
    - It made me do a go-around.
    - It made me brake harder.
    - It had no impact.
- Select the response that best represents your option of this marking:

Figure 4. Photographs of field evaluation markings: (*a*) typical installation; (*b*) placement with no displaced thresholds; (*c*) view of markings without displaced thresholds (insert: Google Earth Imagery 2017); (*d*) view of markings with displaced thresholds (Google Earth Imagery 2017).

- The midpoint marking has significant value. It should be required on all GA runways.
- The midpoint marking has potential value. It should be encouraged where appropriate.
- The value of the midpoint marking is unknown. It should be evaluated in more detail before use.
- The midpoint marking has limited value. Its use should be restricted.
- The midpoint marking has no value or is potentially problematic. It should not be used anywhere.

The first non-demographic question asked whether the pilot noticed the numeric ½ midpoint marking on the runway. This was followed by a question that asked whether this had been the first time they had noticed the marking at that airport. Of the 97 that responded to this question, 47 responded that they had noticed it and it was their first time seeing the marking. Twenty-six pilots indicated they noticed it, but it was not the first time they had noticed it at this airport. Twenty-four pilots did not notice the marking at all.

The next two questions were multiple-choice questions that asked whether the numeric ½ midpoint marking made a difference in the landing and whether the marking had a safety value. The results of these questions are summarized in Tables 5 and 6. These results show that about one-sixth (17%) of the respondents indicated that the runway midpoint marking helped the pilot make a

Table 5. Summary of Responses Regarding Impact on Landing Operations

	Response frequency at each airport (percent of responses in parentheses)						
Multiple-choice response	O61	CFD	CTK	N81	OBI	IWS	Totals
It helped me make a more informed landing maneuver	0	4	0	0	1	10	15 (16.7%)
It confused me	0	5	0	0	1	0	6 (6.7%)
It made me do a go-around	0	1	0	0	0	1	2 (2.2%)
It made me brake harder	0	1	0	0	0	0	1 (1.1%)
It had no impact	1	12	10	3	9	32	67 (74.4%)
Totals	1 (1.1%)	23 (25.6%)	10 (11.1%)	3 (3.3%)	11 (12.2%)	43 (47.8%)	91 (100%)

Note: O61 = Cameron Airpark; CFD = Coulter Field Airport; CTK = Ingersoll Airport; N81 = Hammonton Municipal Airport; OBI = Woodbine Municpal Airport; IWS = West Houston Airport.

Table 6. Summary of Responses Regarding Safety Value of Midpoint Marking

	Response frequency at each airport (percent of responses in parenthesis)						
Multiple-choice response	O61	CFD	CTK	N81	OBI	IWS	Totals
The midpoint marking has significant value. It should be required on all GA runways.	0	3	1	0	1	9	14 (15.7%)
The midpoint marking has potential value. It should be encouraged where appropriate.	0	11	4	2	3	26	46 (51.7%)
The value of the midpoint marking is unknown. It should be evaluated in more detail before use.	0	5	2	0	3	5	15 (16.9%)
The midpoint marking has limited value. Its use should be restricted.	1	1	3	1	2	0	8 (9.0%)
The midpoint marking has no value or is potentially problematic. It should not be used anywhere.	0	4	0	0	2	1	7 (7.9%)
Totals	1 (1.1%)	24 (27.0%)	10 (11.2%)	3 (3.4%)	11 (12.4%)	41 (46.1%)	90 (100%)

Note: O61 = Cameron Airpark; CFD = Coulter Field Airport; CTK = Ingersoll Airport; N81 = Hammonton Municipal Airport; OBI = Woodbine Municpal Airport; IWS = West Houston Airport.

more informed landing decision. Almost three-quarters of the respondents (73.3%) indicated that the marking had no impact on the landing maneuver. Even so, about two-thirds of the pilots (67.4) felt the marking has significant (16%) or potential (52%) safety value. Another 17% felt that the marking needs further evaluation, whereas 16% felt the marking had limited (9.0%) or no (6.7%) value. Overall, these results indicate a generally favorable response for a runway midpoint marking.

Of the responses to the impact on landing operations (Table 5), the "it confused me" response is of concern. The researchers examined the other responses for the six pilots that indicated they were confused by the marking. Some reported confusion can be expected because the FAA had not informed the aviation community of this prototype marking. There were no trends related to weather, plane type, or pilot hours. When asked about the potential safety value, three of the six pilots indicated it had potential value whereas the other three indicated unknown or limited value. The comments from these six pilots were the most revealing. These comments provide some insight into why the marking was confusing. Several indicated that they felt that the marking was an aircraft or some other object on the runway. Similar comments were made by seven of the 39 pilots that offered comments on the response form. Seven of the pilots also commented that, although they liked the idea of the marking, they thought the design of the marking could be improved.

Summary of Field Evaluation

Overall, the field evaluation indicated a general acceptance of the numeric ½ midpoint marking as used in the evaluation, with a potential for a promising future, although the comments in the responses also indicate the potential for pilots to mistake the marking for an aircraft or other object on the runway. Regarding the latter point, such interpretations should be eliminated once FAA promulgates this runway midpoint marking in appropriate literature and Notice to Airmen (NOTAM).

Conclusions and Recommendations

Runway overruns/excursions due to improper landing decisions continue to be a safety issue of concern for the FAA at GA airfields. In this study, researchers developed and evaluated a prototype runway marking that informs pilots of the midpoint of the runway. Of the designs developed and evaluated in the preliminary stages of the project, a numeric ½ midpoint marking was found to have the highest level of understanding. The numeric ½ midpoint marking was further evaluated in a simulation study and also in a field evaluation. The results were favorable, although there is some indication that the marking can be interpreted as an aircraft or other object by pilots. The researchers believe that many of the comments related to an aircraft or object on the runway reflect an observation made at altitude or while in the landing pattern. As the runway midpoint marking is intended to help a pilot that is about to or has just touched down, the confusion with a ghost aircraft appears to have had minimal impact on the intended function of the runway midpoint marking.

Based upon the results obtained from the research, there appears to be merit in the use of a numeric ½ midpoint marking to reduce runway overruns/excursions. Unfortunately, the number of responses received during the field evaluation portion of the project was somewhat limited. With slightly less than a hundred responses to the field evaluation, the ability to draw definitive conclusions about the effectiveness and potential safety value of the numeric ½ midpoint marking is limited.

Acknowledgments

This research project was supported by the FAA Partnership to Enhance General Aviation Safety, Accessibility and Sustainability research program. The authors would like to recognize the project contact Holly Cyrus of the FAA, for her efforts on this project. The authors also acknowledge the airports where the field evaluations were conducted: Cameron Airpark (O61), Cameron Park, California; Coulter Field Airport (CFD), Bryan, Texas; Hammonton Municipal Airport (N81), Hammonton, New Jersey; Ingersoll Airport (CTK), Canton, Illinois; West Houston Airport (IWS), Houston Texas; and Woodbine Municipal Airport (OBI), Woodbine, New Jersey.

Author Contributions

The authors confirm contribution to the paper as follows: study conception and design: GH and BD; data collection: GH and BD; analysis and interpretation of results: GH and BD; draft manuscript preparation: GH and BD. All authors reviewed the results and approved the final version of the manuscript.

The Standing Committee on Aircraft/Airport Compatibility (AV070) peer-reviewed this paper (18-01258).

The findings and conclusions in this report are those of the authors and do not necessarily represent the views of the Federal Aviation Administration. This document does not constitute FAA policy. Although the FAA sponsored this project, it neither endorses nor rejects the findings of this research.

Article

TRR
JOURNAL OF THE TRANSPORTATION RESEARCH BOARD

Transportation Research Record
2018, Vol. 2672(23) 117–127

DOI: 10.1177/0361198118794293
journals.sagepub.com/home/trr
SAGE

Rutting Evaluation of Hot and Warm Mix Asphalt Concrete under High Aircraft Tire Pressure and Temperature at National Airport Pavement and Materials Research Center

Navneet Garg[1], Hasan Kazmee[2], Lia Ricalde[2], and Timothy Parsons[2]

Abstract

The aircraft industry's propensity to increase the range of aircraft results in increased gross weight and tire pressure. Therefore, performance assessment of airfield asphalt concrete mixes under such high tire pressure and extreme environmental exposure becomes imperative for sustainable design implementation. Moreover, a lack of scientific information on the field performance has been a major roadblock in the adoption of green technologies like warm mix asphalt (WMA). An accelerated pavement testing study was undertaken at the Federal Aviation Administration's National Airport Pavement and Materials Research Center to evaluate the rutting performance of hot and warm asphalt mixes at two different binder grades. Six test lanes were constructed – four outdoors and two indoors, each encompassing three different test sections. The constructed test sections were subjected to different combinations of tire pressure, temperature, and environmental exposure using a custom-designed airport heavy vehicle simulator. Both field performance and laboratory characterization tests indicated that WMA exhibited comparable performance to hot mix asphalt. Environmental aging was found to be conducive to the curing and enhanced rutting performance of polymer-modified WMA.

Transportation agencies across the United States have been embracing sustainable green technologies like warm mix asphalt (WMA) to reduce fuel consumption and the carbon footprint during asphalt concrete (AC) production and paving (*1–3*). Currently, the Federal Aviation Administration (FAA) does not have a standard specification for WMA's use on airfields. Due to the limited use and lack of full-scale performance test data, scientific information on the performance of airfield WMA is not adequate. On the other hand, aircraft manufacturers' inclination to include more wide-body aircraft like the Airbus A-380, A-350 and Boeing B-787 into commercial fleets would potentially increase aircraft gross weights and wheel loads leading to significantly high tire pressures. As a result, in 2014 the FAA Advisory Circular 150/5335-5C for airport pavement strength was revised to accommodate 1.75-MPa (254-psi) tire pressure (*4*). Extreme environmental conditions like high temperatures in addition to the high tire pressure may accelerate the damage progression in airfield asphalt pavements. Performance of hot mix asphalt (HMA) and WMA under such conditions is yet to be studied in detail. Several transportation agencies have conducted accelerated pavement testing to investigate the performance of WMA technologies and reported comparable performance to its HMA counterparts (*5–7*).

Using accelerated pavement testing (APT) Mejías-Santiago et al. compared three different WMA technologies with a control HMA for airfield pavements subjected to fighter aircrafts (*7*). The constructed test sections were trafficked with a moving wheel load of 156 kN (35 kip) with 2.3-MPa (325-psi) tire pressure at a controlled temperature of 43°C (109°F). The selected wheel load and tire pressure resembled those of a typical F-15E fighter aircraft; commercial aircraft are much heavier in weight. An APT study was undertaken at the FAA's National Airport Pavement and Materials Research Center

[1]Federal Aviation Administration, William J. Hughes Technical Center, Airport Technology R&D Branch, Atlantic City, NJ
[2]Applied Research Associates, Inc., Egg Harbor Township, NJ

Corresponding Author:
Address correspondence to Navneet Garg: Navneet.Garg@faa.gov

Figure 1. (*a*) FAA heavy vehicle simulator for airports at the National Airport Pavement and Materials Research Center; (*b*) single wheel assembly with a 280-kN (63-kip) load; and (*c*) instrumented heating panels in the HVS-A.

(NAPMRC) to study HMA and WMA behavior under simulated commercial aircraft tire load and high tire pressure. Unlike the previous study (*7*), temperature was varied among the test sections such that the influence of temperature on HMA and WMA rutting accumulation could be established.

Objective and Scope

The primary objective of this paper is to compare the rutting performances of selected airfield HMA and WMA with two different binder grades under the influence of high tire pressure and temperature. This paper provides a brief overview of the airport heavy vehicle simulator (HVS-A), the design of the pavement test sections, the traffic test plan, observations from rutting performances of the AC mixes during traffic tests, and laboratory characterization results from high temperature indirect tensile (IDT) strength and Hamburg wheel tracking tests.

Accelerated Pavement Testing Setup

NAPMRC and HVS-A

Conventional laboratory test protocols often revolve around the stress and distress conditions generally encountered under highway traffic. These stress conditions simulated in the lab are significantly lower than those experienced by airport pavements. Heavy vehicle simulation tests with representative aircraft wheel load and tire pressure can fill the gap between the laboratory test limitations and actual airport pavement stress conditions. Also, the damage progression can be accelerated by replicating slow aircraft speed. The FAA procured a state of the art HVS-A in 2013 for the recently built (2015) NAPMRC to conduct APT on sustainable AC materials (*8*). Figure 1*a* shows the sixth generation HVS-A custom-designed and built by Dynatest, Inc., equipped with enhanced control systems and larger wander capability in comparison to the previous generation HVSs. The HVS-A is capable of applying both bidirectional and unidirectional wheel loads using a single wheel [maximum wheel load of 100,000 lb (444.8 kN)] or dual wheel gear [maximum wheel load of 50,000 lb (222.4 kN)]. The single wheel assembly used in the current study is shown in Figure 1*b*. Figure 1*c* shows the heaters and insulation panels installed around the base rail with which the HVS-A can heat the pavement surface up to a temperature of 67°C (150°F) and maintain it throughout the test duration.

Considering the wide variation in aircraft gear assemblies and operating conditions, airfield pavements encounter wider lateral distribution of wheel passes compared with highway traffic. Previous studies indicate that the airport pavements typically have normally distributed wander patterns (*9*, *10*). A specific wander sequence was employed during the traffic tests with HVS-A. APT with the HVS-A consisted of two different stages, namely response tests and traffic tests.

The response tests involved 22 spatially distributed wheel passes over nine wander tracks at three different load levels [89.0, 133.5, and 177.9 kN (20, 30, and 40 kips)]. The primary objective for the response tests was to check the functionality of the embedded sensors

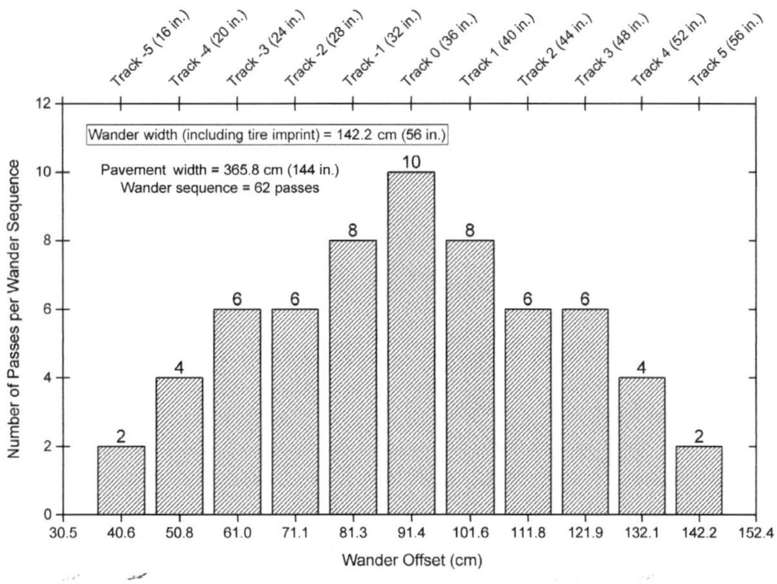

Figure 2. Distribution of wheel passes in each wander sequence.

in the constructed test sections, and study HMA/WMA behavior under different wheel loads. The traffic tests involved a normally distributed repetitive wander sequence with 62 wheel passes over the same number of wander tracks for response testing. The distribution of wheel passes over the transverse width of the pavement sections in the current study is shown in Figure 2. The wander tracks are spaced at a distance of 10.2 cm (4 in.) in which Track 0 indicates the centerline of the pavement test section. The number of assigned wheel passes increases as the wheel track approaches the centerline, almost creating a normally distributed wander pattern.

Design of Full-Scale Test Sections

Figure 3*a* shows the plan view of the designed test sections including AC type, test temperature, and test tire pressure. The surface layers of test lanes 1 and 5 were constructed with polymer-modified PG 76-22 WMA; whereas, lanes 3 and 6 were paved with HMA with the same binder grade. The AC layers of lanes 2 and 4 were constructed with WMA and HMA, respectively with the same binder grade of PG 64-22. Each of the test lanes were then subdivided into three different test sections: North, Center, and South sections. The test temperature of both the North and South sections was maintained at 49°C (120°F) at a depth of 5 cm (2 in.) below the pavement surface; in the meantime, the test surface temperature in the Center sections was kept constant at 32°C (90°F), also 5 cm (2 in.) below the pavement surface. FAA thickness design standard FAARFIELD uses a fixed elastic modulus of 1,379 MPa (200 ksi) for the HMA layer, which corresponds to the modulus at 32°C (90°F). The HVS-A testing was carried out with a 273-kN (61,300-lb) bidirectional single wheel load moving at a speed of 5 km/h (3 mph). To differentiate the effect of high tire pressure, the North and South sections were subjected to 1.75 and 1.45 MPa (254 and 210 psi), respectively. Tire pressure in the Center section was maintained at 1.75 MPa (254 psi) to evaluate the effect of temperature between the North and Center test sections.

As shown in Figure 3*b*, the design thicknesses for AC, P-209 granular base and P-154 type granular subbase were selected to be 5, 12 and 12 in. (127, 305, and 305 mm), respectively. The pavement structure was designed using FAA pavement thickness design standard FAARFIELD so that the base, subbase, and subgrade failure does not occur and distresses are limited to the surface layer. This way, an objective evaluation of the AC mix types would be possible under extreme temperature and tire pressure. The job mix formula for WMA was based on the FAA approved P-401 HMA specification with Type 1 aggregate gradation for the surface mix. A chemical additive known as Evotherm 3G was used to reduce the production and compaction temperature for warm mixes used in this study. Evotherm 3G is capable of reducing the production temperature by 33 to 45°C.

Rut Depth Measurement

According to ASTM EE1703, "rut depth" is defined as the maximum measured perpendicular distance between

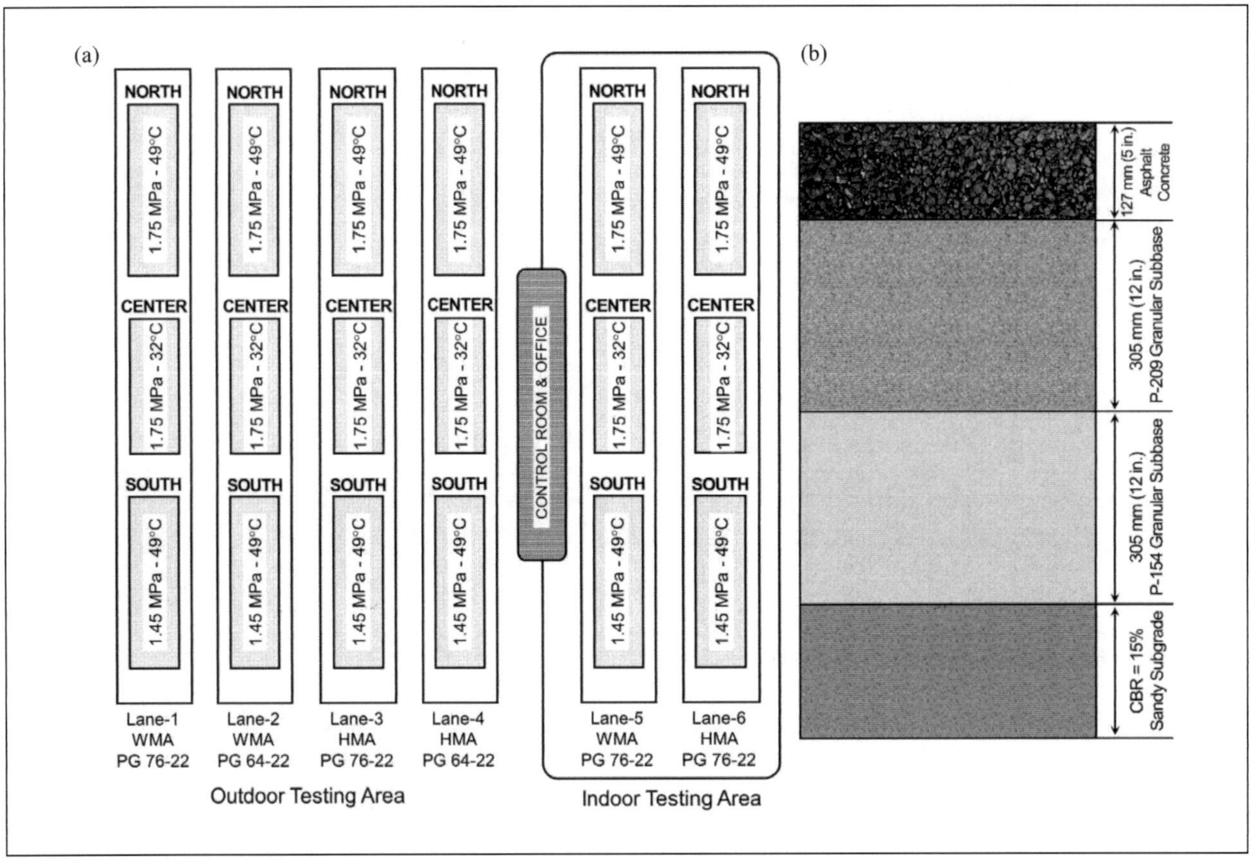

Figure 3. (*a*) Plan view of the full-scale test sections; and (*b*) cross-sectional view of the designed test sections.

the bottom surface of the straightedge and the contact area of the gauge with the pavement surface at a specific location. Despite widespread use of the above terminology, this method of rut measurement has been highly debated and does not quantify the shear induced dilative behavior. In this study, the depth of surface rut was defined as the maximum orthogonal distance between the theoretical true profile and maximum depression at a respective wheel pass. The theoretical true (zero) profile at a specific longitudinal station in a full-scale test section corresponds to the elevation measurements in the transverse direction taken at a certain interval prior to the beginning of HVS-A testing. Figure 4*a* illustrates the adopted definition of rut depth in this study. The US Army Corps of Engineers pavement distress manual adopted by the FAA lists three different severity level for asphalt surface rutting and the high severity level is designated by a rut depth exceeding 25.4 mm (1 in.) (*11*). The rutting failure criterion for this study was set to be the limiting threshold of 25.4 mm (1 in.).

Figure 4*b* illustrates the operational procedure of the on-board laser profiler (built into the HVS-A). The assembly consists of two lasers mounted on the loading carriage at a transverse distance of 183 cm (72 in.) and approximately 76 cm (30 in.) above the actual pavement surface.

Operational Procedure of the Laser Profiler

As shown in Figure 4*b*, the profilers take one full set of measurement along the longitudinal span [11.18 m (36 ft. 8 in.)] of the HVS-A by recording surface elevations every 10.2 cm (4 in.). Over the span of 11.18 m (36 ft. 8 in.), the laser profilers create one row of data with 110 data points. As the loading frame finishes taking one set of measurement along the longitudinal direction, the laser profilers shift 25.4 mm (1.0 in.) from their respective locations in the transverse direction. Thus, the distance between two successive measurement lines at the 4.37-m (14 ft 4-in.) station becomes 12.7 mm (0.5 in.), as shown in Figure 4. Due to the 12.7 mm (0.5 in) interval in the transverse direction at the corresponding station, the 145th row of data for the left laser corresponds to the 1st row of data for the right laser.

The last pass of the left laser overlaps with the first pass of the right laser. Such an overlap of the lasers is used to adjust the left and right lasers to use the same baseline for elevation during the calculation of rutting.

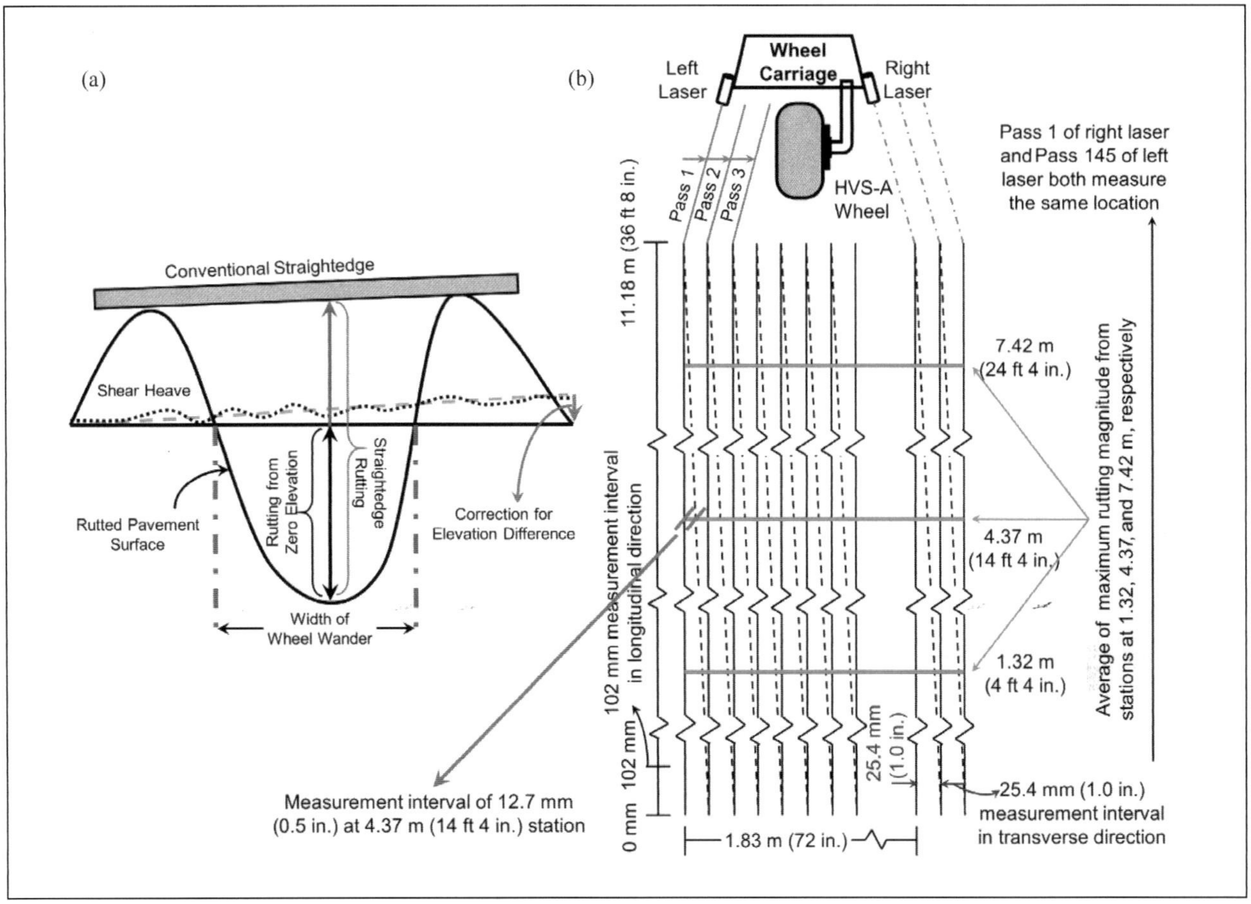

Figure 4. (*a*) Adopted definition of rut depth; and (*b*) operational procedure of HVS-A laser profiler.

Functionality Check, Data Collection, and Analysis

Prior to actual trafficking and subsequent rut measurements, a functionality check is conducted on the laser profiler. Three 3.7-m (12-ft) long 5- × 10-cm (2- × 4-in.) wooden beams are placed in the transverse direction at approximate longitudinal distances of 1.3, 4.4, and 7.4 m (52, 172, and 292 in.), respectively. Readings from the HVS-A profiler are then checked to identify whether the lasers exhibit a spike in those locations accordingly.

Once the functionality check has been completed, surface elevation of the entire test section is recorded using the laser profiler that corresponds to the data for "0" pass. As shown in Figure 4, any difference in elevation between the leftmost and the rightmost point along the stretch of transverse width is corrected with the assumption of a linear slope to achieve a true horizontal profile at "0" pass. The same magnitude of correction was also applied to the data generated for subsequent passes of HVS-A.

Pavement surface profile data are recorded in an increment of a full wander sequence. One full wander sequence in the current study consists of 62 passes of the wheel, with the transverse wheel position varying ± 61 cm (24 in.) from the centerline of the test section. Surface profile data were collected periodically at a multitude of 62 passes such as 62, 124, 186, and so forth.

As the initial rate of rutting is significantly higher than that at later passes, surface elevation is also recorded at a more frequent interval within the first 62 passes. Profile data at "0" passes are then subtracted from the surface profile data collected for the assigned number of passes to evaluate the variation in elevation owing to the subjection of the moving wheel load. The magnitude of rutting at a given number of passes is determined based on the maximum depression or deformation from the theoretical zero transverse profile, as shown in Figure 4.

Rut Performance during Traffic Tests

Table 1 summarizes the number of passes on the flexible pavement test sections to reach the failure criterion and the end of trafficking. Both PG 64-22 test lanes with WMA and HMA were found to be the worst performing test lanes with regrds to the number of passes to failure

Table 1. Number of Passes to Failure and End of HVS-A Trafficking in Full-Scale Test Sections

Number of passes		To failure criterion			End of trafficking		
Tire pressure (MPa)		1.75	1.45	1.75	1.75	1.45	1.75
Temperature (°C)		49	49	32	49	49	32
Lane no.	Mix type	North	South	Center	North	South	Center
1	WMA PG 76-22	601	876	na	2,914	3,906	5,022
2	WMA PG 64-22	166	248	2,962	496	930	5,022
3	HMA PG 76-22	861	767	na	3,968	3,534	5,022
4	HMA PG 64-22	146	281	2,616	620	992	5,022
5	WMA PG 76-22	470	801	na	1,488	2,356	5,022
6	HMA PG 76-22	667	924	na	2,418	2,604	5,022

Note: *Wheel Load = 273 kN (61,300 lb).

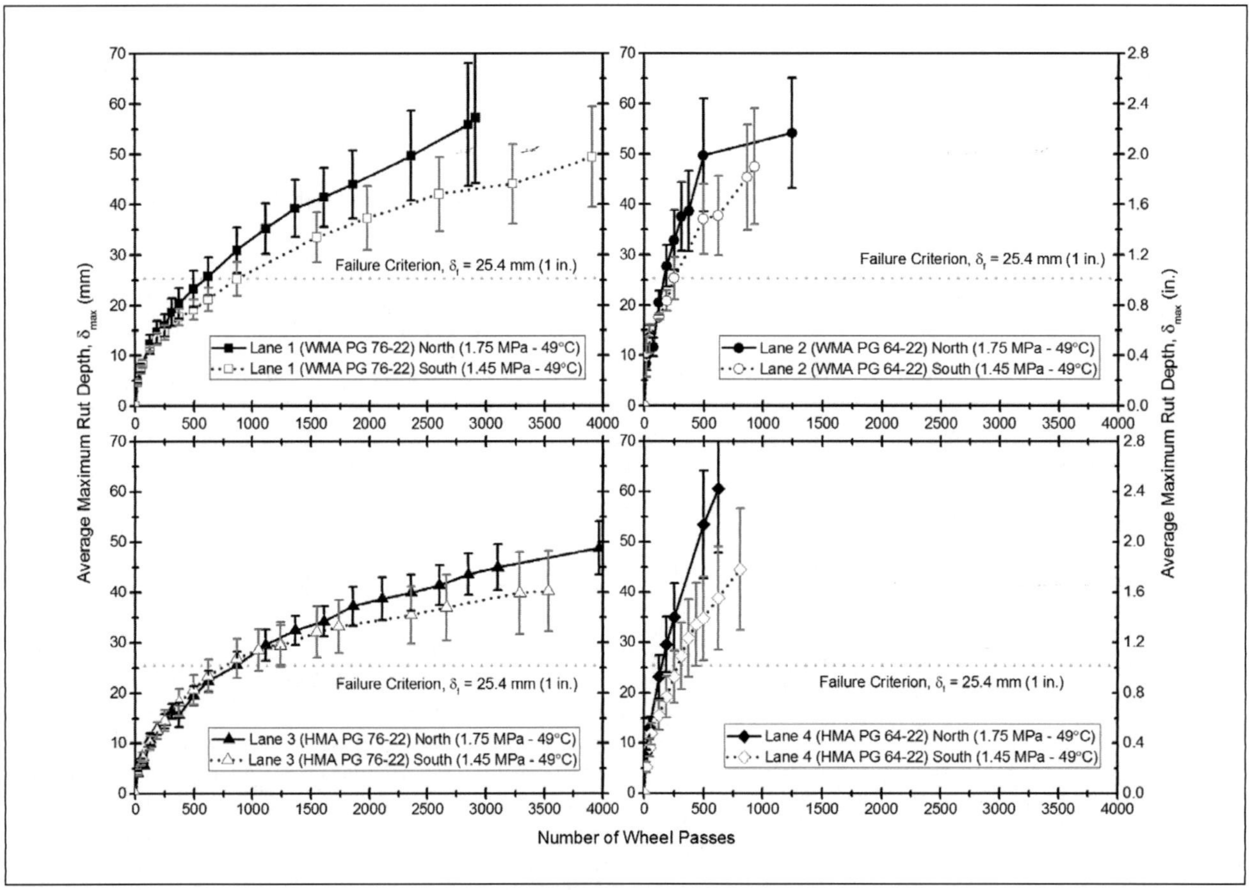

Figure 5. Effect of tire pressure, binder grade, and mix type on asphalt concrete rutting performance.

regardless of the tire pressure and temperature conditions. PG 76-22 HMA performed slightly better than the respective WMA test sections. Furthermore, the outdoor test sections with PG 76-22 asphalt binder outperformed the indoor sections with the same binder grade. All of the Center sections were trafficked to 5,022 passes. Among all the test sections, Lane 4 North with PG 64-22 HMA survived the lowest number of passes. Further details on the effect of tire pressure, temperature, mix type and environmental aging have been presented in the following subsections.

Effect of Tire Pressure, Binder Grade, and Mix Type

Figure 5 shows the average maximum rut depths in North and South sections of lanes 1 through 4 with increasing number of passes. The effect of tire pressure was more prominent at the later stages of HVS-A testing. For instance, Lane 4 North and South sections survived about 146 and 281 passes before reaching the failure threshold of 25.4 mm (1 in.). However, as the testing was continued beyond the failure criterion, the rate of rutting accumulation in the corresponding test sections diverged. Lane 4 (HMA PG 64-22) North section, which was subjected to a higher tire pressure accumulated 51 mm (2 in.) of rutting at approximately 460 passes. On the other hand, the South section with low tire pressure exhibited 45 mm (1.8 in.) of rutting at 806 passes. Lane 2 sections constructed with PG 64-22 warm mix showed similar rutting behavior under the influence of varying tire pressures.

As expected, lanes 2 and 4 North sections with PG 64-22 binder grade survived the lowest number of wheel passes to reach the failure criterion. The effect of binder grade was evident in both North and South sections. For example, the Lane 1 North section constructed with PG 76-22 warm mix withstood approximately 3.6 times more passes than the Lane 2 PG 64-22 warm mix North section. On a similar note, an additional 486 passes were needed for Lane 3 South (HMA PG 76-22) sections to reach the designated failure criterion in comparison with the Lane 4 South (HMA PG 64-22) section.

The effect of mix type was only present in the PG 76-22 mixes. As shown in Figure 5, rutting trends exhibited by PG 64-22 hot and warm closely tracked one another under the two different tire pressures. Similarly, the PG 76-22 WMA exhibited comparable rutting resistance with respect to the corresponding HMA. The Lane 3 North section with PG 76-22 HMA survived approximately 260 more passes to reach failure with respect to the Lane 1 North section with PG 76-22 WMA. Under lower tire pressure 1.45 MPa (210 psi), both North and South sections appeared to rut in a similar magnitude. However, as the HVS-A test was continued beyond the failure criterion, the WMA section (Lane 1 South) withstood approximately 30% fewer passes before reaching 38 mm (1.5 in.) of rutting in comparison with its corresponding HMA (Lane 3 South) counterpart.

Effect of Pavement Temperature

Pavement temperature had a more dominant influence on rutting progression in comparison with tire pressure. Figure 6 illustrates the effect of pavement temperature on rutting accumulation in North and Center sections of lanes 1 through 4 with increasing numbers of passes. Both North and Center sections were subjected to similar tire pressure 1.75 MPa (254 psi). North and Center sections were maintained at the constant temperatures of 49 and 32°C (120 and 90°F), respectively.

Figure 6 shows that the Lane 4 (PG 64-22 HMA) North section failed at approximately 146 passes; whereas the Lane 4 Center section survived about 2,616 passes to reach the failure criterion of 25.4 mm (1 in.). The PG 64-22 WMA sections in Lane 2 exhibited a similar rutting accumulation trend. Notably, the PG 64-22 WMA performed slightly better than the PG 64-22 HMA at low temperature 32°C (90°F) surviving another 346 wheel passes prior to reaching the failure threshold.

Under moderate temperature [32°C (90°F)], higher quality mixes with PG 76-22 binder grade accumulated up to only 20 mm (0.8 in.) of rutting after approximately 5,000 passes. However, the Lane 1 (WMA PG 76-22) Center section consistently accumulated a higher magnitude (not very significant though) of rutting than the Lane 3 (HMA PG 76-22) Center section, which reached 18 mm (0.7 in), contrary to the observation in rutting trends of PG 64-22 WMA and HMA sections.

Effect of Curing and Environmental Aging

Figure 7 illustrates the effect of curing and environmental aging on rutting behavior in the constructed test sections. The outdoor sections with both hot and warm mix types outperformed the indoor sections under high tire pressure. This implies that curing and environmental aging contribute to the stiffening of asphalt mixes and thereby increase the resistance against rutting.

PG 76-22 HMA performed slightly better than PG 76-22 WMA as shown in Figure 7. At the initial stage, all of the indoor and outdoor sections exhibited similar rates of rutting progression. However, when the constructed test sections with PG 76-22 AC were tested beyond the limiting criterion, the rutting trends became considerably different in the cases of the outdoor and indoor sections.

Despite satisfactory performances by the PG 76-22 HMA and WMA mixes, temperature was found to be the most influential factor dictating the magnitude and rate of rutting accumulation. None of the Center sections subjected to 32°C (90°F) of heat reached the failure criterion until the end of HVS-A trafficking at 5,022 passes. Conversely, both high and low tire pressure sections at higher temperatures failed below 1,000 passes regardless of the mix type and environmental exposure. Thus, the conducive effect of environmental aging was superseded by the effect of temperature.

Post-Traffic Material Characterization

Previous research studies indicate that the rutting progression in airfield AC mixes can be linked to the high temperature IDT strength of the corresponding mixes

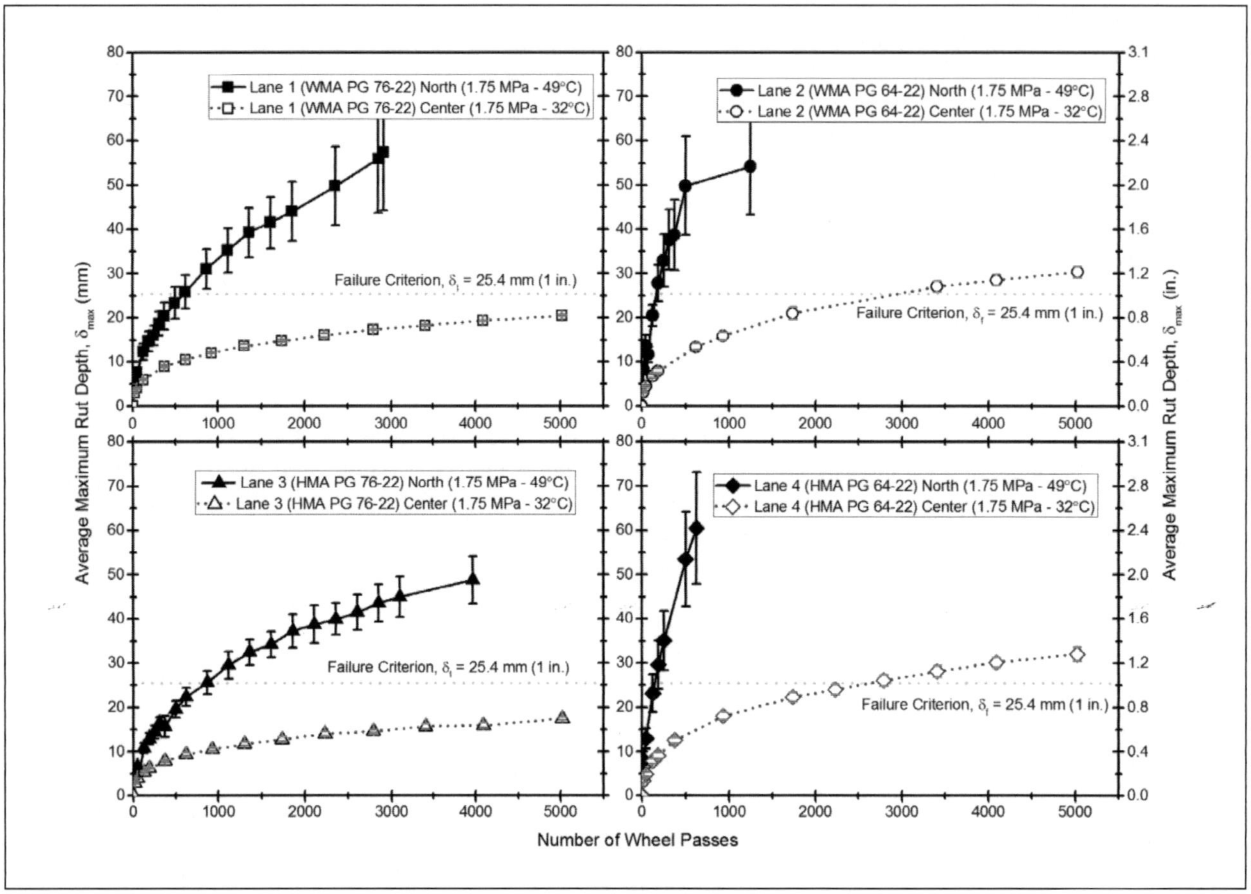

Figure 6. Effect of temperature on asphalt concrete rutting performance.

(*12*). IDT strength tests were conducted on both field cores and laboratory compacted specimens in accordance with the ASTM D6931 standard specification. The test temperature was selected to be 40°C instead of 25°C, the only modification to the aforementioned test procedure. Hamburg wheel tracking tests were also conducted on the field cores following the ASHTO T324-16 standard specification (*2, 3*).

Hamburg Wheel Tracking Test

Figure 8 shows the accumulation of permanent deformation in the field cores during the Hamburg wheel tracking tests. As lanes 1 and 5 had the same PG 76-22 WMA surface course and lanes 3 and 6 had the respective HMA surfacing, representative field cores for those two mix types were collected from lanes 1 and 6 only. As expected from the field observations, PG 76-22 mixes exhibited a substantially low magnitude of rutting compared with their PG 64-22 counterparts. All of the specimens underwent a certain magnitude of consolidation indicative of post-compaction. At the end of 20,000 passes in submerged conditions, both of the PG 76-22 mixes accumulated approximately 6.5 mm (1/4 in.) of rutting. The effect of moisture susceptibility was evident from the comparison of rutting progression in PG 64-22 HMA and WMA field cores. From the very beginning of the test, Lane 2 PG 64-22 WMA field cores exhibited a very steep creep slope compared with that of Lane 4 PG 64-22 HMA field cores. Even beyond the stripping inflection point, the strip slope for PG 64-22 WMA was found to be approximately 2.5 times the strip slope for the corresponding HMA. The PG 64-22 hot and warm mixes reached the failure criterion of 12.7 mm (0.5 in.) at 13,300 and 3,900 wheel passes, respectively.

High Temperature Indirect Tensile (IDT) Strength Test

Figure 9 shows the IDT strength of the field cores and laboratory compacted specimens. A large variation in air voids was noticed in the extracted cores. Notably, all of the laboratory compacted specimens had approximately 3.5% air voids as dictated by the airfield AC job mix formula. The average air voids in the outside lanes (lanes 1

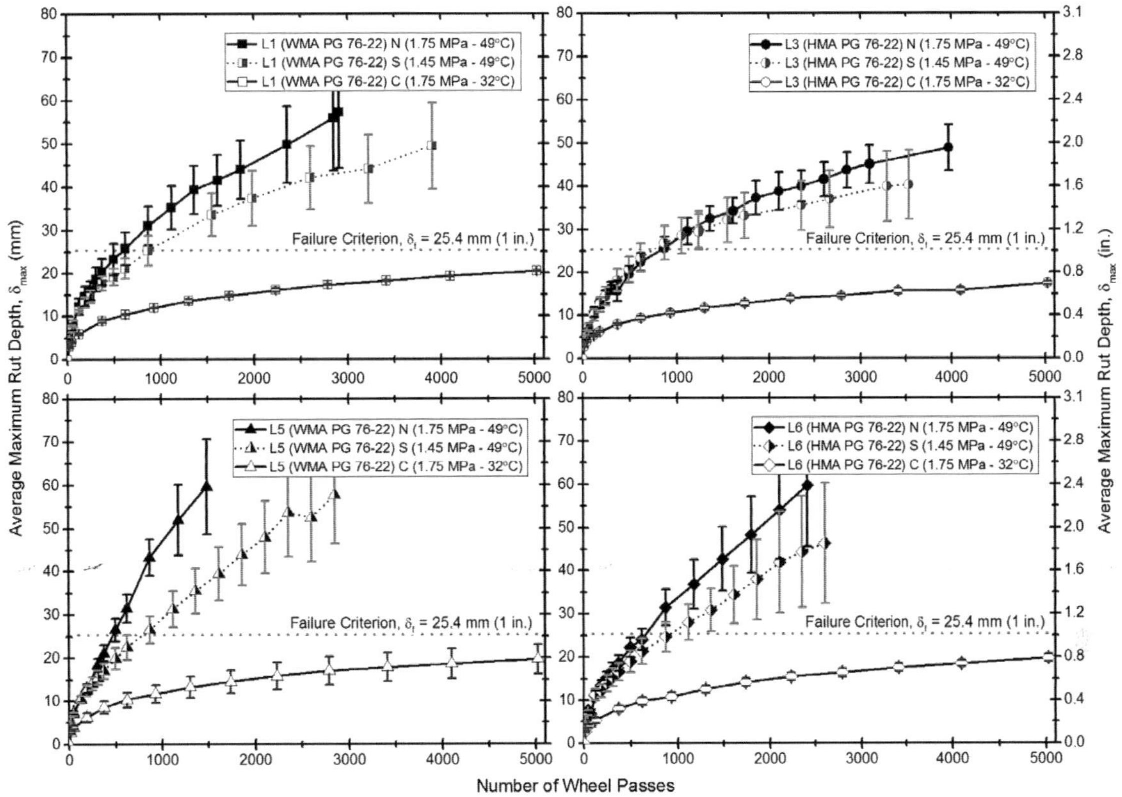

Figure 7. Effect of curing and environmental aging on asphalt concrete rutting performance.

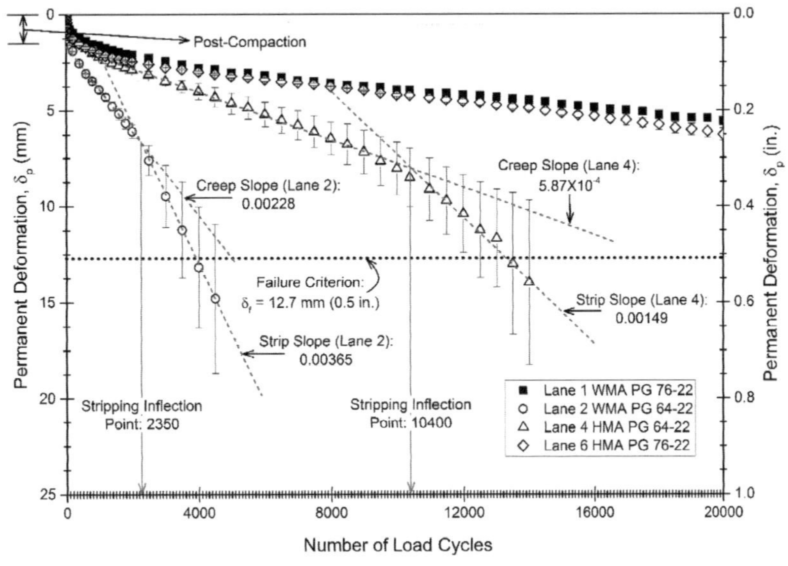

Figure 8. Hamburg wheel tracking tests on field cores.

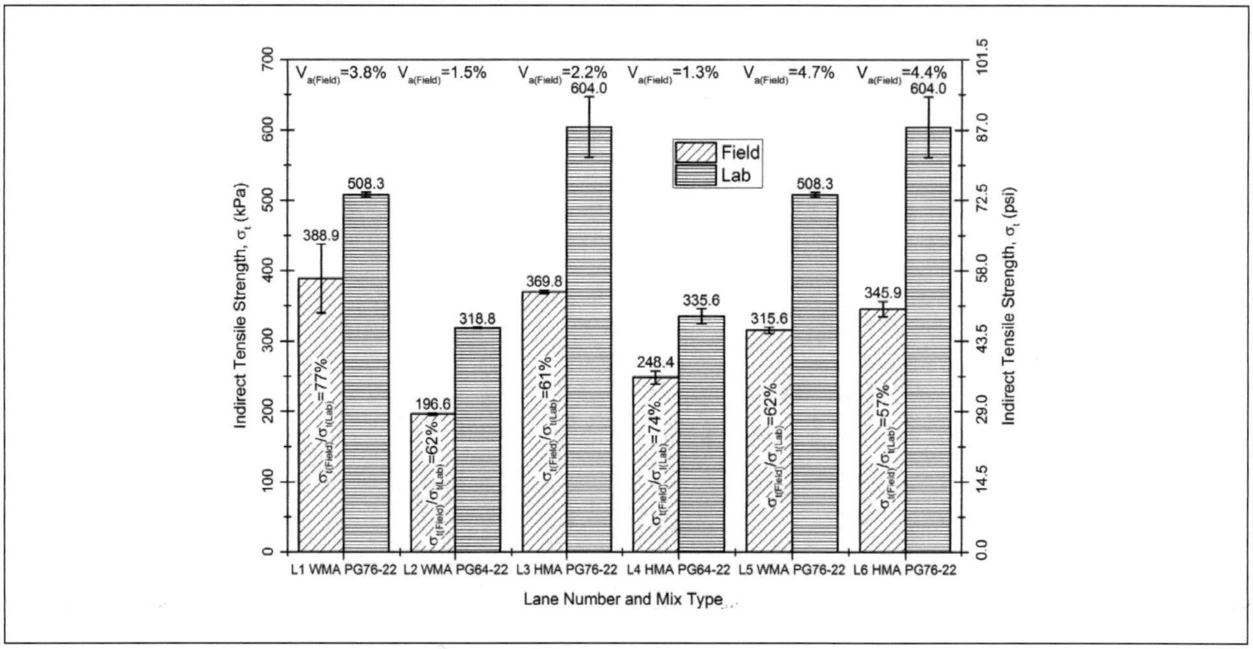

Figure 9. Indirect tensile strength of field cores and laboratory compacted specimens.

through 4) were found to be lower than those in the inside lanes. Despite such a discrepancy, both lanes 3 and 6 field cores exhibited a similar magnitude of tensile strength. Conversely, Lane 1 PG 76-22 field cores had higher tensile strength than Lane 5 PG 76-22 field cores. Environmental exposure might have contributed to enhanced curing and strength gain in the Lane 1 PG 76-22 WMA cores. In accordance with the poor rutting performance during APT and Hamburg wheel tracking tests, Lane 2 PG 64-22 WMA field cores exhibited the lowest tensile strength among all the mixes, followed by the Lane 4 HMA cores with similar binder grade. However, the ratio of field to laboratory strength was found to be the lowest in the case of Lane 6 PG 76-22 HMA specimens. Such an anomaly can be attributed to a higher than designed percentage of air voids in the corresponding field cores.

Conclusion

This paper summarizes the rutting performances of airfield AC mixes under the influence of high tire pressure and temperature. Six different test lanes were constructed with four different AC mixes encompassing HMA and WMA with two different binder grades. Each of the test lanes were subdivided into three different test sections: North, South, and Center. While trafficking, the North, South, and Center sections were subjected to 1.75, 1.45, and 1.75 MPa (254, 210, and 254 psi) of tire pressure, respectively. The temperatures in those three sections were maintained at 49, 49, and 32°C (120, 120, and 90°F), respectively at 50 mm (2 in.) below pavement surface. Rutting performances observed during the APT were examined against the laboratory characterization test results. Based on the key observations, the following conclusions can be drawn:

i. The WMA with polymer-modified binder (PG 76-22) exhibited comparable performance to the HMA in terms of rutting progression;
ii. Rutting performance of AC was more sensitive to temperature than tire pressure and environmental aging;
iii. Polymer-modified binder significantly improves the rutting performances of both HMA and WMA concrete;
iv. The effect of tire pressure had a significant impact on mixes with PG 64-22 binder grade, and had minimal impact on the rutting accumulation of higher quality PG 76-22 AC mixes;
v. The effect of high tire pressure under high temperature showed significant rutting for all the pavements and mixes tested;
vi. Center section pavements under the effect of high tire pressure and normal temperature [32°C (90°F)] presented low rutting compared with North and South section pavements;
vii. PG 64-22 WMA was susceptible to moisture-induced rutting damage and exhibited low IDT strength according to the laboratory characterization tests;

viii. Mixes with polymer-modified binder (PG 76-22) accumulated a minimal amount of rutting compared with the neat PG 64-22 binder mixes during the Hamburg wheel tracking tests, confirming the findings from heavy vehicle simulation tests.

An advanced pavement analysis tool called PANDA-AP (Pavement Analysis using Non-linear Damage Approach for Airports) is currently being developed. It implements mechanistic-based models that account for damage and rutting not only in the viscoelastic and viscoplastic AC layer but also in unbound pavement layers (including base, subbase, and subgrade). It uses the Drucker-Prager CAP model for unbound materials. Users will be able to compare the performance of different asphalt mixes and pavement structures. Results from NAPMRC and the National Airport Pavement Testing Facility are being used to calibrate PANDA-AP.

Acknowledgments

The work described in this paper was supported by the FAA Airport Technology Research and Development Branch. The authors acknowledge the support of Murphy Flynn and Wilfredo Villafane of FAA, David Traverzo, Henry Fermin, Jeffery Stein, Dennis Bataille, Monir Haggag, and Matthew Willson of Systems Research and Applications Corporation (CSRA), John Gerke, Christopher Tomlinson, and Russell Knieriem of Applied Research Associates, Inc. (ARA), and Peter Millar of Dynatest for their support during the construction and testing of pavements.

Author Contributions

The authors confirm contribution to the paper as follows: study conception and design: NG; data collection: NG, HK, LR, TP; analysis and interpretation of results: NG, HK, LR, TP; draft manuscript preparation: NG, HK, LR, TP. All authors reviewed the results and approved the final version of the manuscript.

References

1. Prowell, B. D., G. C. Hurley, and B. Frank. *Warm-Mix Asphalt: Best Practices*. National Asphalt Pavement Association, Lanham, Md, 2012.
2. Hurley, G. C., B. D. Prowell, and M. Huner. *Evaluation of Aspha-Min Zeolite for Use in Warm Mix Asphalt (NCAT Report 05-04)*. National Center for Asphalt Technology, Auburn, Ala., 2005.
3. Hurley, G. C., and B. D. Prowell. *Evaluation of Evotherm(R) for Use in Warm Mix Asphalt (NCAT Report 06-02)*. National Center for Asphalt Technology, Auburn, Ala., 2006.
4. Federal Aviation Administration. *Standardized Method of Reporting Airport Pavement Strength - PCN (Advisory Circular No. 150/5335-5C)*. Office of Airport Safety & Standards - Airport Engineering Division, Federal Aviation Administration, Washington, D.C., 2014.
5. Jones, D., R. Z. Wu, B. Tsai, C. Barros, and J. Peterson. Accelerated Loading, Laboratory, and Field Testing Studies to Fast-Track the Implementation of Warm Mix Asphalt in California. In *Advances in Pavement Design through Full-scale Accelerated Pavement Testing* (Jones, D., ed.), Taylor & Francis Group, London, 2012.
6. Baek, C., Y.-C. Suh, S. Yang, J. Lee, and S. Kwon. An Evaluation of Warm-Mix Asphalt Performance Using Accelerated Pavement Testing in Korea. *KSCE Journal of Civil Engineering*, Vol. 18, No. 7, 2014, pp. 2120–2127. https://doi.org/10.1007/s12205-014-1366-5.
7. Mejías-Santiago, M., J. Doyle, and J. Rushing. Accelerated Pavement Testing of Warm-Mix Asphalt for Heavy-Traffic Airfields. *Transportation Research Record: Journal of the Transportation Research Board*, 2014. 2456: 11–20.
8. Flynn, M., N. Garg, W. Villafane, and D. Traverzo. Design and Engineering Challenges in the Development of the FAA's Full Scale Accelerated Pavement Test Facility NAPMRC. *Proc., 5th International Conference on Accelerated Pavement Testing*, San Jose, Costa Rica, 2016, Springer, pp 37–52.
9. HoSang, V. A. *Field Survey and Analysis of Aircraft Distribution on Airport Pavements*. Report No. FAA-RD-74-36. Federal Aviation Administration, Washington, D.C., 1975.
10. Rufino, D., J. Roesler, E. Tutumluer, and E. Barenberg. Wander Patterns for Commercial Aircraft at Denver International Airport. *Proc., 27th International Air Transportation Conference*, ASCE, 2001, Chicago, IL, pp. 158–170.
11. U.S. Army Corps of Engineers. *Asphalt Surfaced Airfields: Paver Distress Identification Manual*. Construction Engineering Research Laboratory, Champaign, Ill., 2009.
12. Li, Q., N. Garg, and M. Haggag. Evaluation of Rutting Performance of Airport Pavement Hot-Mix Asphalt Using Gradation-Based Method and Laboratory Testing. Presented at 94th Annual Meeting of the Transportation Research Board, Washington, D.C., 2015.

The Standing Committee on Aircraft/Airport Compatibility (AV070) peer-reviewed this paper (18-03566).

The contents of this paper reflect the views of the authors, who are responsible for the facts and accuracy of the data presented within. The contents do not necessarily reflect the official views and policies of the FAA. The paper does not constitute a standard, specification, or regulation.

Article

TRR
JOURNAL OF THE TRANSPORTATION RESEARCH BOARD

Transportation Research Record
2018, Vol. 2672(23) 128–136

DOI: 10.1177/0361198118773186
journals.sagepub.com/home/trr
SAGE

Evaluation of Asphalt Drainable Base Structural Performance for Flexible Airport Pavement Design

Qiang Li[1], Benjamin Mahaffay[2], and Jeffrey Gagnon[2]

Abstract

In 2014, the Federal Aviation Administration's (FAA) National Airport Pavement Test Facility (NAPTF) completed construction of its flexible pavement test section, Construction Cycle 7 (CC7). Among the objectives of CC7 was to study the performance of an asphaltic drainable base pavement section compared with a P-209 aggregate base pavement section under full-scale airport loading. This research addresses the pre-traffic material characterization, full-scale traffic test data analysis and post-traffic test data analysis of these two pavement sections in dry conditions. This paper will discuss the preliminary results from laboratory testing and full-scale traffic tests for material properties which may be used in design and construction specifications to improve FAA pavement design software (FAARFIELD).

The Federal Aviation Administration (FAA) operates a state-of-the-art, full-scale pavement test facility dedicated solely for airport pavement research, known as the National Airport Pavement Test Facility (NAPTF). Located at the William J. Hughes Technical Center near Atlantic City, New Jersey, the $21million test facility was constructed in April 1999 and consists of a 900 ft (274.3 m) long by 60 ft (18.3 m) wide fully enclosed test area with fully embedded instrumentation and a dynamic computerized data acquisition system. (*1, 2*). The main purpose of the facility is to collect data through accelerated pavement testing (APT) to update the latest FAA design procedures and advisory circulars to foster a safe and efficient air transportation system. APT is conducted with the 1.3 million pound, fully-programmable, rail-based national airport pavement test vehicle (NAPTV) consisting of two loading carriages capable of simulating up to six-dual wheel gear configuration per carriage at a capacity up to 75,000 lb per wheel. The NAPTV is capable of various gear configurations with the ability to simulate today's heaviest long-range, wide body jet airliners (*3*).

During the recent construction cycle seven (CC7), six flexible pavement test sections (LFP-1, LFP-2, LFP-3, LFP-4, LFC-5, and LFS-6) were constructed on the north side of the testing section. LFP-1 through LFP-4 were designed as perpetual asphalt concrete pavement test sections (asphalt concrete thickness ranging from 8 to 15 in.) (*4*), LFC-5 was a conventional flexible pavement (P-209 crushed aggregate base course, and aggregate subbase), and LFS-6 was a conventional pavement utilizing 8 in. (203 mm) of an asphaltic treated drainable base in lieu of the typical P-209 crushed stone base, placed over 29 in. (737 mm) of a P-154 subbase course. A CL-CH soil classification material (DuPont clay) with a CBR value of 5-6 was used as subgrade. Final constructed pavement cross-sections of LFC-5 and LFS-6 are shown in Figure 1.

It is well documented that excessive water within the pavement structure leads to early distresses and premature structural and functional pavement failures (*5–8*). Therefore, it is a critical design consideration to provide a structurally adequate subsurface drainage layer within the pavement structure. The primary objective of this research paper is to evaluate the structural performance of a flexible pavement using an asphaltic treated drainable base material under full-scale airport loading APT conditions.

[1]SRA International, Inc., Linwood, NJ
[2]Federal Aviation Administration, Airport Technology R&D Branch, William J. Hughes Technical Center, Atlantic City International Airport, NJ

Corresponding Author:
Address correspondence to Qiang Li: Qiang.Li@gza.com

Table 1. Summary of Parameters Used for HWD Data Analysis

Parameter	Formula	Unit	Parameter's objective
ISM	ISM=F/D0 F = vertical dynamic load D0 = deflection at the center sensor	lbf/mils	Characterize the stiffness of all structural layers
AREA	AREA = 6*(D0 + 2*(D2 + D3 + D4 + D5 + D6) + D7)/D0	in.	Reflect the structural response of the whole pavement structure
AUPP	AUPP = 0.5*(11*D0-2(D2 + D3 + D4 + D5 + D6)–D7)	mils	AUPP is correlated to maximum tensile strain at bottom of AC layer

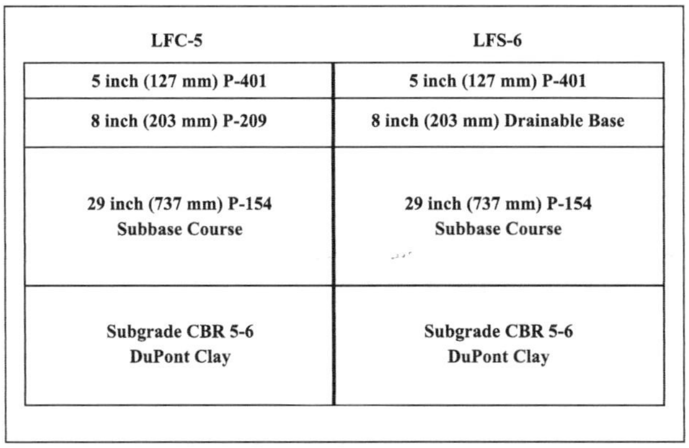

Figure 1. CC7 north side pavement profile cross sections.

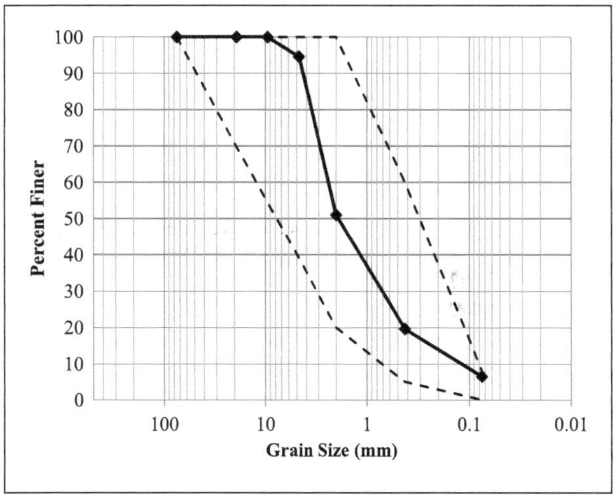

Figure 2. Subbase base gradation (dashed lines represent FAA P-154 gradation limits).

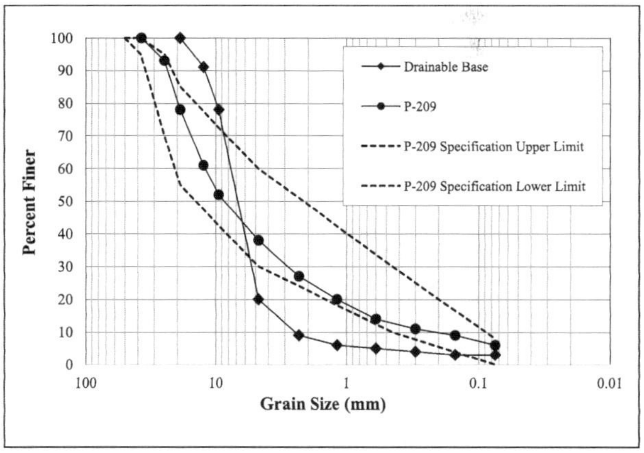

Figure 3. Drainable base and granular base gradation (dashed lines represent FAA P-209 gradation limits).

Material Placement and Properties

The P-154 subbase course was a granular argillite material purchased from a local supplier conforming to the FAA Advisory Circular AC 150/5370-10G (9). Figure 2 shows the P-154 particle size distribution with FAA P-154 gradation limits. The material was placed in approximately 10 in. compacted lifts to a final thickness of 29 in. to the desired compaction effort of greater than 95% of the dry density. Sand cone testing yielded an average of 4.1% moisture (dry of 7.6% optimum) with an average dry density of 136.18 lb/ft^3.

The P-209 crushed aggregate base course was clean crushed calcium carbonate (limestone) aggregate purchased from a local aggregate supplier conforming to the FAA Advisory Circular AC 150/5370-10G (9), as shown in Table 1 and Figure 3. The material was placed in a single lift with a paver and compacted to a final thickness of 8 in. (203 mm) to the desired compaction effort of greater than 98% of the dry density. Sand cone testing yielded an average of 2.7% moisture content (dry of the 5.7% optimum) with a dry density of 156.83 lb/ft^3.

The asphaltic drainable base material was an open graded aggregate mix of calcium carbonate (limestone) aggregate purchased from a local aggregate supplier, and the gradation is shown in Figure 3. The mix was designed at a 3% asphalt content using 64-22 grade asphalt resulting in a density of 129 lb/ft^3. The material was placed with an asphalt paver and compacted to a final thickness of 8 in. (203 mm) exhibiting approximately 23% air voids and a permeability averaging 1,476 ft^3 per day.

Table 2. Summary of PSPA and LWD Results for CC7 Unbounded Materials

		Subgrade	Subbase	Base
PSPA seismic modulus	Minimum (ksi)	15	47	78
	Maximum (ksi)	34	135	168
	Mean (ksi)	20.87	81.77	114.16
	Standard deviation (ksi)	3.99	19.50	21.57
	CV	19.1%	23.8%	18.9%
LWD modulus, ksi	Minimum (ksi)	4.1	16.5	26.0
	Maximum (ksi)	8.9	33.2	35.2
	Mean (ksi)	6.3	24.1	30.4
	Standard deviation (ksi)	1.3	3.7	2.6
	CV	21%	15%	8%

The asphalt concrete material used for surface layer met P-401 specifications for hot mix asphalt (*9*). PG76-22 asphalt binder was used in the asphalt concrete with an asphalt content of 6% and air voids of 3.4%.

Pre-Traffic Material Characterization

Portable seismic pavement analyzer (PSPA) is a nondestructive testing method based on wave propagation theory. The PSPA is set on the test surface and a source initiates highly-repeatable seismic waveforms that are recorded by three accelerometers. These waveforms are transferred back to the computer for analysis. The modulus of the soil materials can be determined based on compression or shear wave velocity. The modulus value obtained from PSPA is a low-strain elastic modulus. A more detailed explanation on theory and equipment can be found elsewhere (*10–13*).

A KUAB Model 240 heavy weight deflectometer (HWD) was utilized to perform the HWD tests. The standard procedure for the CC7 north side HWD tests has been established with the following parameters: the segmented 12 in. loading plate, a pulse width of 30 msec, and four drop heights consisting of a 36 kips "seating drop" followed by impact loads of 12, 24, and 36 kips (*14*). The peak loads and deflections are recorded for all four drops along with air and pavement surface temperatures. The deflections are measured at the center of load plate (D0), at 12 in. offset (D2), at 24 in. offset (D3), at 36 in. offset (D4), 48 in. offset (D5), 60 in. offset (D6), and 72 in. offset (D7). An additional sensor is placed at 12 in. offset (D1) on the front of loading plate.

In this study, impact stiffness modulus (ISM), AREA and area under the pavement profile (AUPP) were used to characterize the shape of deflection basin (Table 1) (*14–18*). AREA denotes the normalized area of a vertical slice taken through a deflection basin. AUPP has a promising relationship with the tensile strain at bottom of the asphalt concrete layer (*19*).

Light weight deflectometer (LWD) is a portable version of the HWD. LWD measures the deflection from a falling weight and estimates the modulus.

A series of PSPA and LWD tests were performed on the top lift of subgrade, subbase and base to characterize the material properties and to evaluate the uniformity of unbound material. PSPA and HWD tests were conducted on top of the asphalt concrete layer to estimate the uniformity of pavement structure before full-scale accelerated traffic tests.

Table 2 summarizes the PSPA test results for CC7 subgrade, subbase and subgrade. PSPA seismic moduli are roughly twice the unconfined low-strain resilient modulus, and this relationship is independent of the type of material used. In the absence of any laboratory testing data, this relationship can be used to convert seismic modulus to resilient modulus (*13*). The results appear to indicate the coefficient of variation (CV) of seismic moduli is around 20%, therefore the unbounded material used in CC7 are fairly uniform. The LWD test results of CC7 subgrade, subbase and subgrade are summarized in Table 2. The LWD test results also show that the unbounded material used in CC7 are fairly uniform.

HWD and PSPA tests were conducted to evaluate uniformity of the pavement before the traffic tests. There are nine test locations in each section. The ISM was calculated based on HWD data to evaluate the uniformity. ISM is correlated to the stiffness of all structural layers. Since drainable base is stiffer than granular base, the mean ISM of LFC-5 is lower than LFS-6. Table 3 represents the statistical analysis results. CV of the six test sections is between 3% and 7%. HWD test results appear to indicate that the pavement structures are fairly uniform. The PSPA test results were shown in Table 3. PSPA tests measure the seismic moduli of P-401. Two sections used the same mix design, so the mean value of the two sections is close to each other. The CV for the two sections is less than 8%.

Table 3. Summary of P-401 Uniformity Test Results

	Section	LFC-5	LFS-6
HWD	ISM max (lbf/mils)	1247.2	1548.9
	ISM min (lbf/mils)	1035.1	1366.1
	ISM mean (lbf/mils)	1108.8	1462.3
	Standard deviation (lbf/mils)	66.2	62.4
	CV	6.0%	4.3%
PSPA	Seismic modulus max (ksi)	3580	3740
	Seismic modulus min (ksi)	2770	2870
	Seismic modulus mean (ksi)	3203.9	3282.2
	Standard deviation (ksi)	233.7	193.3
	CV	7.3%	5.9%

Full-Scale APT

Full-scale traffic tests of the pavement structure simulating actual loading conditions were performed to monitor pavement performance. Due to traffic tests being performed at the indoor NAPTF, all were performed in dry conditions. The pavement sections were trafficked using the NAPTV over a 12-month period from 2015 to 2016 totaling 23,364 passes. During traffic testing the NAPTV simulates realistic aircraft wander by varying the lateral position of the carriage to simulate a normal distribution of aircraft traffic. The wander pattern used for this study consists of 66 vehicle passes, arranged in 9 wander positions (or tracks). The wander positions and sequences were chosen to simulate a normal distribution of aircraft traffic with a standard deviation of 30.5 in. (representing the current design condition for airport taxiways) (*2*). The NAPTV carriage was configured to a six-wheel gear configuration (3-D) at 55,000 lb per wheel. During the trafficking tests, the PSPA and HWD tests were performed periodically at four predetermined nondestructive test locations, while rut depth measurements and crack maps were also developed to monitor pavement performance.

PSPA Tests

PSPA tests were performed periodically during trafficking at the four predetermined locations. The PSPA data was analyzed and the temperature corrected seismic modulus in the non-traffic area is shown in Figure 4*a*. This figure shows that the temperature correction used is an effective way to reduce the effect of temperature. The temperature corrected PSPA seismic moduli data collected in the traffic area during full-scale traffic tests is shown in Figure 4*b*. Results show the decreasing rate of seismic modulus is lower in the asphaltic drainable base pavement section (LFS-6) than the aggregate base pavement section (LFC-5), which shows the LFS-6 section deteriorates at a slower rate than the LFC-5 section.

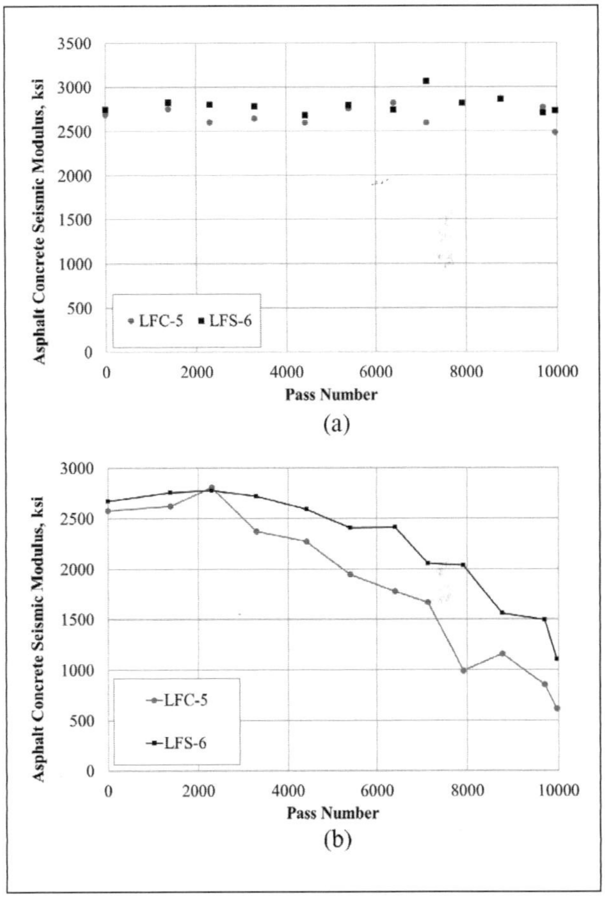

Figure 4. PSPA data analyses showing seismic moduli in (*a*) non-traffic area and (*b*) traffic area.

HWD Tests

HWD tests were performed periodically during trafficking at the same four predetermined locations. The HWD data was used to calculate ISM, AREA and AUPP utilizing temperature corrected deflections. These deflection parameters were plotted versus pass number as shown in Figure 5. ISM is an indicator of the entire pavement

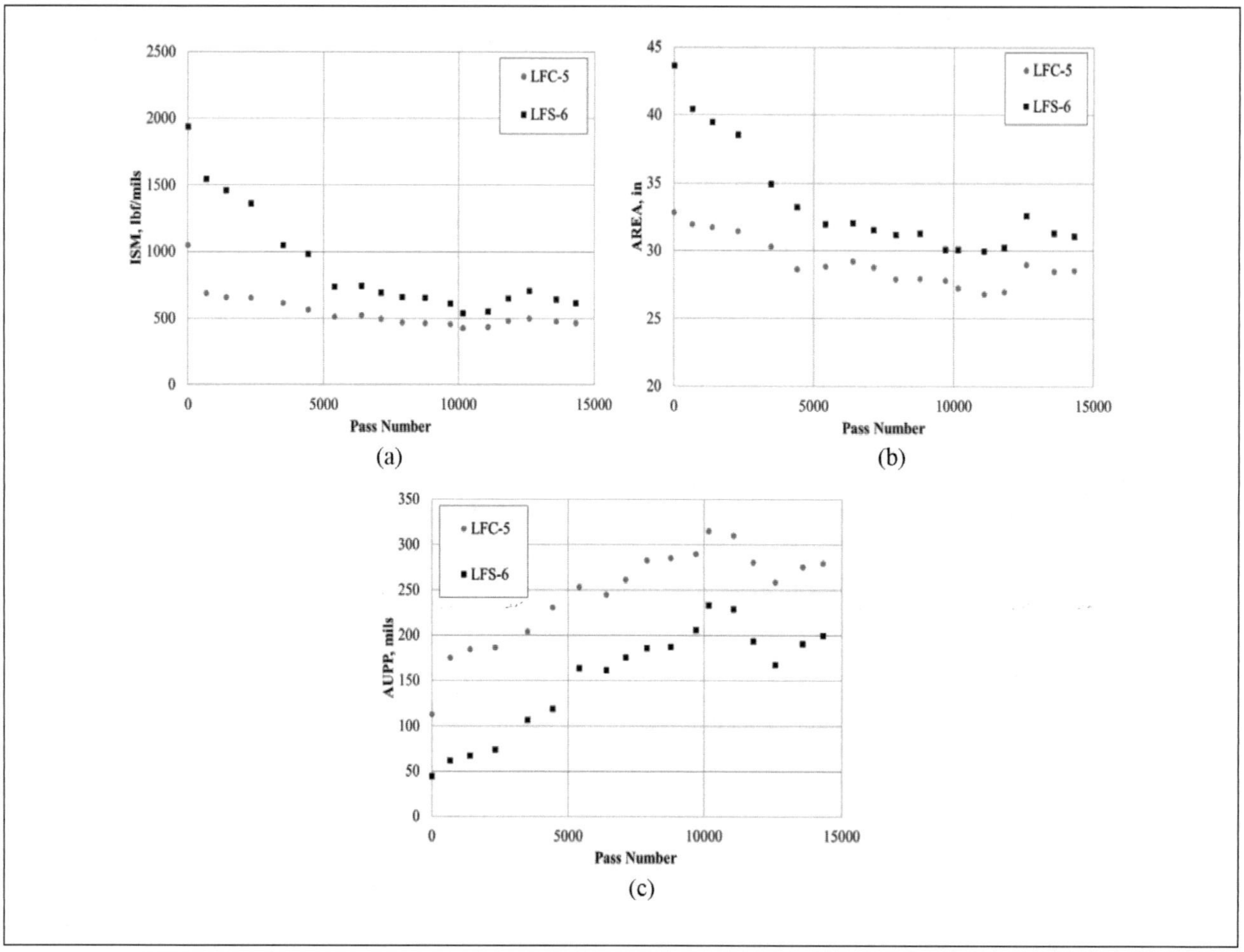

Figure 5. HWD data analyses for (*a*) ISM, (*b*) AREA, and (*c*) AUPP.

structure stiffness. Figure 5*a* shows LFS-6 has higher ISM and less stiffness reduction than LFC-5. AREA also reflects the structural response of the whole pavement structure. The AREA results show a similar trend to ISM. Considering the drainable base is stiffer than granular base, higher AREA was observed in LFS-6 as shown in Figure 5*b*. AUPP is related to the maximum tensile strain at the bottom of the hot-mix asphalt (HMA) surface layer. The maximum tensile strain at bottom of the HMA surface layer will increase when aircraft traffic deteriorates and produces micro-cracks or macro-cracks in pavement. Figure 5*c* shows lower tensile strains in the asphaltic drainable base pavement section (LFS-6) than the P-209 aggregate base pavement section (LFC-5) thus, the degradation rate due to traffic for LFS-6 is slower than LFC-5.

Pressure Cell Measurement

Vertical stress on top of the subgrade materials is a critical parameter to predict subgrade rutting. Pressure Cells (PCs) were placed on the top of subgrade surface to measure the vertical stress under the same traffic load. PC 5 and PC 6 were installed in LFC-5 and LFS-6 respectively. The maximum vertical stress of each wander was plotted versus pass number in Figure 6. PC 5 was damaged after 2,868 passes. The result appears to indicate that subgrade vertical stress in LFC-5 is higher than LFS-6 during traffic. It is because the drainable base distributes the load more effectively than P-209 granular base. The significant decrease of vertical stress around 2,300 passes is caused by the decrease of the pavement temperature.

Maximum Rut Depth and Crack Mapping

The maximum rutting depth was periodically measured using a 16 ft (4.88 m) straight edge, and crack mapping was conducted during trafficking. The rut depth and pass number were plotted and show rutting depth increases as pass number increases, as shown in Figure 7. At approximately 5,000 passes the P-209 crushed aggregate base

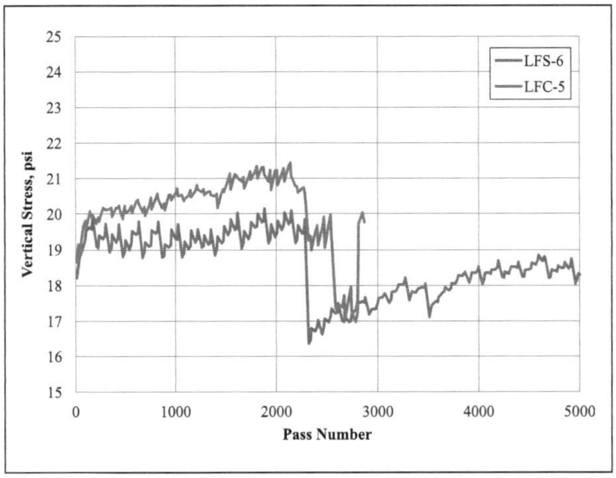

Figure 6. PC measurement during traffic test.

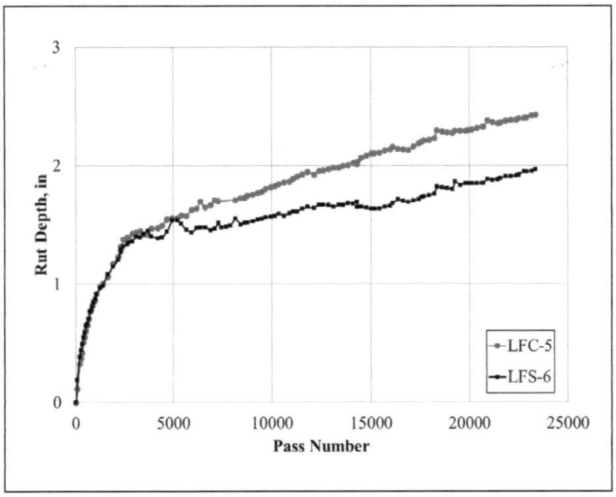

Figure 7. Maximum rutting depth.

(LFC-5) pavement section begins to show higher rut depth and have an increased rutting rate compared to the asphaltic drainable base pavement section (LFS-6).

Figure 8 shows the final crack map after 23,364 passes and it illustrates less longitudinal and transverse cracks were observed in asphaltic drainable base pavement area (LFS-6). The decreased cracking observed in the LFS-6 test area lines up with the lower tensile strains observed from the AUPP results. Therefore, asphaltic drainable base pavement section not only provides higher rutting resistance, but also decreased fatigue cracking than a typical crushed granular base pavement section.

Post-Traffic Trench Test

CC7 traffic testing was completed in December 2016. Trenching provides a viable option to examine failed test

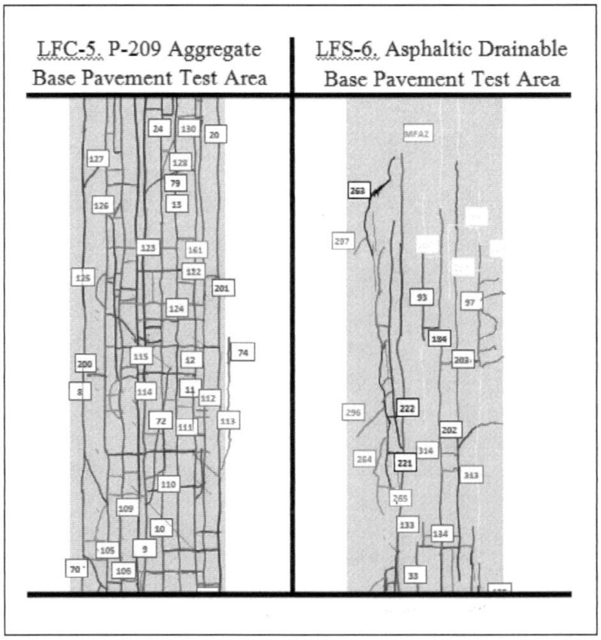

Figure 8. Crack map.

items, confirm the contribution of each pavement layer to the total rutting, and retrieve and preserve samples for laboratory material characterization. Using a laser straight line, the face of each trench was measured to get the profile of each layer of the pavement (Figure 9). Trench profiles showed most of the observed rutting was attributed by the subbase P-154 layer for both sections. The P-154 layer thickness reduction in LFC-5 is higher than in LFS-6. The LWD test and nuclear density test were performed on top of the P-154 surface. The results (Figures 10 and 11) show that LWD modulus and nuclear density in the traffic area is higher than in non-traffic area. The LWD modulus and nuclear density in LFC-5 are higher than in LFS-6. The reason is the drainable base can distribute the traffic load more effectively than the P-209. Under the same traffic load, the vertical stress on top of the P-154 surface in section LFC-5 is higher than in LFS-6.

Summary

The research findings include:

- Pre-traffic testing using nondestructive methods demonstrates the CC7 pavement structure sections were uniform. The PSPA and ISM uniformity test results show the CV between the asphalt drainable base pavement section (LFS-6) and the P-209 crushed aggregate base pavement section (LFC-5) was less than 8%. According to ISM testing, the

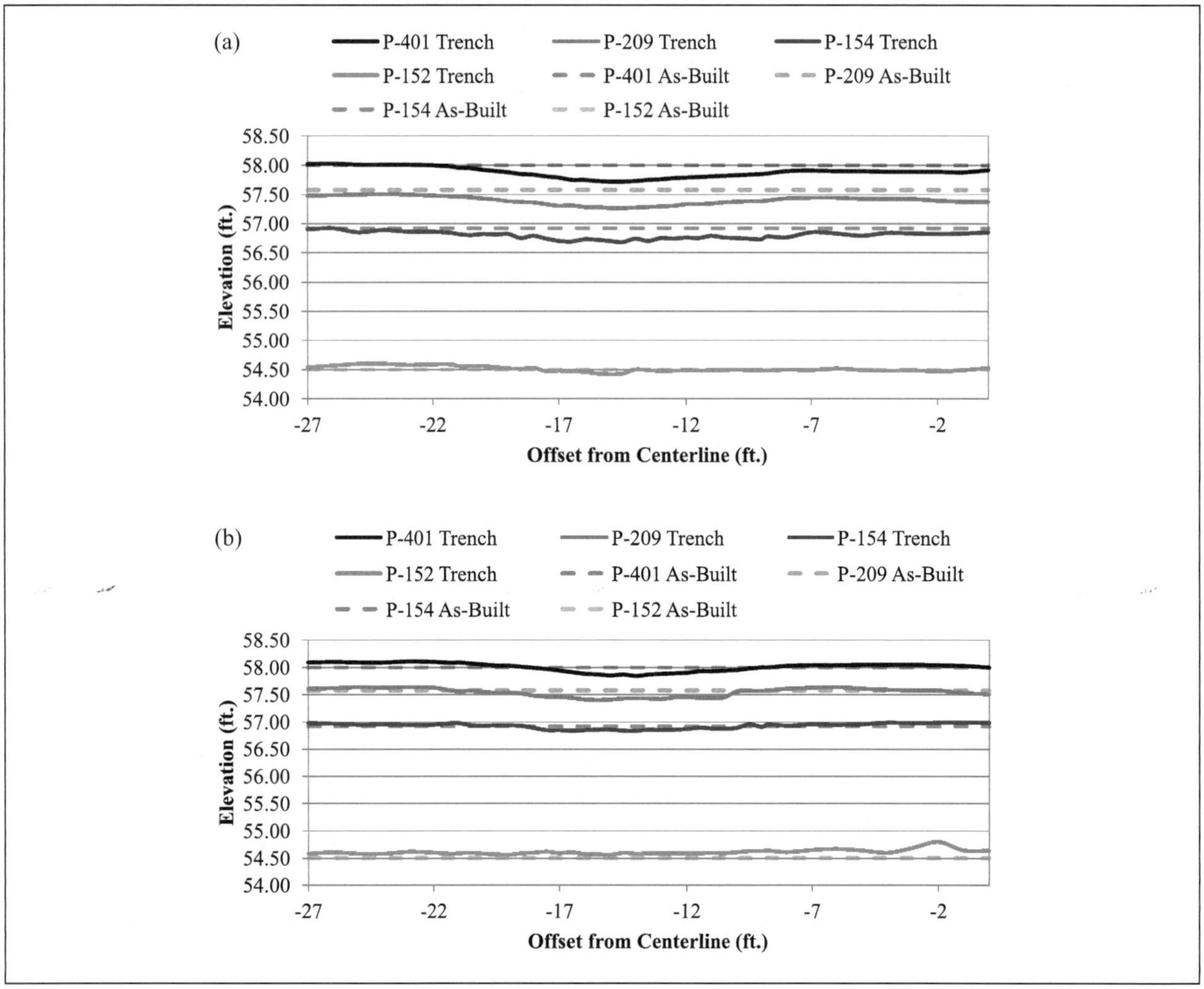

Figure 9. Trench pavement layer profile for (*a*) LFC-5 and (*b*) LFS-6.

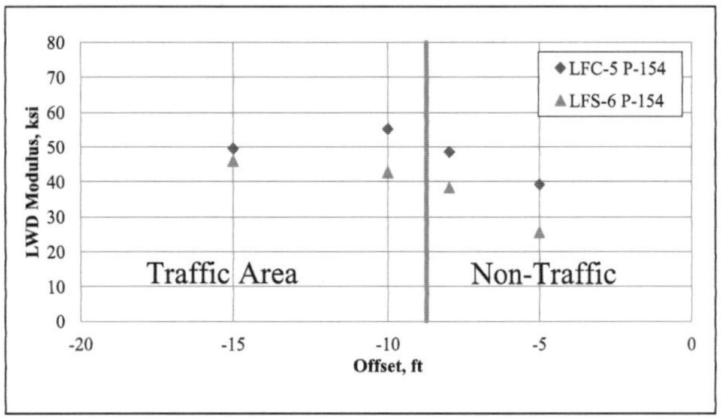

Figure 10. LWD test on top of P-154 surface.

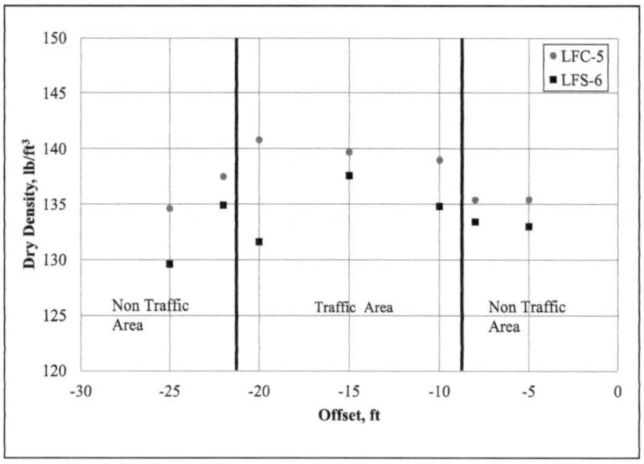

Figure 11. Nuclear density test on top of P-154 surface.

asphaltic drainable base pavement section (LFS-6) yielded a higher stiffness than the P-209 aggregate base pavement section (LFC-5).

- Analyses of PSPA and HWD data using ISM and AUPP methods from full-scale traffic testing using the NAPTV indicate the rate of deterioration for the asphaltic drainable base (LFS-6) pavement section was lower than the P-209 crushed aggregate base pavement section (LFS-5).
- Rut depth measurement and crack mapping from the full-scale traffic testing proved the asphaltic drainable base pavement section (LFS-6) demonstrated higher rutting resistance properties and decreased fatigue cracking than the P-209 crushed aggregate base pavement section (LFC-5).
- Trench profiles showed most of the observed rutting was attributed by the subbase P-154 layer for both sections. The P-154 layer thickness reduction in LFC-5 is higher than LFS-6. The asphalt drainable base gave greater structural benefit by distributing the traffic load in the lower layers more effectively than P-209.

In conclusion, this research demonstrates the structurally superior asphaltic drainable base course layer designed within flexible airport pavements can provide a structural benefit and help to increase pavement life and performance.

It is recommended that the FAA continue to incorporate additional treated drainable base materials, with a potential to incorporate saturated conditions in the upcoming construction cycles. The goal is to incorporate drainable base layers into FAA design software (FAARFIELD) and to update design and construction specifications for use in flexible and rigid airport pavement designs.

Acknowledgments

The work described in this paper was supported by the FAA Airport Technology Research and Development Branch. The authors would like to acknowledge the support of Navneet Garg and Murphy Flynn (FAA), and SRA staff—Renaldo Aponte, Jeff Stein, Henry Fermin, and Carlos Rosado.

Author Contributions

The authors confirm contribution to the paper as follows: study conception and design: Li, Mahaffay, Gagnon; data collection: Li, Mahaffay; analysis and interpretation of results: Li, Mahaffay, Gagnon; draft manuscript preparation: Li, Mahaffay, Gagnon. All authors reviewed the results and approved the final version of the manuscript.

References

1. Gervais, E. L., G. F. Hayhoe, and N. Garg. Towards a Permanent ACN Solution for 6-Wheel Landing Gear Aircraft. *Proc., ASCE Airfield Pavements Specialty Conference*, Las Vegas, Nev., 2003.
2. Garg, N., and G. F. Hayhoe. Asphalt Concrete Strain Responses at High Loads and Low Speeds at the National Airport Pavement Test Facility (NAPTF). *Proc., ASCE Airfield Pavement Specialty Conference - Advancing Airfield Pavements*, Chicago, Ill., 2001.
3. FAA. *FAA Airport Technology Research & Development Branch Home Page*. FAA William J. Hughes Technical Center, Atlantic City Intl. Airport, N.J. http://www.airporttech.tc.faa.gov. Accessed July 25, 2017.
4. Li, Q., C. E. Cary, and N. Garg. Use of Heavy Weight Deflectometer in Monitoring Flexible Pavement Deterioration during Accelerated Pavement Testing. *Proc., NDE/NDT for Highway and Bridges: Structural Materials Technology*, Portland, Ore., 2016.
5. Cheng, D. Z., D. N. Little, R. L. Lytton, and J. C. Holste. Surface Free Energy Measurement of Aggregates and Its Application on Adhesion and Moisture Damage of Asphalt-Aggregate System. *Proc., 9th International Center for Aggregate Research Symposium*, Austin, Tex., 2001.
6. Dawson, A. R., and A. R. Hill. Prediction and Implications of Water Regimes in Granular Bases and Subbases. *Proc., International Symposium on Sub-drainage in Roadway Pavements and Subgrades*, Granada, Spain, 1998, pp. 121–128.
7. Elsayed, A. S., and J. K. Lindly. Estimating Permeability of Untreated Roadway Bases. *Transportation Research Record: Journal of the Transportation Research Board*, 1996. 1519: 11–18.
8. Little, D. N., and I. V. Jones. Chemical and Mechanical Processes of Moisture Damage in Hot-Mix Asphalt Pavements. In *Moisture Sensitivity of Asphalt Pavements - A National Seminar*, Transportation Research Board National Seminar, San Diego, Calif., 2003, pp. 37–70.
9. Federal Aviation Administration, AAS-100, Office of Airport Safety & Standards. *150/5370-10G - Standards for Specifying Construction of Airports*. U.S. Department of Transportation, 2014, https://www.faa.gov/documentLibrary/media/Advisory_Circular/AC-150-5370-10G-updated-201605.pdf.
10. Nazarian, S., M. R. Baker, and K. Crain. *SHRP Report H-375: Fabrication and Testing of a Seismic Pavement Analyzer*. Transportation Research Board of the National Academies, Washington, D.C. 1993.
11. Nazarian, S., D. Yuan, and M. R. Baker. *Rapid Determination of Pavement Moduli with Spectral-Analysis-of-Surface-Waves Method*. Report Research Project 0-1243. Center for Geotechnical and Highway Materials Research, University of Texas at El Paso, El Paso, Tex., 1995.
12. Nazarian, S., D. Yuan, and M. Arellano. Quality Management of Base and Subgrade Materials with Seismic Methods. *Transportation Research Record: Journal of the Transportation Research Board*. 2002. 1786: 3–10.

13. Nazarian, S., R. Williams, and D. Yuan. *A Simple Method for Determining Modulus of Base and Subgrade Materials.* ASTM STP 1437. ASTM, West Conshohocken, Pa., 2003.
14. Li, Q., N. Garg, and M. Haggag. Evaluation of Flexible Airport Pavement Using HWD and PSPA. *Proc., ICPT 9th International Conference on Road and Airfield Pavement Technology*, Dalian, China, 2015.
15. Horak, E. The Use of Surface Deflection Basin Measurements in the Mechanistic Analysis of Flexible Pavements. *Proc., 5th International Conference on the Structural design of Asphalt Pavements*, Ann Arbor, Michi., 1987.
16. Horak, E. *Aspects of Deflection Basin Parameters Used in a Mechanistic Rehabilitation Design Procedure for Flexible Pavements in South Africa.* PhD thesis. Department of Civil Engineering at the University of Pretoria, Pretoria, South Africa, 1988.
17. Xu, B., S. Ranji Ranjithan, and Y. Richard Kim. New Relationships between Falling Weight Deflectometer Deflections and Asphalt Pavement Layer Condition Indicators. *Transportation Research Record: Journal of the Transportation Research Board*, 2002. 1806: 48–56.
18. Horak, E., and S. Emery. Falling Weight Deflectometer Bowl Parameters as Analysis Tool for Pavement Structural Evaluations. *Proc., 22nd Australian Road Research Board (ARRB)*, Brisbane, Australia, 2006, 15pp.
19. Thompson, M. R. *Area Under the Pavement Profile to Predict Strain.* Informal Presentation at FWD Users Group Annual Meeting, Indianapolis, Ind., 1989.

The Standing Committee on Aircraft/Airport Compatibility (AV070) peer-reviewed this paper (18-05233).

The contents of this paper reflect the views of the authors, who are responsible for the facts and accuracy of the data presented within. The contents do not necessarily reflect the official views and policies of the FAA. The paper does not constitute a standard, specification, or regulation.

TRR
JOURNAL OF THE TRANSPORTATION RESEARCH BOARD

Transportation Research Record
2018, Vol. 2672(23) 137–145

DOI: 10.1177/0361198118776521
journals.sagepub.com/home/trr
SAGE

Examining Head-Down Time in Transportation: Case Study in Single-Pilot General Aviation Operations

Sara A. Wilkins[1]

Abstract

The "head-down time" problem refers to the potential inability of an operator (e.g., pilot, car driver, or train driver) to divide his or her attention optimally between the primary visual field (e.g., out the window) and an auxiliary tool (e.g., visual display screen). This problem has become an important safety concern in road, rail, and air traffic environments. To improve transportation safety, researchers are applying technology to minimize head-down time. One of these emerging technologies is a pilot-oriented cognitive assistant called the Digital Copilot. The Digital Copilot is a working prototype that provides contextual information and aviation related reminders to the pilot in a timely manner. However, the effects of using this cognitive assistant on the pilot had not been evaluated. This research uses a cognitive task modeling software, called Cogulator, to model a pilot's expected thoughts and actions and compare them across two different environments: one using the Digital Copilot and one using three different electronic flight bags (EFBs) without cognitive assistance. This study examines how effectively the cognitive assistant can manage mental effort, simplify tasks, and decrease head-down time during preparation for the approach phase of flight for the following five tasks: determine weather frequency, determine communication frequency, review and follow a checklist, determine if the tower is open, and determine the preferred landing runway. Results show that the Digital Copilot provides: time savings in all tasks except for determining weather frequency; savings in head-down time for all tasks; and working memory load savings, or no change, for all tasks.

The "head-down time" problem refers to the potential inability of an operator (e.g., pilot, car driver, or train driver) to divide his or her attention optimally between the head-up primary visual field (e.g., out the window) and a head-down auxiliary tool (e.g., visual display screen) (*1*). This problem has become a safety concern in road, rail, and air traffic environments. In cars and trains, the operator's attention can be diverted from scanning out the window for hazards ahead as a result of in-vehicle controls, displays, onboard information systems, or mobile phones, which force the operator's head down. In general aviation (GA), pilots are reviewing charts, communicating on the radio, and programing navigation equipment while simultaneously flying the airplane. This not only equates to a high level of mental effort, but also limits the pilots' ability to look out the window for traffic, obstructions, or other hazards. This is a greater concern in single-pilot operated aircraft, which are 30% more likely to be involved in an accident than a dual-pilot crew (*2*). There are strong business aviation needs for a safe transportation system that provides speed, efficiency, and convenience for professionals to travel across the country. Although business aircraft accidents do not attract media attention to the extent that airliners do, over the past decade, the number of accidents resulting in fatalities involving private and chartered corporate aircraft is five times more than for airlines (*3*).

Researchers are applying new technologies to help with the head-down problem and thereby improve transportation safety. For example, measures of driving performance in a fixed-base simulator with a National Advanced Driving Simulator (NADS) scenario were superior when driving with the head-up display (HUD) using a Pioneer HUD, installed according to the manufacturer's recommendation, compared with the head-down display (HDD), using a Surface Pro 3 tablet fixed in front of an existing center console display (*4*). This study required drivers to follow a lead vehicle while performing a secondary task "composed of strings of random letters and numbers called 'pseudo-text'" at multiple levels of visual complexity (*4*). In addition, audio displays have been found to benefit navigation performance and mental

[1]The MITRE Corporation, McLean, VA

Corresponding Author:
Address correspondence to Sara A. Wilkins: sawilkins@mitre.org

workload for drivers as shown in a fixed-base high-fidelity simulator that evaluated selected in-vehicle route guidance systems (*5*). HUDs have proven successful in the aviation industry to view navigation, radar, and warning systems on a single screen without taking up any console space, as well as to align information with the outside world, and in the automobile industry to provide information otherwise available on the dashboard, such as speed. Similar benefits may be realized in the rail industry in train cabs to enhance observation, warning, and decision control (*6*).

In an effort to increase GA safety, the National Transportation Safety Board (NTSB) strongly recommends that GA pilots take advantage of applicable training opportunities and new technologies available (*7*). One of these new technologies is a pilot-oriented cognitive assistant called the Digital Copilot. While not a HUD, the Digital Copilot, developed by the MITRE Corporation, is a working prototype cognitive assistant that provides contextual information and reminders to the pilot in a timely manner (*8*). The Digital Copilot accepts spoken commands and automatically provides the pilot with relevant information based on the flight context. Thus, a cognitive assistant may offer similar benefits to a HUD by limiting the pilot's need for head-down time.

In order to maximize cognitive assistance, the Digital Copilot design considers the appropriate modality for presenting information to the pilot. This is directly related to the multiple resource theory, which assumes that humans have a limited set of resources available for mental processes to allocate to a task or combination of tasks. Multiple resource theory and performance predictions are often related to mental effort. That is, regardless of the resources used, humans have a limited cognitive capacity. For example, auditory and visual tasks demand common perceptual resources. When multiple tasks overload demand, multiple resource theory can predict performance failure and provide guidance to restore performance (*9*). For instance, the Society of Automotive Engineers has a recommended practice that describes criteria that in-vehicle tasks should meet in order to be acceptable while driving (*10*). When operating in safety critical environments, such as driving, railway signaling, and flying, mental workload must be kept at a manageable level to ensure appropriate allocation of resources and processes to accomplish tasks.

Because the specific pilot benefits of using the Digital Copilot as a cognitive assistant had not been evaluated, this study used task modeling to compare how effectively it can manage mental effort, simplify tasks, and decrease head-down time during preparation for the approach phase of flight. These task models were built using an open-source software tool called Cogulator, developed by the MITRE Corporation. Cogulator is based on a technique known as goals, operators, methods, and selection rules (GOMS) (*11*). Each task is defined by a goal composed of operators, which are the perceptual, motor, and cognitive processes used to achieve the goal. Cogulator is able to estimate a single-pilot's task times, working memory load, and head-down time using a task decomposition and the information needed to perform that task. This study hypothesized that the Digital Copilot cognitive assistant technologies would decrease task times, head-down time, and mental effort for tasks relative to non-cognitive assistant technologies in the form of EFBs.

Background

The following sub-sections provide background information on head-down time and cognitive modeling in the transportation industry. Specifically, this section focuses on aviation transportation and provides details on Cogulator, the cognitive and task modeling tool used in this study.

Problem Statement: Head-Down Time in Transportation

Head-down time in automobiles, trains, and airplanes increases the risk that the operator may miss critical events. This is more pronounced in stressful situations that require a higher level of mental effort (*12*). In these conditions, a person's field of view and response time are degraded, which has been cited in studies of pilots (*13*), car drivers (*14*, *15*), and air traffic control operators (*16*). One study showed that, in the United States, 13% of vehicle accidents were related to visual distractions (*17*). Other studies have indicated that a mere two seconds without looking at the road is more likely to lead to a crash or near crash (*18*, *19*). Between 1970 and 2004, approximately 34% of road traffic accidents in the United States may have been avoided if a HUD was present that displayed a "particular type of symbology" or behaved in a "particular manner in order to provide the driver with the safety critical information that was required," such as instrument and navigation information (*6*). In trains, head-down time "could result in the failure to detect signals, whistle boards, rear end marking devices, and other conditions that cause derailments or collisions" (*20*). The same report concluded that "increased mental workload and head-down time could also degrade situation awareness and result in speeding, excessive braking, missed radio communications, and poor train handling" (*20*). The NTSB has determined that locomotive engineers have failed to observe and respond to red signals due to visual distractions resulting in devastating accidents (*21*, *22*). Another study suggests Category A Signals Passed at Danger (SPADs) may have been prevented by a HUD indicating the relevant signal information (*6*). Of 80 rail accidents assessed involving fatalities to train occupants between 1967 and 2002, a HUD would have made a positive contribution in approximately 43% of all cases (*6*). However, this is a subjective judgment that presumes "availability of information not currently available and/or displayed in the cab, such as line speed limit information", in addition to head-up instruments, last signal aspect information, night vision, track circuit information, conformal cue for signals, and driver's reminder appliance information (*6*).

NTSB investigators have determined that the probable cause of numerous GA accidents was a result of failure to maintain an adequate visual lookout (*23*). Although portable electronic devices (PEDs) such as cell phones, tablets, and handheld global positioning system (GPS) units are useful, they can lead to more head-down time, limiting a pilot's ability to see other aircraft (*23*). EFBs, displayed on a PED, have the potential to be helpful, but risk negative side effects if not implemented appropriately (*24*). Therefore, HUDs and cognitive assistants have become a popular research area to minimize the amount of time spent switching focus between the primary out the window view and a head-down center console or dashboard. In aviation, HUDs resulted in "faster detection responses in symbology and presentation of traffic, as well as increased flight path tracking accuracy" (*13*). Although pilot procedures are designed to minimize head-down time, as more visual information is presented in-vehicle, such as on moving map displays, increasing head-down time becomes a concern. This is especially problematic in single-pilot GA because there is no second pilot to distinguish the tasks to relay critical information about the aircraft. During the phases of flight that require the most mental effort—traditionally the approach and landing phases—the pilot needs to configure the aircraft, communicate with air traffic control and/or other traffic, manage speed, and align the aircraft appropriately for landing, regardless of a congested airspace or inclement weather. According to the National Business Aviation Association (NBAA), a single pilot can easily become task saturated and overwhelmed in high traffic and low visibility conditions (*25*). Since pilots flying under visual flight rules rely on visual cues for maintaining aircraft orientation, and visual acquisition of other aircraft, obstacles, and landmarks, HUDs and cognitive assistants are particularly advantageous.

Cognitive Modeling in the Transportation Industry

Cognitive modeling is used to simulate human performance and mental task processes in a computerized model. Cognitive models are derived from the basic principles of cognition and their goal is to explain scientifically how the human brain accomplishes complex tasks, such as perceiving, thinking, or making decisions, to produce a logically valid and precisely quantitative prediction. Cognitive modeling has been used in the automotive industry to develop in-vehicle interfaces to minimize distraction and head-down time (*26*) and to assess visual demand of typical data entry tasks in automotive navigation systems for smart phones (*27*). One software application, Distract-R, developed by Dario Salvucci at Drexel University with funding from the National Science Foundation and Ford Motor Company (*28*), enables researchers to prototype interfaces, specify driving tasks and scenarios, and simulate, visualize, and analyze the results of the cognitive model (*29*). Another software application, Adaptive Control of Thought-Rational (ACT-R), developed by the Department of Psychology at Carnegie Mellon University (*30*), has been used to model driver behavior (*31*). In the rail industry, a cognitive task analysis "examined the cognitive demands and activities of locomotive engineers in today's environment and the changes in cognitive demands and activities that are likely to arise with the introduction of new train control technologies" (*32*). In the aviation domain, ACT-R has been used to evaluate the impact of adding a synthetic vision system (SVS) to a commercial airliner cockpit (*33*), model civil aviation pilot skills (*34*), and model possible sources of aviation error (*35*). Stephanie M. Doane and Young Woo Sohn, from the Department of Psychology at Mississippi State University, measured and recorded eye fixations and control movements to predict pilot visual attention using the software application ADAPT, which uses a computational cognitive model to "predict visual information gathering and action planning performance for multiple pilots during simulated flight maneuvers" (*36*). Furthermore, researchers developed an accessible cognitive modeling tool to address the limitations of GOMS modeling "to aid cockpit automation designers in assessing the potential for automation-induced pilot performance problems" (*37*).

This case study is different since it focuses on single-pilot GA operations and examines the specific benefits of using cognitive assistance during preparation for the approach phase of flight. The research used Cogulator, a human performance modeling tool that estimates task times, working memory load, and mental effort (*11*). Cogulator combines elements of different GOMS-style modeling techniques to "enable the researcher to infer the number of steps it will take to accomplish a given task and assess the interaction between task elements and the physical and mental functions" required to perform the task (*38*). To build the model, the user starts with a goal, which is the task, and uses operators, or predefined actions a human can take with an associated time. The method consists of the steps defined by the operators to achieve the goal. Selection rules, or "if–then" statements to describe decision logic, were not used in the models in this study. Cogulator uses the task decompositions to estimate three metrics: working memory load, task time, and head-down time.

Working memory is the cognitive construct in which information is temporarily stored and manipulated in order to complete complex tasks (*39*). It is thought to be "an important driver of mental effort" (*38*). Working memory can pull information from long-term memory in order to make judgments and calculations, and organize elements (*40*). Cogulator measures working memory load by calculating the number of unique pieces of information a person has to juggle concurrently (*38*). Working memory has a limited capacity, which is generally thought to be "seven plus or minus two chunks of information" (*41*). A chunk is a unit of information that is

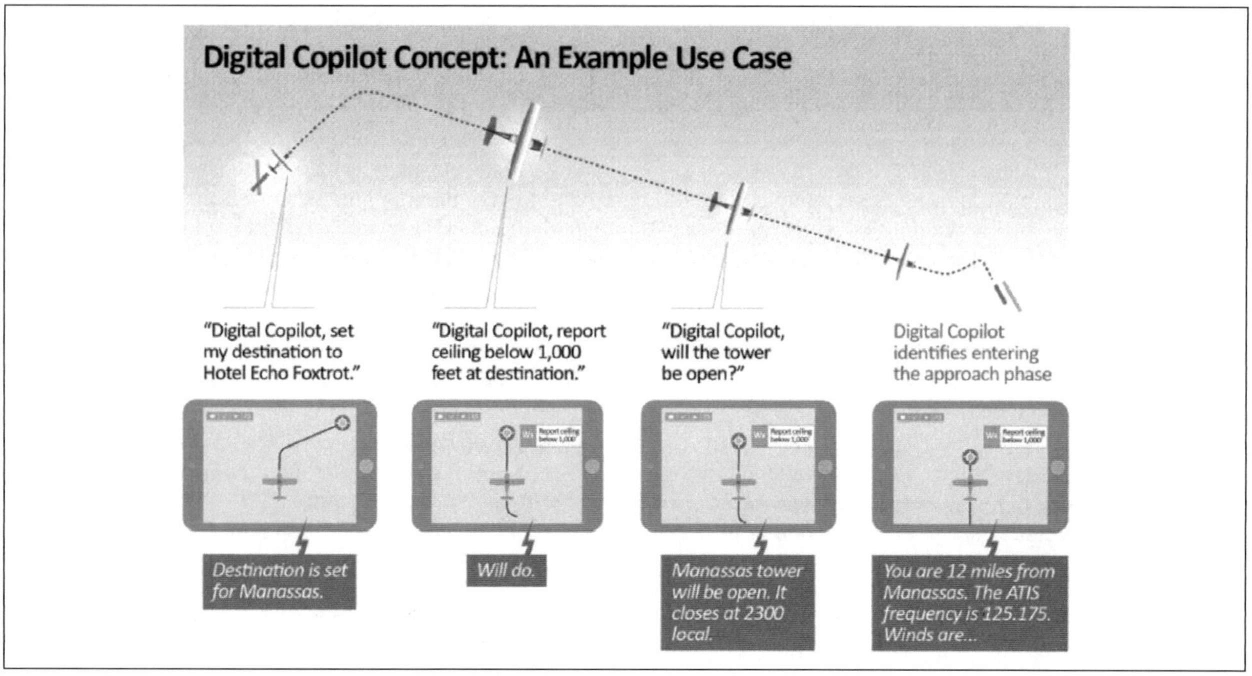

Figure 1. The Digital Copilot concept: an example use case. © 2017 IEEE

formed by using an association in long-term memory to bind multiple elements into a single unit (*40*). Over time, these chunks decay exponentially (*11*). Task time measures "the amount of time that elapsed for a task to be completed" (*38*). Cogulator determines task time by summing the operators necessary to perform a task; that is, the perceptual (i.e., hear and see), cognitive, or motor (i.e., say and hands) resources. This enables the researcher to calculate the head-down time metric by summing the time associated with the "see" operator when the pilot's head is down.

Cognitive Modeling with the Digital Copilot

The Digital Copilot is a working prototype of a pilot-oriented cognitive assistant that provides a set of features "focused on reduction of working memory load through information management" (*8*). The Digital Copilot seeks "to bring some of the benefits of a crew to a single-pilot cockpit by monitoring the aircraft position and determining the factors that influence decisions, such as the phase of flight and pilot intent" (*38*). While not a HUD, the Digital Copilot helps to reduce the pilot's head-down time by providing the pilot with on-demand information or contextual notifications using speech synthesis (*8*). These capabilities can reduce the volume of information the pilot must manage, offload work related to searching for and retrieving information, and cross-check pilot compliance with procedures and clearances (*38*). Figure 1 illustrates an example use case.

The pilot issues commands to the Digital Copilot primarily through speech recognition and the Digital Copilot provides information back to the pilot primarily through speech synthesis. For example, the pilot can request information about the traffic pattern altitude and direction and the Digital Copilot will respond with the information verbally. The Digital Copilot also makes inferences about the pilot's intent and offers cognitive assistance to the pilot in the form of notifications, reminders, and contextual information by aggregating information across various onboard databases as well as Automatic Dependent Surveillance-Broadcast (ADS-B) data (*8*). Although the Digital Copilot does not explicitly ask to verify inferences, the pilot is aware of these inferences because the Digital Copilot includes the inferred facts in its response. If this inference is wrong, the pilot can correct the Digital Copilot in a subsequent request. Figure 2 illustrates how the Digital Copilot processes information.

Method: Building Models using Cogulator

The Digital Copilot has not been tested with actual pilots to determine its effects on mental effort (e.g., workload and task management). Instead, this study uses the Cogulator software to model a typical GA pilot's overall task time, head-down time, and working memory load to evaluate whether these modeled metrics can be reduced while using the Digital Copilot, as compared with using an EFB. The as-is mode of

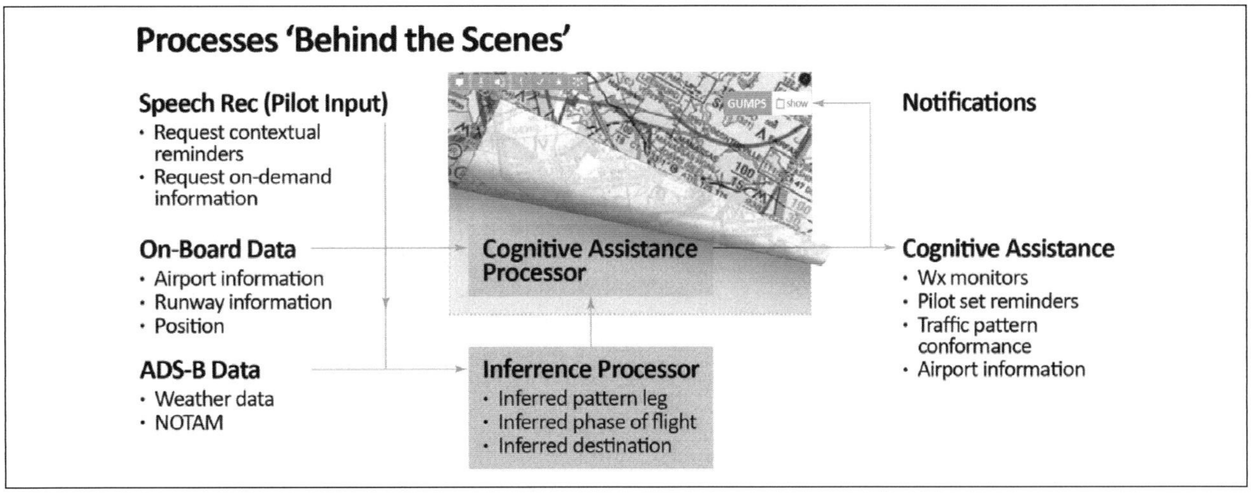

Figure 2. Digital Copilot: information processes. © 2017 IEEE

Table 1. Task Comparison Between the Digital Copilot and Tested EFBs

Capability	Digital copilot	Tested EFBs
Determine weather frequency	• On-demand assistance • Pilot uses speech recognition to retrieve frequency • Infers desired frequency is set for destination • Information provided verbally and textually	• Pilot retrieves frequency using a series of touchpad entries • Information provided textually
Determine communication frequency	• Contextual information • Frequency automatically provided when approaching the tower • Information provided verbally and textually	• Pilot retrieves frequency using a series of touchpad entries • Information provided textually
Review and follow a checklist	• On-demand assistance • Pilot uses speech recognition to request the reading of the checklist • Information provided verbally and textually • Items checked when complete	• Pilot retrieves the checklist using a series of touchpad entries • Information is provided verbally and textually or only textually • Pilot checks off items when complete
Determine if tower is open	• On-demand assistance • Pilot uses speech recognition to request information • Reply based on time estimate to the destination and the air traffic control tower schedule • Information provided verbally and textually	• Pilot looks up information using a series of touchpad entries; in some cases, pilot converts Zulu time to local time • Information is provided textually • Pilot determines if tower is open based on estimated arrival time
Determine preferred landing runway	• Contextual information • Infers preferred runways based on weather data and current route • Pilot retrieves runway by zooming in at the destination airport	• Pilot looks up preferred runway using a series of touchpad entries or pilot does mental math to determine preferred runway • Information provided textually

operation does not use the Digital Copilot and assumes the pilot uses an EFB for information management. Three commonly used EFB applications were considered to "ensure a valid comparison among different user interface-related variances" (*38*). The to-be mode of operation assumes the pilot uses the Digital Copilot as a form of cognitive assistance supplementary to an EFB. In both modes of operation, the study assumes "the actor is an experienced pilot and the models do not consider errors during the tasks" (*38*).

The researcher used task decomposition in order to model the cognitive, perceptual, and motor behaviors in Cogulator for the following five tasks: determine weather frequency,

Table 2. Results Comparing Digital Copilot with Minimum EFB

	Digital copilot				Minimum EFB			
Task	Task time (seconds)	Head-down (seconds)	Proportion head-down (%)	Memory items	Task time (seconds)	Head-down (seconds)	Proportion head-down (%)	Memory items
Determine weather frequency	5.8	0.6	10%	2	4.3	1.9	44%	2
Determine communication frequency	3.2	0	0%	2	7.7	2.9	38%	3
Review and follow a checklist	24.3	7.2	30%	13	38.1	36.5	96%	13
Determine if tower is open	6.5	0.6	9%	2	9.7	7.4	76%	4
Determine preferred landing runway	2.6	1.1	42%	2	6.4	4.4	69%	3

determine communication frequency, review and follow a checklist, determine if the tower is open, and determine the preferred landing runway. These tasks are described in greater detail in Table 1. These tasks were chosen because they are common tasks for pilots during preparation for the approach phase of flight and they are available in the three EFBs and the Digital Copilot. Because three different EFBs were used in this study, the minimum EFB metrics refer to the fastest task times and fewest memory chunks modeled of the EFBs considered in this study.

Results

The following section explains the model outputs and analysis of the results.

Model Outputs

The results are summarized by comparing the Digital Copilot metrics with the minimum EFB metrics (Table 2). Of the five tasks modeled, the results show that a pilot's use of the Digital Copilot is estimated to result in time savings in all tasks except for determining weather frequency; in head-down time savings for all tasks; and in working memory load savings or no change for all tasks. The motivation for this research is to improve the safety of single-pilot GA operations, however, this work has not yet attempted to translate these improvements into any safety impact assessment. The results presented here may help inform studies on that broader safety topic.

Determining the weather frequency is estimated to take 1.5 seconds longer—but requires only one-third as much head-down time—when using the Digital Copilot compared with the minimum EFB. The Digital Copilot takes more time because shifting the eyes from looking out the window to reading the frequency off the EFB screen is faster than aurally receiving the frequency. While the "hear" operator takes longer than the "read" operator, "it is helpful to unload tasks from the eyes to the ears so the eyes can support tasks focused on flying the airplane" (*38*). Literature suggests that an airframe-referenced auditory display has utility and pilot acceptance in GA operations (*42*). A reduction in head-down time could reduce a pilot's vulnerability to external hazards, such as other aircraft or terrain (*43*).

A pilot's ability to determine the tower frequency using the Digital Copilot compared with the minimum EFB is estimated to take less than half the amount of time, saving 4.5 seconds, and, more notably, involves no head-down time for the pilot. This contextual notification eliminated the time needed for a pilot to look up the information and relieved the pilot of mental workload associated with remembering to complete this task (*38*).

A pilot's ability to review and follow a checklist is estimated to save about one-third of the time, or over 13 seconds, and more than five times the savings in head-down time, or over 29 seconds, using the Digital Copilot compared with the minimum EFB. While this task does not show any appreciable differences in working memory requirements, pilots using the Digital Copilot can request on-demand assistance for reminders. For example, a reminder to complete the "Before Landing Checklist" would result in a reduction in mental effort, although it does not change the working memory requirements of completing the checklist (*38*).

A pilot's ability to determine tower hours using the Digital Copilot is estimated to take about one-third less time, saving over 3 seconds, and uses less than one-twelfth the amount of head-down time, saving almost 7 seconds.

A pilot's ability to determine the preferred runway using the Digital Copilot is estimated to take about less than half the amount of time, saving almost 4 seconds, and use one-third less head-down time, saving over 3 seconds.

The number of unique memory items were either the same or less using the Digital Copilot compared with the minimum EFB. Because the Digital Copilot uses contextually-based information to determine the communication frequency, it requires, at a minimum, one less piece of information compared with an EFB. For example, the Digital Copilot automatically states the communication frequency for the pilot when the airplane is about 20 miles away from the arrival airport so the pilot only needs to remember the arrival airport and the communication frequency, whereas the minimum EFB also requires the pilot to remember to check the communication frequency. In order to determine if the tower is open, the Digital Copilot requires the pilot to store the arrival airport and the answer to the question, whereas the minimum EFB requires the pilot to store the arrival airport, the tower hours, the current time, and the answer to the question. Determining the preferred runway using the Digital Copilot requires the pilot to recall the arrival airport and the preferred runway direction, whereas the minimum EFB requires the pilot to recall the arrival airport and store the preferred runway and the preferred direction as two different items. For example, first the pilot stores "16L-34R" as an option, and then determines the preferred end, storing "16L" separately. The remaining two tasks, determine weather frequency and review and follow a checklist, have the same unique memory items.

Discussion

Pilots can use EFBs to conduct many tasks related to flight operations, and embedding cognitive assistant technologies within this software may increase their functionality to support pilots further. However, the ways in which humans use new technology is a recognized problem. Studies on human performance affected by technologies such as cognitive assistants are evaluated based on mental workload, situation awareness, complacency, and skill degradation (*44*). Cognitive assistant technologies may lead to increased safety and enhanced pilot performance since non-flying related tasks could be completed by the cognitive assistant, thereby reducing the pilot's workload. The ability to infer the pilot's intent based on his or her actions helps manage cognitive mismatches because the technology is able to detect inconsistencies between the pilot's actions and the operational task requirements (*45*). A cognitive assistant may reduce the amount or change the type of pilot training required. Younger pilots may benefit more from a cognitive assistant since they are able to adapt to advanced technology faster due to greater familiarity with computers (*46*). When the cognitive assistant is working as intended, the pilot may be able to perform other tasks more efficiently, with less head-down time, in order to focus on flying.

However, in the event of a malfunction, misinterpretation, or misinformation, the results could be catastrophic. Reduced situational awareness may result if the cognitive assistant has incomplete or inaccurate information (*44*). Because the cognitive assistant may not be completely reliable, there is a chance of increased workload to fix an error (*44*), or pilot workload may be shifted, rather than reduced (*45*). Flight schools may have to focus on how to manage cognitive assistant technologies to mitigate this (*45*). Other unintended consequences, such as distraction, may cause spatial disorientation (*47*). When pilots are distracted, they have a tendency not to recognize hazards (*47*). A pilot may encounter a data transmission error, technological glitch, unauthorized tampering with the back-end system architecture, loss of service, or a dead battery, all of which would prevent him or her from using the cognitive assistant as intended. If the cognitive assistant is highly reliable, but not completely reliable or robust, then pilots may "relax their vigilance for off-nominal behavior, missing the occasional time that there is an undetected problem" (*44*). In especially high workload situations, the pilot may be more reliant on the cognitive assistant, and may not realize that the information presented could be incorrect (*44*). In these circumstances, if pilots became too reliant on the technology, they would be in danger, especially if they experience skill degradation due to lack of use of certain skills (*44*). In order to maximize benefit and minimize risk, there needs to be a method for integrating this technology into the cockpit and testing it.

New aviation technology must also consider the appropriate modality for presenting information to the pilot in order to maximize cognitive assistance while minimizing head-down time and potential modality interference. According to the multiple resource theory, humans have a limited set of resources available for mental processes to allocate to a task or combination of tasks. These resources include visual, auditory, cognitive, and psychomotor (*48*, *49*). Overload occurs when task demand exceeds the resource, which may cause the quality of performance of primary or secondary tasks to decline (*50*). Thus, it can be more advantageous for time-shared tasks to use different resources for these purposes. As described by Wickens, "all other things being equal (i.e., equal resource demand or single task difficulty), two tasks that both demand one level of a given dimension (e.g., two tasks demanding visual perception) will interfere with each other more than two tasks that demand separate levels of the dimension (e.g., one visual, one auditory task)" (*51*). In other words, cross-modal time sharing is preferred over intra-modal time sharing (*51*). Therefore, it is beneficial that the Digital Copilot cognitive assistant uses the auditory channel to communicate with the pilot. However, it can be difficult to understand multiple auditory messages all at once, so a visual backup component is useful for when the pilot is communicating on the radio with air traffic control or other air traffic.

Conclusion

As cognitive assistants like the Digital Copilot become more capable, researchers need to continue to test their features to determine the effects in terms of task times, head-down time, and mental effort. Potential research includes examining features in other phases of flight, those related to flying under instrument flight rules, and comparing results with a traditional paper flight bag (*38*). This study did not include traditional paper flight bags because it seemed more likely that a pilot using a PED with an EFB would be more open to and interested in using a cognitive assistant for aviation, compared with a pilot who flies with less modern technology. Moreover, according to a 2013 study by the Aircraft Owners and Pilots Association, nearly 75% of respondents used "portable/mobile" devices for aviation applications (*52*). Nevertheless, this would be a noteworthy comparison. In addition to measuring the specific pilot benefits, the safety benefits of using the Digital Copilot are also being quantified using methods of human reliability analysis and probabilistic risk assessment to quantify human errors and accident sequences across all phases of flight. Results show that the Digital Copilot has the potential to reduce the accident rate by about 20% (*53*). Similarly, as more cognitive assistance tools become available, they need to be evaluated to assess their effects on pilots. Although these cognitive assistants may offer the potential for safety benefits, future studies need to examine how they may introduce new dangers through distraction or misinformation.

Although cognitive modeling is already well established in the road, rail, and air traffic environments, the working memory aspect of Cogulator is a new feature that can be applied across the transportation domain to measure mental workload. Aside from counting the unique memory items as a metric for working memory load, Cogulator can average the amount of information that must be handled to complete the task, inform when the user is at maximum capacity, and show the exponential decay of the memory items over time. Other studies should examine this metric more closely as part of an evaluation of human interaction with emerging technologies within the transportation domain. Cogulator has a wide range of applicability in the transportation domain to continue to assess the head-down time problem across industries and evaluate the human factors of new technologies.

References

1. Hillburn, B. *Head-Down Time in Aerodrome Operations: A Scope Study*. Center for Human Performance Research, 2004.
2. National Business Aviation Association. *2017 NBAA Top Safety Focus Area*. https://www.nbaa.org/ops/safety/top-safety-focus-areas/2017/. Accessed July 22, 2017.
3. Pandey, B. Business Aviation: Air Safety: Safety in Business Aviation. *SP's AirBuz*, 2017.
4. Smith, M., L. J. Gabbard, and C. Conley. Head-Up vs. Head-Down Displays: Examining Traditional Methods of Display Assessment While Driving. *Proc., 8th International Conference on Automotive User Interfaces and Interactive Vehicular Applications*, Ann Arbor, MI, 2016.
5. Srinivasan, R., and P. Jovanis. Effect of Selected In-Vehicle Route Guidance Systems on Driver Reaction Time. *Human Factors*, Vol. 39, No. 2, 1997, pp. 200–215.
6. Davies, K., J. Hinton, P. Thorley, and D. Buse. *Feasibility of Head-Up Displays in Driving Cabs*. Rail Safety and Standards Board, Bristol, 2007.
7. National Transportation Safety Board. *NTSB 2016 Most Wanted Transportation Safety Improvements: Prevent Loss of Control in Flight in General Aviation*. National Transportation Safety Board, Washington, D.C., 2016.
8. Estes, S. L., K. J. Burns, J. R. Helleberg, K. M. Long, M. E. Pollack, and J. L. Stein. *Digital Copilot: Cognitive Assistance for Pilots*. Association for the Advancement of Artificial Intelligence, 2016.
9. Wickens, C. D. Multiple Resources and Mental Workload. *Human Factors*, Vol. 50, No. 3, 2008, pp. 449–455.
10. Green, P. Estimating Compliance with the 15-Second Rule for Driver-Interface Usability and Safety. Presented at the Human Factors and Ergonomics Society 43rd Annual Meeting, Santa Monica, California, 1999.
11. MITRE Corporation. *Cogulator: A Primer*. http://cogulator.io/primer.html. Accessed June 13, 2017.
12. Easterbrook, J. A. The Effect of Emotion on Cue Utilization and the Organisation of Behavior. *Psychological Review*, Vol. 66, No. 3, 1959, pp. 183–201.
13. Fadden, S., P. M. Ververs, and C. D. Wickens. Cost and Benefits of Head-Up Display Use: A Meta-Analytic Approach. *Proc., Human Factors and Ergonomics Society Annual Meeting*, Chicago, IL, 1998.
14. Crundall, D. E., and G. Underwood. Effects of Experience and Processing Demands on Visual Information Acquisition in Drivers. *Ergonimics*, Vol. 41, No. 4, 1998, pp. 448–458.
15. Mack, A., and I. Rock. Inattentional Blindness. *PSYCHE*, Vol. 5, No. 3, 1999. http://journalpsyche.org/files/0xaa59.pdf
16. Willems, B., and E. Stein. *Controller Workload and Visual Scanning During Simulations Containing Scripted Midair Conflicts and Varied Task Load*. Federal Aviation Administration, Washington, D.C., 2001.
17. Burns, P., and T. Lansdown. *E-Distraction: The Challenges for Safe and Usable Internet Services in Vehicle*. Transport Research Laboratory, United Kingdom, 2001.
18. Zwahlen, H. T., C. C. Adams, and D. P. DeBals. Safety Aspects of CRT Touch Panel Controls in Automobiles. *Proc., Second International Conference on Vision in Vehicles II*, Nottingham, United Kingdom, 1988.
19. Redelmeier, D. A., and R. J. Tibshirani. Association Between Cellular-Telephone Calls and Motor Vehicle Collisions. *New England Journal of Medicine*, Vol. 336, No. 7, 1997, pp. 453–458.
20. Department of Transportation Federal Railroad Administration. *Emergency Order to Restrict On-Duty Railroad Operating Employees' Use of Cellular Telephones and Other Distracting Electronic and Electrical Devices*. Department of Transportation Federal Railroad Administration, Washington, D.C., 2008.
21. National Transportation Safety Board Office of Public Affairs. *NTSB Determines Engineer's Failure to Observe and Respond to Red Signal Caused 2008 Chatsworth Accident; Recorders in Cabs Recommended*, 2010. https://www.ntsb.gov/news/press-releases/Pages/NTSB_Determines_Engineers_Failure_to_Observe_and_Respond_to_Red_Signal_Caused_2008_Chatsworth_Accident;_Recorders_in_Cabs.aspx. Accessed July 21, 2017.

22. National Transportation Safety Board Office of Public Affairs. *Train Engineer Distracted by Cell Phone Cited as Probable Cause of Train Collision*, 2003. https://www.ntsb.gov/news/press-releases/Pages/Train_Engineer_Distracted_by_Cell_Phone_Cited_as_Probable_Cause_of_Train_Collision.aspx. Accessed July 21, 2017.
23. National Transportation Safety Board. *See and Be Seen: Your Life Depends on It*, July 2015. https://pacs.ou.edu/media/filer_public/8f/df/8fdfcaa8-95aa-4836-88b4-2837a7fe89cb/2015-julyntsbseeandbeseen.pdf. Accessed July 21, 2017.
24. Abbott, K. *Some Lessons Learned About Pilots and Flight Deck Automated Systems*. HindSight. Vol. 20, pp. 38–40, Winter 2014. Accessed July 14, 2017.
25. National Business Aviation Association. *Award-Nominated NBAA Video Anchors NBAA's Single-Pilot Safety Resources*. June 17, 2016. https://www.nbaa.org/ops/safety/single-pilot/20160617-award-nominated-nbaa-video-anchors-nbaa-single-pilot-safety-resources.php. Accessed July 21, 2017.
26. Purucker, C., F. Naujoks, A. Prill, and A. Neukum. Evaluating Distraction of In-Vehicle Information Systems While Driving by Predicting Total Eyes-Off-Road Times with Keystroke Level Modeling. *Applied Ergonomics*, Vol. 58, 2017, pp. 543–554.
27. Quaresma, M. *Assessment of Visual Demand of Typical Data Entry Tasks in Automotive Navigation Systems for iPhone*. LEUI Laboratory of Ergonomics and Usability of Interfaces in Human-Technology Systems, Rio de Janeiro, Brazil, 2012.
28. Distract-R. http://cog.cs.drexel.edu/distract-r/about.php. Accessed October 26, 2017.
29Salvucci, D. D. Rapid Prototyping and Evaluation of In-Vehicle Interfaces. *ACM Transactions on Computer Human Interaction*, Vol. 16, No. 2, 2009, p. 33.
30. ACT-R. http://act-r.psy.cmu.edu/. Accessed October 26, 2017.
31. Salvucci, D. D. Predicting the Effects of In-Car Interfaces on Driver Behavior Using a Cognitive Architecture. *CHI*, Vol. 3, No. 1, 2001, pp. 120–127.
32. Roth, E., and J. Multer. *Technology Implications of a Cognitive Task Analysis for Locomotive Engineers*. U.S. Department of Transportation, Cambridge, MA, 2006.
33. Byrne, M. D., A. Kirlik, M. D. Fleetwood, D. G. Huss, A. Kosorukoff, R.-S. Lin, and C. S. Fick. A Closed-Loop, ACT-R Approach to Modeling Approach and Landing With and Without Synthetic Vision System (SVS) Technology. Presented at the Human Factors and Ergonomics Society 48th Annual Meeting, New Orleans, LA, 2004.
34Xue, H.-J., J.-F. Pang, Y.-C. Luan, and Y.-Q. Tie. Computational Cognitive Modeling of Civil Aviation Pilots. *Proc., International Conference on Automatic Control and Artificial Intelligence*, Xiamen, China, 2012.
35. Byrne, M. D., and A. Kirlik. Using Computational Cognitive Modeling to Diagnose Possible Sources of Aviation Error. *The International Journal of Aviation Psychology*, Vol. 15, No. 2, 2005, pp. 135–155.
36. Doane, S. M., and Y. W. Sohn. ADAPT: A Predictive Cognitive Model of User Visual Attention and Action Planning. *User Modeling and User-Adapted Interaction*, Vol. 10, No. 2, 2000, pp. 1–45.
37. Gil, G.-H., and D. B. Kaber. An Accessible Cognitive Modeling Tool for Evaluation of Pilot–Automation Interaction. *The International Journal of Aviation Psychology*, Vol. 22, No. 4, 2012, pp. 319–342.
38. Wilkins, S. A. Examination of Pilot Benefits from Cognitive Assistance for Single-Pilot General Aviation Operations. *Proc., 36th Digital Avionics Systems Conference (DASC)*, IEEE, 2017.
39. Baddeley, A. The Episodic Buffer: A New Component of Working Memory? *Trends in Cognitive Science*, Vol. 4, No. 11, 2000, pp. 417–423.
40. Wickens, C. D. and Hollands J. G. *Engineering Psychology and Human Performance*. Upper Saddle River, NJ, Prentice Hall, 2000.
41. Miller, G. A. The Magical Number Seven Plus or Minus Two: Some Limits on Our Capacity for Processing Information. *Psychological Review*, Vol. 63, No. 2, 1956, pp. 81–97.
42. Navqi, M. H., A. J. Wigdahl, and R. J. Ranaudo. Utility of an Airframe Referenced Spatial Auditory Display for General Aviation Operations. *International Society for Optics and Photonics*, Vol. 7327, 2009, p. 73270G.
43. Gore, B. F. Man–machine Integration Design and Analysis System (MIDAS) v5: Augmentations, Motivations, and Directions for Aeronautics Applications. In *Human Modelling in Assisted Transportation* (P. Cacciabue, M. Hjälmdahl, A. Luedtke, and C. Riccioli, eds.), Springer, Milano, 2011.
44. Frost, S., K. Goebel, and J. Celaya. *A Briefing on Metrics and Risks for Autonomous Decision-Making in Aerospace Applications*. Infotech@Aerospace, Garden Grove, California, 2012.
45. Baxter, G., D. Besnard, and D. Riley. Cognitive Mismatches in the Cockpit: Will They Ever Be a Thing of the Past? *Applied Ergonomics*, Vol. 38, No. 4, 2007, pp. 417–423.
46. Wiener, E. L. *Human Factors of Cockpit Automation: A Field Study of Flight Crew Transition*. NASA Ames Research Center, Moffett Field, CA, 1985.
47. Small, R. L., A. M. Fisher, J. W. Keller, and C. D. Wickens. A Pilot Spatial Orientation Aiding System. *Proc., AIAA 5th Aviation, Technology, Integration, and Operations Conference (ATIO)*, Arlington, VA, 2005.
48Wickens, C. D. *Engineering Psychology and Human Performance*. Charles E. Merrill, Columbus, OH, 1984.
49. Horrey, W. J., and C. D. Wickens. Multiple Resource Modeling of Task Interference in Vehicle Control, Hazard Awareness and In-Vehicle Task Performance. *Proc., Second International Driving Symposium on Human Factors in Driver Assessment, Training, and Vehicle Design*, Iowa City, Iowa, 2003.
50. Green, P. *Driver Interface/HMI Standards to Minimize Driver Distraction/Overload*. University of Michigan Transportation Research Institute, Ann Arbor, MI, 2008.
51Wickens, C. D. Multiple Resources and Performance Prediction. *Theoretical Issues in Ergonomics Science*, Vol. 3, No. 2, 2002, pp. 159–177.
52. Aircraft Owners and Pilots Association (AOPA). *General Aviation Publication Reader Preference Study: A Media & Market Study Conducted Among FAA Certified Pilots*. Erdos & Morgan, Syosset, NY, 2013.
53. Burns, K., Bonaceto, C., Estes, S., & Helleberg, J. Evaluating the Operational Safety of a Digital Copilot. In The 2017 AAAI Fall Symposium Series: Technical Reports, Palo Alto, California, 2017.

The Standing Committee on Aviation Security and Emergency Management (AV090) peer-reviewed this paper (18-01450).

Article

TRR
JOURNAL OF THE TRANSPORTATION RESEARCH BOARD

Transportation Research Record
2018, Vol. 2672(23) 146–157

DOI: 10.1177/0361198118792337
journals.sagepub.com/home/trr
SAGE

Modeling Airline Cost Pass-Through within Regional Aviation Markets

Bojun Wang[1], Aidan O'Sullivan[1], Lynnette Dray[1], and Andreas W. Schäfer[1]

Abstract

Studies assessing the impact of market-based environmental policies in aviation rely on various scenarios of airline cost pass-through, because there is little empirical evidence with respect to the impacts of airline costs on airfares. Instead, the costs effect has been indirectly measured by proxy variables such as distance, fuel price, and aircraft sizes. This paper provides empirical evidence of airline cost pass-through by developing an airfare model that explicitly captures airline operating costs. Using a feasible generalized two-stage least squares (FG2SLS) approach, we obtained coefficients of airline fuel costs per passenger, non-fuel costs per passenger, and non-fuel costs per flight modeling for seven world regions (20 region-pair markets). A comparison of the estimated cost pass-through elasticities conducted across regional markets suggests that airlines may respond to the cost increases differently, depending on the cost types and the markets they operate in. Based on the estimated coefficients, we systematically evaluate the potential impacts of introducing a carbon tax policy within two major regional markets with distinct cost pass-through elasticities.

To achieve carbon-neutral growth in international aviation, the International Civil Aviation Organization has introduced a market-based measures (MBMs) scheme. The Carbon Offsetting and Reduction Scheme for International Aviation (CORSIA), which is planned to be introduced in 2021, will effectively increase the airlines' fuel costs (*1*). Airlines will respond to the higher fuel costs by adjusting technology, operations, and air fares. However, depending on the elasticity of demand relative to supply, higher fares in a competitive market may result in lower sales and market shares. Therefore, an airline has to strike a balance between recovering cost increases and maintaining its market share. Cost pass-through rates will manifest this balance. Additionally, as airline operating costs consist of several components, an airline could potentially be more vulnerable to changes in one cost component than another. For example, an increase in fuel costs may have greater impact on airlines than the same increase in aircraft landing costs. Thus, it would be interesting to assess the potential pricing responses of airlines to changes in different operating costs under competition.

Cost pass-through can be measured either by the absolute price increase resulting from an absolute increase of costs, known as the absolute cost pass-through, or by the percentage change in price for a percentage change in marginal cost, known as the cost pass-through elasticity (*2*). Using either measure, previous empirical studies have mainly assessed cost pass-through in the oil industry, the wholesale electricity industry, and the exchange rates to import prices, etc. (*2*). However, there is little empirical evidence with respect to airline cost pass-through (*3*), and thus the economic impact of MBMs on airline pricing behavior remains largely unclear. As a result, most of the studies that aim to assess the extent to which MBMs could lead to aviation emissions reduction had to assume rates of cost pass-through. For instance, both Anger and Kohler (*4*) and Scheelhaase et al. (*5*) assumed that the allowance cost of the European Emissions Trading Scheme (EU ETS) would be fully passed on to passengers and concluded that increases in airfares would be small because of a low carbon price. In evaluating impacts of the EU ETS on airline networks, Albers et al. used two scenarios of 35% and 100% cost pass-through (*6*). Similarly, Meleo et al. assumed three cost pass-through scenarios: 0%, 50%, and 100%, in their study on the direct costs increases in the Italian airline market (*7*). The widely-adopted scenario-based cost pass-through in

[1]Air Transportation Systems Laboratory, UCL Energy Institute, University College London, London, UK

Corresponding Author:
Address correspondence to Bojun Wang: bojun.wang.13@ucl.ac.uk

previous research suggests that a solid understanding of the behavior change in airline pricing is at the heart of evaluating economic impacts of MBMs on the aviation sector. Therefore, more empirical research is needed on this subject.

Estimating the cost pass-through requires the estimation of airfares in a competitive environment. How airlines set fares has been studied extensively, albeit with a focus on the U.S. domestic airline market. A review of the literature reveals findings on several factors that affect airfares.

Passenger demand was found to be positively correlated with fares, if the demand impact outweighs the economies of density effect (*8*, *9*), and negatively correlated otherwise (*10*). Airline competition also has significant impacts on airfares, and has thus far received the most extensive discussion in literature, since the deregulation of the airline industry in the U.S. in 1978 (*11*, *12*). Studies discovered that, after deregulation, airlines with market power tend to charge higher fares (*8*, *13*, *14*). In these studies, market structure has been measured by the level of market concentration using the Herfindahl–Hirschman Index (HHI). Higher market concentration indicates relatively low competition, and vice versa. Another critical indicator of market competition is the presence of low-cost carriers (LCCs). Fares are found to be drastically depressed if a LCC enters a market (*15*, *16*). And Brueckner et al. further demonstrated that the presence of LCCs will affect airfares in not only local airport-pair markets but also neighboring markets (*17*).

Studies have also found significant yet inconclusive impacts of flight delays on airfares. Forbes showed that every one-minute increase in delay leads to a \$1.42 reduction in fares (*18*). In contrast, Zou and Hansen, who estimated the delay effects separately for non-stop and one-stop routes, concluded that delays will result in higher fares (*10*). Although flight frequency is relatively less often included in previous fare models, it also has significant effects on prices. According to Chi and Koo, higher flight frequencies could result in either higher or lower fares, depending on whether the effect of greater frequency on costs is larger than its effect on demand (*19*). Finally, although the impact of operating costs on airfares has been widely acknowledged, few studies have modeled airline operating costs explicitly. Instead, in most cases, distance, fuel price, and aircraft sizes have been used as proxy variables of airline costs. When controling for demand, frequency, and competition effects, longer origin–destination (OD) distance, higher fuel price, and larger aircraft are associated with higher fares (*8*, *10*, *19*). However, these proxy variables cannot capture changes in specific airline costs, and thus are not useful to quantify how much of an airline's costs burden is passed through to passengers via airfares, which is the focus of this research.

Based on this literature review, this paper aims to evaluate empirically airline cost pass-through for seven world regions (20 region-pair markets) and to compare the pass-through elasticities across different airline markets. An airfare model that explicitly captures airline operating costs as well as other key influencing factors is developed and estimated for each of these regional markets.

Our research makes several contributions. First, we provide empirical evidence on airline cost pass-through for future research which otherwise would have to rely on presumed cost pass-through rates when evaluating the economic impacts of MBMs on the aviation sector. Second, having estimated the fare model at a global scale, our results have implications for both developed and developing airline markets. This is particularly important to regions beyond the U.S. domestic market, where aviation emissions are projected to grow more rapidly over the next 20 to 30 years (*20*) yet have not been explored in the current body of airline pricing studies. Finally, coefficients estimated from our model could be used to evaluate potential impacts of MBMs such as CORSIA on airline pricing behavior, but also to help policy makers to design other aviation emissions reduction policies. Notably, our airfare model is a core component of the updated Aviation Integrated Model AIM2015 (*21*).

The next section of this paper describes the data underlying this work, the model specification, and the three key operating cost variables, namely: fuel costs per passenger, non-fuel costs per passenger, and non-fuel costs per flight. The model estimation and estimated coefficients are then discussed in the third section. Based on the coefficient estimates, a carbon tax policy scenario is evaluated within two regional markets with distinct cost pass-through elasticities. Using the AIM2015 model, the scenario analysis compares the system-wide impacts of increased airline costs on airfare, demand, and aviation CO_2 emissions. The final section offers conclusions.

Data and Empirical Model

This section presents the airfare model developed in this research. We first describe the datasets used to construct the model variables. The specification of the model then follows and we conclude with a detailed discussion of the three key operating costs variables.

Data

Data describing airfares, passenger demand, market shares, flight frequency, and route characteristics are either directly obtained, or constructed from the Sabre Market Intelligence database (*22*). Fleet data is obtained from FlightGlobal and is used to derive aircraft type by

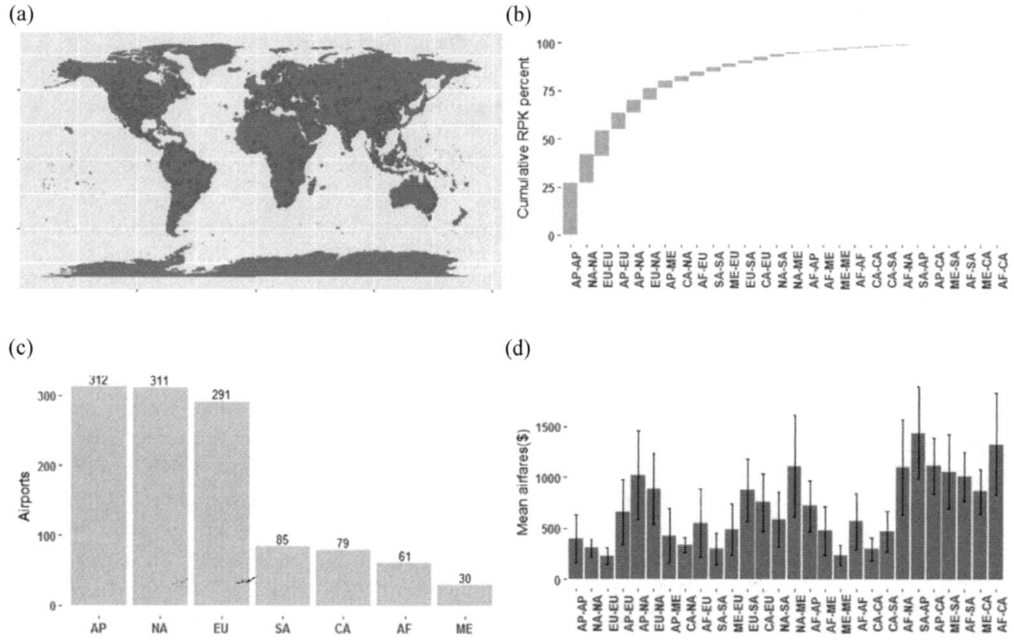

Figure 1. Descriptive summary of datasets: (*a*) global coverage of the datasets; (*b*) global RPK percentage by regional markets; (*c*) number of airports by region; (*d*) mean airfare with standard deviation by regional markets.

segment (*23*). Aircraft are grouped into nine different size classes, based on the Sustainable Aviation aircraft categories (*24*). En-route and airport landing charges by size class are provided by the RDC airport charges database (*25*).

The seven world regions included in this study are: Africa (AF), Asia and Pacific (AP), Central America (CA), Europe (EU), ME (Middle East), North America (NA), and South America (SA). Our fare model uses cross-sectional data for the year 2015, with itineraries connecting different OD region pairs grouped into intra-regional markets (e.g. EU–EU) and inter-regional markets (e.g. NA–EU). We weighted airfares (including taxes) by the number of passengers paying different observed prices based on booking classes, and aggregated ticket prices of all airlines operating on the same route as the annual weighted average price. The unit of observation is a unique route between an OD airport pair, connected by a maximum of four flight segments. To ensure a robust model estimation, routes with very low demand are removed. We restrict low-traffic routes to those with a share of the total OD passengers on a given city pair below 5%, and annual passengers fewer than 52 (1 passenger per week) in intra-regional markets or 520 (10 passengers per week) in inter-regional markets.

Figure 1 describes four key aspects of the cleaned data. Our data covers all continents over the world (*a*). The largest five markets in terms of revenue passenger kilometers (RPK) are AP–AP, NA–NA, EU–EU, AP–EU, and AP–NA, and the top 20 region-pair markets out of 28 account for 90% of the global total RPK (*b*). The share of RPK is closely linked to the total number of airports available in each region (*c*). As 78% of global airports are located in AP, NA, and EU, markets connecting these regions account for the largest proportion of the global RPK. Finally, from (*d*) we can see that overall fares are higher in inter-regional markets than in intra-regional markets (also with greater variation in fares), and the highest average fares are found in those smallest markets potentially because of the limited supply.

Model Specification

Informed by the literature reviewed earlier, we formulate fare as a function of several factors that have demonstrated significant pricing impacts. These factors are grouped in an overview equation in Equation 1 as *Cost, Demand, Competition*, and *OD Country Fixed Effects*, where m, k, n are origin, connecting, and destination airport(s), respectively:

Table 1. Definition of the Fare Model Variables

Variable	Definition
Fare	The weighted average route fare (including taxes), calculated as the observed price weighted by the number of passengers paying this price by booking class. Fares are aggregated across all airlines on the same route.
FuelCostPerPax	The sum of average fuel costs per passenger of all flight segments used on the given itinerary. Flight segment total fuel costs are fuel costs of all aircraft operating on the given segment.
NonFuelCostPerFlt	The sum of average non-fuel per flight costs of all flight segments used on the given itinerary.
NonFuelCostPerPax	The sum of average non-fuel per passenger costs of all flight segments used on the given itinerary.
LegMeanHHI	OD route-level Herfindahl–Hirschman Index (HHI). The HHI is calculated as the sum of squared market shares of all airlines in respective markets. Route HHI is measured by the geometric mean of segment HHIs covered by the route. A market contains all routes that connect the OD metropolitan areas, capturing competition at both local and adjacent airport pairs.
AirportMeanHHI	Endpoint airport-level market concentration measured by the average HHIs at origin and destination airports. A market here is defined to contain all airlines that depart from the origin airport, and all airlines that arrive at the destination airport.
CUIMean	Average airport capacity utilization index of the endpoint airports, calculated as the ratio of average aircraft movements during 16-hours daytime operating at a given airport over the airport's declared capacity. This variable is used to measure the effect of delays exogenously.
Freq	Annual total flight frequency on a given route.
Pax	Annual total number of passengers on a given route.
RouteShare	Ratio of OD passengers on a given route over total OD passengers between the OD city pair.
Nlegs	The total number of flight segments used by a given route. Route characteristics that affect the demand.
HubsPass	The total number of hubs used by a given route. Route characteristics that affect the demand.
OriginCountry	Dummy variables of origin country. OriginCountry equals 1 if flying from country O (O = 2, 3, 4...).
DestCountry	Dummy variables of destination country. DestCountry equals 1 if flying to country D (D = 2, 3, 4...).

$$(Fare)_{mkn} = f(Cost_{mkn}, Demand_{mkn}, Competition_{mkn}, CountryFE_{OD}) \quad (1)$$

Fares are determined by the complex interactions between supply and demand, where supply is expressed mainly via airline costs (but also flight frequency). Competition, which not only often acts as the equilibrium-shifter but also has significant influence on airline cost pass-through (*3*), is included. The unobserved effects of endpoint countries that may affect airline pricing differently (e.g. taxes on fares) are captured by country fixed effects. Following this rationale, the airfare model is specified in Equation 2 as follows:

$$\begin{aligned}\ln(Fare)_{mkn} = {} & \beta_0 + \beta_1 \ln(FuelCostPerPax)_{mkn} + \beta_2 \ln(NonFuelCostPerPax)_{mkn} \\ & + \beta_3 \ln(NonFuelCostPerFlt)_{mkn} + \beta_4 \ln(Freq)_{mkn} + \beta_5 \ln(Pax)_{mkn} \\ & + \beta_6 \ln(RouteShare)_{mkn} + \beta_7 \ln(CUIMean)_{mkn} + \beta_8 \ln(LegMeanHHI)_{mkn} \\ & + \beta_9 \ln(AirportMeanHHI)_{mkn} + \beta_{10} \ln(Nlegs)_{mkn} + \beta_{11} \ln(HubsPass)_{mkn} \\ & + \sum_{O=2,3,4...} \delta_O Origin\,Country_O + \sum_{D=2,3,4...} \theta_D DestCountry_D + \varepsilon_{mkn}\end{aligned} \quad (2)$$

where m, n, and k denote origin, destination, and connecting airport(s), respectively; O and D denote origin country and destination country, and ε is the random error. Definitions of the fare model variables are provided in Table 1. As mentioned earlier, the costs pass-through can be measured by either the absolute pass-through rate or the pass-through elasticity. In this study, we measure the pass-through by elasticity using a log-log model specification, which allows us to compare the percentage changes of airfares given the same percentage of increase in different costs types. Notably, we do not include any dummy variable for LCCs, as they mainly operate in limited markets (e.g. U.S. and EU). Therefore, including this variable does not significantly improve the model's performance when estimating other world regions.

Airline Operating Costs Variables

As discussed in the literature review, airline operating costs have been largely measured by proxy variables such as distance, fuel price, and dummy variables for aircraft sizes in previous airfare models (*8*, *10*, *19*). Such proxy variables cannot directly quantify the effects of changes in airline costs to airfares. In contrast, this fare model includes three operating costs variables that have this capacity. As shown in Figure 2, we first categorized all flights on a flight segment into nine aircraft size classes

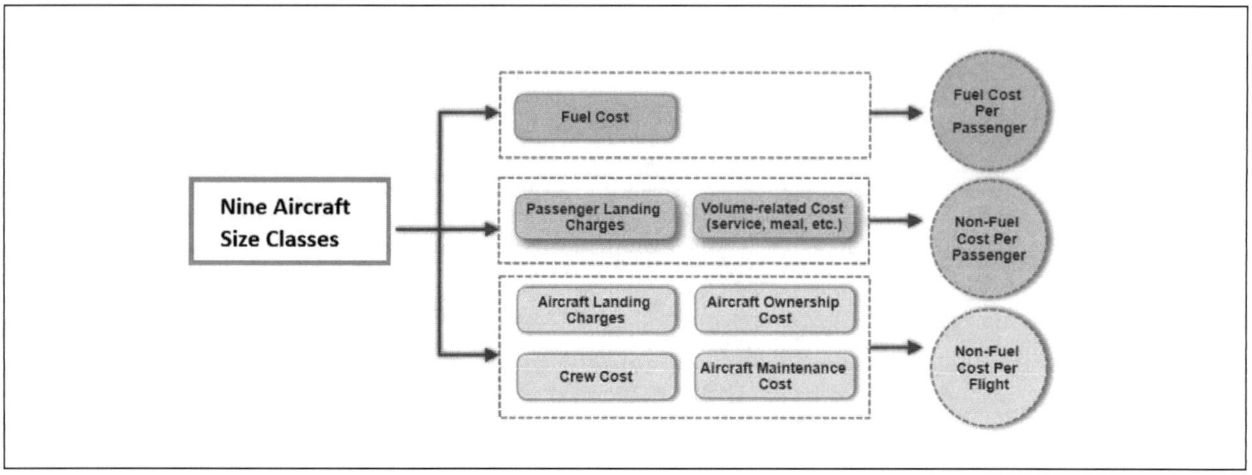

Figure 2. Procedure for constructing airline operating costs variables.

based on the Sustainable Aviation categories (*24*). Each aircraft size class is associated with seven different operating cost components, which are either calculated by the AIM2015 Direct Operating Cost model (*25*) or derived from the RDC airport charges database (*26*) as input to calculate total operating costs of all flights on a given segment.

The seven operating cost components are grouped into three categories: fuel costs, non-fuel passenger costs, and non-fuel flight costs. Fuel costs and non-fuel costs are distinguished because fuel costs is generally the single largest cost component that has shown great fluctuations over the past 15 years (*3*), whereas other cost components are relatively stable over time. Thus, fuel cost is the main source of volatility in airline total costs and needs to be measured separately. Additionally, fuel costs and non-fuel passenger costs are divided by passenger demand as costs per passenger because their total segment costs are determined by the enplaned passenger numbers. For instance, fuel costs of a full flight will certainly be higher than that of an empty flight with the same aircraft size due to heavier weight and more fuel consumption. In contrast, non-fuel flight costs is averaged by flight frequency because each aircraft's flight-based costs (Figure 2) are fixed and do not change with the number of enplaned passengers. Equations 3 to 5 show how the three costs variables are derived, as follows:

In Equation 3, $FuelC_{f,l}$ represents the fuel costs of an aircraft in size class f flying on segment l, and $Freq_{f,l}$ denotes the annual total flight frequency of aircraft in this size class on segment l. The total fuel costs of all aircraft on the segment l are then averaged by the annual total passengers of this segment (Pax_l). Finally, the fuel cost per passenger on itinerary *mkn* is the sum of fuel costs per passenger of all segments (*Legs*) on *mkn*.

Similarly, in Equation 4, $(PaxLanC + VolC)_{f,l}$ is the total non-fuel passenger costs associated with one passenger, that is, passenger landing charges and volume-related costs (Figure 2) on aircraft size f flying over segment l; and $Pax_{f,l}$ is the annual total passengers on aircraft size f and segment l. Lastly, in Equation 5 the non-fuel flight costs is the sum of flight landing charges (*FltLandC*), crew costs (*CrewC*), maintenance costs (*MtnC*), and ownership costs (*OwnC*) (Figure 2) for aircraft size f and on segment l. The total non-fuel flight costs for the segment is averaged by total flight frequency over all size classes on segment l ($Freq_l$).

Model Estimation Results and Discussion

In this section, the model estimation is briefly described, followed by interpretation and discussion of the coefficients of the three costs variables estimated from the

$$(FuelCostPerPax)_{mkn} = \sum_{l \in Legs} \frac{\sum_{f,l} FuelC_{f,l} Freq_{f,l}}{Pax_l} \quad \forall Legs \in mkn \quad (3)$$

$$(NonFuelCostPerPax)_{mkn} = \sum_{l \in Legs} \frac{\sum_{f,l} (PaxLndC + VolC)_{f,l} Pax_{f,l}}{Pax_l} \quad \forall Legs \in mkn \quad (4)$$

$$(NonFuelCostPerFlt)_{mkn} = \sum_{l \in Legs} \frac{\sum_{f,l} (FltLndC + CrewC + MtnC + OwnC)_{f,l} Freq_{f,l}}{Freq_l} \quad \forall Legs \in mkn \quad (5)$$

world's top 20 region-pair markets. The section concludes with an analysis of the system-wide impacts of introducing a carbon tax in two regional markets with distinct cost pass-through elasticities on demand, fares, and CO_2 emissions, using the AIM2015 model.

Model Estimation

Airfare models are complicated by endogeneity bias arising from the demand effects from fares, that is, a change in demand by a change in airfares. From Equation 2, this potentially affects five right-hand side variables, namely Pax, RouteShare, Freq, LegMeanHHI, and AirportMeanHHI. The number of OD passengers on a given route is clearly endogenous as changes in fares also affect passenger demand. RouteShare, defined as the share of total OD passengers on this city pair using a given route, may be endogenous because it is a function of OD demand (Pax). Similarly, HHIs are also potentially endogenous, given that an airline's market share, which is input to calculate HHIs, is expected to be a function of the price it charges (*13, 19*). Flight frequency may be endogenous because increases in frequency will have lower per flight costs, thus lowering fares, which in turn attracts more demand, resulting in a change in frequency. After conducting the traditional Breusch-Pagan test (*27*) and the Hausman test (*28*), we found that the null hypotheses of homoskedasticity and exogeneity can be rejected at the 0.1% level in this model, indicating that heteroskedasticity exists and the five variables are endogenous.

To correct for the endogeneity and heteroskedasticity bias, we estimate the model using a feasible generalized two-stage least squares (FG2SLS) procedure, with lagged Pax, RouteShare, LegMeanHHI, AirportMeanHHI, and Freq in year 2014 as instrumental variables (IVs). The estimation procedures are: (a) estimate OLS residuals from the reduced-form equation; (b) regress the log of the squared residuals over all the exogenous variables (including the IVs); (c) estimate the error variance from the fitted values in step (b); (d) apply 2SLS with the dependent variable, the explanatory variables, and all the IVs divided by the estimated error variance (*29, 30*).

Results and Discussion

Due to space constraints, Table 2 reports the coefficients of only the three key operating costs variables estimated from the top 20 regional-pair markets, which account for 90% of the global RPK (Figure 1*b*). Overall, out of the 60 coefficient estimates across the 20 airline markets, only 10 are not statistically significant. The other 50 estimated coefficients have high statistical significance at least at the 1% level, and all have the expected positive signs. This demonstrates that airlines do pass increases in fuel costs, non-fuel passenger costs, and non-fuel flight costs on to passengers through higher airfares.

Importantly, the coefficients tend to vary in magnitude depending on the specific type of costs that airlines pass through and the particular regional market in which they operate. Additionally, we found fairly large values for the first stage *F*-statistics of the added IVs, suggesting that the chosen IVs are sufficiently strong. The adjusted R^2 values range from 0.520 to 0.938, indicating that our model explains a significant proportion of the variance in airfares of the 20 markets. Next, the cost pass-through elasticities for each costs type will be interpreted and discussed for intra- and inter-regional markets, respectively.

Among the seven intra-regional markets, six coefficients prove to be statistically significant for the fuel costs variable at the 0.1% level. Airlines in AP–AP are the most responsive to changes in fuel costs, with an elasticity between 0.36 and 0.39. The relatively high elasticity in AP–AP can be explained by the fact that fuel costs account for a larger share of total airline costs, due to a wider geographical coverage (i.e., ranging from Russia to Australia), compared with other intra-region markets. This is followed by SA–SA (0.20–0.28), EU–EU (0.23–0.25), and AF–AF (0.20–0.27), which have very similar pass-through elasticities within 95% confidence intervals. NA–NA is slightly less elastic to fuel costs changes; for every 10% increases in fuel costs, fares in NA–NA will increase by 1.9% to 2.1%. The elasticity with statistical significance is the lowest in CA–CA (0.14–0.20), which shows only about half of the fuel costs effects compared with AP–AP (0.36–0.39). The only statistically insignificant coefficient is found in ME–ME, suggesting that increases in fuel costs do not have an airfare impact in ME–ME. This can be attributed to the fact that the majority of routes in this region are operated by Gulf national flag carriers that have considerably cheaper fuel costs due to the region's proximity to oil production and refining facilities, leading to lower supply chain costs (*31*).

All seven estimates for the non-fuel flight costs are statistically significant at the 0.1% level. Sharply contrasting with the result from the fuel costs estimation, ME–ME is found to be the most responsive to changes in non-fuel flight costs. For each 10% increase in these costs, the mean of fares in ME–ME will increase by 5.5% to 7.7%. Once again, this result demonstrates that with their considerably cheaper fuel costs, airlines in the ME-involved markets are the most vulnerable to changes in the other big component of airline operating costs, i.e. the flight-based costs. In contrast, NA–NA and EU–EU are the least elastic to non-fuel flight costs changes, with the estimated coefficients both between 0.15 and 0.16. Between the highest and the lowest elasticities, SA–SA ranks the

Table 2. Feasible Generalized Two-Stage Least Squares (FG2SLS) Estimation Results[a]

Results for the seven intra-regional markets

Cost variables	Market	Coef.	Std. error	Pass-through	Obs.	Adj.R^2
ln(FuelCostPerPax)	AP–AP	0.375***	0.008	0.36–0.39	29,880	0.854
	SA–SA	0.238***	0.018	0.20–0.28	3,519	0.859
	EU–EU	0.237***	0.006	0.23–0.25	43,263	0.532
	AF–AF	0.233***	0.021	0.20–0.27	2,802	0.838
	NA–NA	0.200***	0.004	0.19–0.21	70,429	0.574
	CA–CA	0.168***	0.015	0.14–0.20	2,190	0.838
	ME–ME	0.007	0.053	Not significant	626	0.864
ln(NonFuelCostPerFlt)	ME–ME	0.662***	0.063	0.55–0.77	626	0.864
	SA–SA	0.361***	0.019	0.32–0.40	3,519	0.859
	AF–AF	0.270***	0.021	0.24–0.30	2,802	0.838
	AP–AP	0.214***	0.008	0.20–0.23	29,880	0.854
	CA–CA	0.163***	0.014	0.14–0.19	2,190	0.838
	NA–NA	0.154***	0.003	0.15–0.16	70,429	0.574
	EU–EU	0.154***	0.005	0.15–0.16	43,263	0.532
ln(NonFuelCostPerPax)	SA–SA	0.584***	0.051	0.48–0.68	3,519	0.859
	AF–AF	0.276***	0.046	0.20–0.36	2,802	0.838
	AP–AP	0.156***	0.013	0.13–0.18	29,880	0.854
	NA–NA	0.152***	0.009	0.14–0.17	70,429	0.574
	EU–EU	0.069***	0.007	0.06–0.08	43,263	0.532
	ME–ME	0.146	0.255	Not significant	626	0.864
	CA–CA	–0.046	0.031	Not significant	2,190	0.838

Results for the thirteen inter-regional markets

Cost variables	Market	Coef.	Std. error	Pass-through	Obs.	Adj.R^2
ln(FuelCostPerPax)	AP–EU	0.339***	0.017	0.31–0.37	12,313	0.798
	EU–SA	0.321***	0.038	0.25–0.39	2,623	0.662
	CA–NA	0.232***	0.010	0.21–0.25	8,487	0.534
	EU–NA	0.217***	0.018	0.18–0.25	11,193	0.520
	NA–SA	0.181***	0.041	0.12–0.24	2,366	0.700
	AP–NA	0.162***	0.025	0.11–0.21	6,302	0.567
	AF–EU	0.151***	0.023	0.11–0.19	4,496	0.877
	CA–EU	0.141***	0.033	0.08–0.20	1,754	0.654
	EU–ME	0.118***	0.035	0.07–0.17	3,286	0.775
	AF–AP	0.056**	0.022	0.01–0.10	1,910	0.866
	AP–ME	0.004	0.018	Not significant	3,579	0.865
	NA–ME	0.011	0.053	Not significant	1,399	0.651
	AF–ME	–0.009	0.022	Not significant	1,093	0.938
ln(NonFuelCostPerFlt)	AF–ME	0.406***	0.025	0.36–0.45	1,093	0.938
	AF–AP	0.367***	0.023	0.33–0.41	1,910	0.866
	EU–ME	0.335***	0.033	0.29–0.38	3,286	0.775
	NA–ME	0.325***	0.069	0.20–0.45	1,399	0.651
	CA–EU	0.278***	0.041	0.20–0.35	1,754	0.654
	NA–SA	0.278***	0.038	0.22–0.34	2,366	0.700
	AP–NA	0.274***	0.028	0.22–0.33	6,302	0.567
	AP–ME	0.266***	0.016	0.24–0.30	3,579	0.865
	AF–EU	0.259***	0.020	0.22–0.29	4,496	0.877
	EU–SA	0.222***	0.044	0.15–0.30	2,623	0.662
	EU–NA	0.172***	0.021	0.13–0.21	11,193	0.520
	AP–EU	0.110***	0.015	0.08–0.14	12,313	0.798
	CA–NA	0.075***	0.011	0.06–0.09	8,487	0.534
ln(NonFuelCostPerPax)	AP–NA	0.253***	0.014	0.23–0.28	6,302	0.567
	AP–EU	0.183***	0.016	0.16–0.21	12,313	0.798
	AF–EU	0.168***	0.024	0.12–0.21	4,496	0.877
	EU–SA	0.149***	0.026	0.10–0.20	2,623	0.662
	EU–ME	0.124***	0.032	0.07–0.18	3,286	0.775
	NA–ME	0.114**	0.038	0.04–0.19	1,399	0.651

(continued)

Table 2. (continued)

Results for the seven intra-regional markets						
Cost variables	Market	Coef.	Std. error	Pass-through	Obs.	Adj.R^2
	AF–AP	0.111**	0.044	0.03–0.19	1,910	0.866
	EU–NA	0.080***	0.016	0.05–0.11	11,193	0.520
	CA–NA	0.072***	0.018	0.04–0.11	8,487	0.534
	AP–ME	−0.038	0.032	Not significant	3,579	0.865
	CA–EU	−0.039	0.034	Not significant	1,754	0.654
	NA–SA	−0.072	0.038	Not significant	2,366	0.700
	AF–ME	−0.056	0.061	Not significant	1,093	0.938

***Significant at the 0.1% level. **Significant at the 1% level. *Significant at the 5% level.
[a]Markets ranked in descending order by their pass-through elasticity within each cost variable.

second most elastic market yet only has an elasticity (0.32–0.40) slightly more than half of that of ME–ME; elasticities of the third to the fifth markets are AF–AF (0.24–0.30), AP–AP (0.20–0.23), and CA–CA (0.14–0.19), respectively.

Five out of seven intra-region markets for the non-fuel passenger costs have statistically significant estimates, with ME–ME and CA–CA as exceptions. SA–SA, which was found to be the second most elastic market to changes in both fuel and non-fuel flight costs, proves to be the most sensitive market to this costs type (0.48–0.68). This is followed by AF–AF having a pass-through elasticity between 0.20 and 0.36 within 95% confidence intervals. Coefficients of AP–AP (0.13–0.18) and NA–NA (0.14–0.17) are not statistically significantly different within 95% confidence intervals. EU–EU has the lowest elasticity with statistical significance, with every 10% increase in non-fuel per passenger costs leading to fare increase of just 0.6% to 0.8%.

We now turn to the coefficients of the 13 selected inter-regional markets. Given that each of the inter-regional markets connects two different OD regional markets with distinct characteristics, the estimated cost pass-through elasticities could provide additional insights.

Out of 13 estimates for the fuel costs variable, 10 have statistically significant coefficients at the 0.1% level. The three statistically insignificant coefficients all concern the ME-involved markets, namely AP–ME, NA–ME, and AF–ME. This is similar to our findings in the intra-region market estimates, and again demonstrates that the ME-involved markets are generally unaffected by fuel cost increases. Out of the 10 statistically significant estimates, AP–EU is the inter-regional market most sensitive to fuel costs, with estimated pass-through elasticity of about 0.31 to 0.37. EU–SA follows closely with an estimated coefficient between 0.25 and 0.39 within 95% confidence intervals. Two groups of markets are found to have very similar elasticities. One group contains CA–NA (0.21–0.25) and EU–NA (0.18–0.25), and the other group includes NA–SA (0.12–0.24), AP–NA (0.11–0.21), AF–EU (0.11–0.19), and CA–EU (0.08–0.20). The markets least elastic to fuel costs changes are EU–ME (0.07–0.17) and AF–AP (0.01–0.10).

Similar to the results found in the intra-regional markets, the non-fuel flight costs coefficients of all 13 inter-regional markets are statistically significant at the 0.1% level. This finding provides an important implication that policy interventions in increasing non-fuel flight costs will have the most widespread significant effects on airline pricing behavior globally. Nevertheless, the effects still differ significantly in magnitude between the 13 markets. Consistent with our findings from the intra-region markets' coefficients, the ME-involved markets are the most responsive to changes in non-fuel flight costs. Out of the four markets with the highest elasticities to this costs type, three are the ME-involved markets. In particular, average fares in AF–ME will increase by 3.6% to 4.5% as a result of a 10% increase in non-fuel flight costs. AF–AP follows closely after AF–ME with an elasticity between 0.33 and 0.41. EU–ME and NA–ME have slightly lower coefficients at 0.29 to 0.38 and 0.20 to 0.45, respectively. In addition, coefficients of CA–EU, NA–SA, AP–NA, AP–ME, and AF–EU all range from 0.20 and 0.35 within 95% confident intervals. The market with the lowest elasticity to this cost type is CA–NA; with 10% increases in the per flight costs, fares will only increase by 0.6% to 0.9%.

Lastly, four out of 13 coefficients for the non-fuel per passenger costs are statistically insignificant. Together with the two statistically insignificant coefficients found in the intra-market results, this costs type has six out of the 10 statistically insignificant estimates across all the 60 estimated coefficients. In addition, the values of cost pass-through elasticities of the non-fuel per passenger costs are overall smaller than those of the other two operating costs categories. For instance, AP–NA has the highest elasticity but only at 0.23 to 0.28, which is significantly lower than the most fuel costs-sensitive inter-region market, AP–EU (0.31–0.37), and the most flight costs-sensitive market AF–ME (0.36–0.45). The remaining eight statistically significant estimates of non-fuel passenger costs are all below 0.20, with the lowest elasticity at 0.04 to 0.11 found in CA–NA. Thus we can conclude that, overall, increases in non-fuel per passenger costs have the smallest impact on airline pricing behavior.

There are two potential reasons to explain this result. First, the weight of non-fuel passenger costs is generally the smallest in an airline's total operating costs. For instance, by our calculation, in year 2015 the weights of three costs types in total segment costs for flight segment LHR–PEK are: fuel costs (41%), non-fuel flight costs (53.3%), and non-fuel passenger costs (5.7%). Therefore, increases in non-fuel passenger costs are less likely to have a large impact on airlines' pricing. Secondly, the literature has demonstrated that increases in the firm-level costs are much less passed onto passengers than the sector-wide cost changes (*2*, *3*). As a major part of the non-fuel passenger costs, the volume-related costs (in-flight services, meals, etc.) clearly belong to the firm-level costs. Thus, our empirical results confirm that firm-level cost pass-through is indeed less than sector-wide cost pass-through.

Overall, there is statistical evidence that airlines do pass a significant proportion of their operating costs on to passengers across different regional markets. However, the commonly assumed 100% cost pass-through is not supported by our empirical results. Rather, most of the estimated pass-through elasticities are below 0.5, after controling for the key supply, demand, and competition effects on airfares in the model. Additionally, we found that increases in airline non-fuel flight costs will significantly impact airline pricing across all regional markets. In contrast, increases in airline fuel costs are unlikely to affect most of the ME-involved markets, as the fuel costs of many Gulf national flag airlines in these markets are significantly cheaper (*31*). Finally, changes in non-fuel passenger costs have the least impact on airlines and the impacts are only significant in 14 markets out of 20.

To conclude this section, we compare the potential impacts of an emissions reduction policy that increases airline fuel costs on airfares, demand, and CO_2 emissions in two major regional markets. The fare model developed in this work is a core component of the updated aviation systems model AIM2015 (*21*), which predicts the system-wide impacts of aviation technology, operational, and policy scenarios. Using the AIM2015, we simulate a carbon tax policy over three baseline scenarios. For sources of these baseline scenarios see Dray et al, (*21*). Specifically, the three baseline scenarios are:

1. The worst-case scenario: low GDP growth to 2050, high oil prices, and pessimistic technology adoption (late availability date, high cost, low benefit)
2. The mid-case scenario: mid GDP growth to 2050, central oil prices, and mid-range technology adoption
3. The best-case scenario: high GDP growth to 2050, low oil prices, and optimistic technology adoption (early availability date, low cost, high benefit)

Two regional markets with distinct cost pass-through elasticities are selected, i.e. AP–AP and EU–EU, where a carbon tax was introduced in 2015 at \$36/tCO_2 which will linearly reach \$150/tCO_2 by 2050. Notably, this is a relatively high carbon tax scenario, compared with the highest carbon tax in the EU ETS to date at \$36 (*32*). Figure 3 depicts first the key scenario inputs of GDP growth and oil prices, followed by the scenario projections for both regional markets.

According to Figure 3, demand in terms of RPK from 2015 to 2050 could increase the most under the best-case scenario in both markets (graphs a.1 and a.2). In AP–AP, total RPK under the best case is projected to increase by 494%, double the increase under the mid-case scenario, and 3.9 times that of the worst-case scenario. Demand in EU–EU follows a similar trend, with an increase under the best-case scenario by 353%, 1.9 times the mid-case scenario, and 5.3 times the worst-case scenario. The strong demand growth of the best case is partially boosted by the projected low fares. According to graphs b.1 and b.2, in both markets average fare per RPK turns out to be the lowest under the best-case scenario and the highest under the worst-case scenario. In addition, within each baseline scenario, fares fluctuate as fuel prices change (input.1) yet the fluctuation is more significant in AP–AP than in EU–EU. Finally, driven by the projected strong growth in demand, the total direct CO_2 emissions could increase the most under the best-case scenario (AP–AP 444% and EU–EU 294%), despite an optimistic assumption on low-carbon technological improvements.

Having compared how the baseline scenarios could change in future without any emissions reduction policy interventions, now we assess whether a high carbon tax would make any difference to demand, fare, and CO_2 emissions in the two markets. Importantly, given that the fuel cost pass-through elasticity is higher in AP–AP (0.36–0.39) than in EU–EU (0.23–0.25) (Table 2), fares in AP–AP are expected to increase more compared with

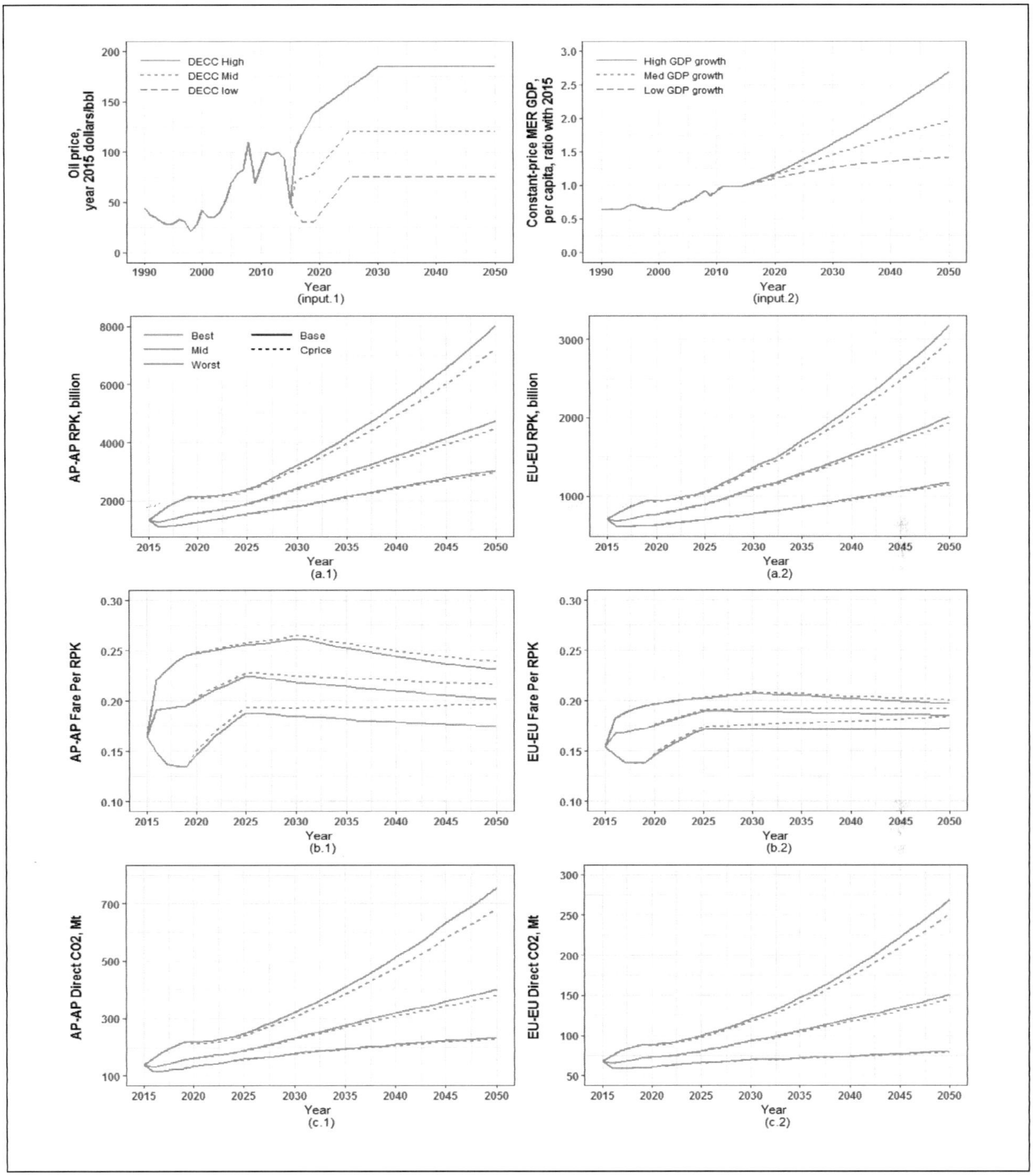

Figure 3. AIM2015 projections on demand, average fare, and direct CO_2 emissions in AP–AP and EU–EU regional airline markets.

fares in EU–EU, after the introduction of the carbon tax. Consistent with our expectations, by 2050 fare per RPK in AP–AP (Figure 3, b.1) could be 12.5% (the best case), 7.1% (the mid case), and 3.4% (the worst case) higher than those of the no-intervention projections, whereas fare per RPK in EU–EU is projected to increase by just 6.9% (the best case), 3.8% (the mid case), and 1.8% (the worst case), respectively. Our results prove that airlines would adjust airfares based on their fuel cost elasticities in different regional markets.

As a result of the increased airfares, by 2050 demand in AP–AP (Figure 3, a.1) could reduce by 9.7% (the best case), 5.7% (the mid case), and 2.8% (the worst case), respectively. Impacts on EU–EU demand are smaller due to the smaller increases in prices, with the post-policy RPK decrease by 6.2% (the best case), 3.8% (the mid case), and 2.0% (the worst case), respectively, by 2050. The associated direct CO_2 emissions follow a similar trend. Compared with the no-intervention projections, emissions under the best-case scenario could drop by 9.6% in AP–AP and by 6.6% in EU–EU by 2050 if the carbon tax was introduced. Emissions reduction under the mid-case scenario is projected to be 5.4% in AP–AP and 3.7% in EU–EU. The worst-case scenario would be least affected, with emissions only decreased by 2.5% in AP–AP and 1.8% in EU–EU.

Conclusions

The research presented in this paper shows that airlines operating in different regional aviation markets may have distinct pricing responses to market-based measures (MBMs) by explicitly modeling the cost pass-through of fuel costs per passenger, non-fuel costs per passenger, and non-fuel costs per flight. It contributes to the field by enabling future research to have more certainty on the potential cost pass-through of airlines when evaluating the economic impacts of MBMs to aviation. To our best knowledge, this is the first study that empirically estimates airline cost pass-through under competition at a global scale.

The coefficient estimates indicate that increases in airline fuel costs may have the greatest impacts on the intra-AP regional market and the inter-regional market connecting Asia Pacific and Europe. In contrast, the ME-involved markets are unlikely to be affected by the fuel costs increases because fuel costs are considerably cheaper for the Gulf legacy carriers. Thus in order to reduce emissions in the ME-involved markets, market-based policies need to increase not fuel costs but other airline operating costs in these markets. Changes in non-fuel flight costs are found to have significant impacts on all regional markets, although it would be more difficult to design MBMs that can affect the largely-fixed flight-based operating costs. Increases in non-fuel per passenger costs would have the lowest impact on airfares as these are largely influenced by increases in firm-level costs and also have the smallest share in airlines' total operating costs.

Our results also provide some useful insights into the potential impacts of introducing a carbon tax on airline pricing in different regional markets. The application of the aviation systems model AIM2015 using the estimated cost pass-through elasticities suggests that such a policy intervention may have important and different effects on airfares, demand, and CO_2 emissions. Without any policy interventions, the growth of aviation CO_2 emissions can only be drastically slowed down in a low GDP and high oil price future, as the result of a low growth in demand. In contrast, the introduction of a high carbon tax proves to increase airfares, leading to a nearly 10% reduction in CO_2 emissions in the most fuel costs-sensitive market, AP–AP, even if GDP grows rapidly and oil prices are low. However, the effects of a carbon tax are likely to be smaller in markets with lower fuel-cost elasticity. This finding also implies that air transport passengers in the fuel-costs elastic markets could pay higher airfares compared with passengers in the less elastic markets under a global carbon tax policy. Lastly, given that the ME-involved markets are not affected by carbon tax, airlines in these markets could obtain large windfall profits due to both their unaffected fuel costs and increased airfares.

Acknowledgments

This work is supported by EPSRC grant EP/M027031/1. We are grateful for their support. The first author also personally acknowledges the Chinese Scholarship Council (CSC) for the scholarship provided to his PhD.

Author Contributions

The authors confirm contribution to the paper as follows: study conception and design: Bojun Wang, Aidan O'Sullivan; data collection: Bojun Wang, Lynnette Dray; analysis and interpretation of results: Bojun Wang, Aidan O'Sullivan, Lynnette Dray, Andreas W. Schäfer; draft manuscript preparation: Bojun Wang. All authors reviewed the results and approved the final version of the manuscript.

References

1. ICAO (International Civil Aviation Organization). *Carbon Offsetting and Reduction Scheme for International Aviation (CORSIA)*, 2016. http://www.icao.int/environmental-protection/Pages/market-based-measures.aspx. Accessed November, 2016.
2. RBB Economics. *Cost Pass-through: Theory, Measurement, and Potential Policy Implications*, 2014.
3. Koopmans, C., and R. Lieshout. Airline Cost Changes: To What Extent Are They Passed Through to the Passenger? *Journal of Air Transport Management*, Vol. 53, 2016, pp. 1–11.
4. Anger, A., and J. Kohler. Including Aviation Emissions in the EU ETS: Much Ado About Nothing? A Review. *Transport Policy*, Vol 17, 2010, pp. 38–46.
5. Scheelhaase, J., W. Grimme, and M. Schaefer. The Inclusion of Aviation into the EU Emission Trading Scheme: Impacts on Competition Between European and Non-European Network Airlines. *Transportation Research Part D*, Vol. 15, 2010, pp.14–25.

6. Albers, S., J. A. Bühne, and H. Peters. Will the EU-ETS Instigate Airline Network Reconfigurations? *Journal of Air Transport Management*, Vol. 15, 2009, pp. 1–6. http://dx.doi.org/10.1016/j.jairtraman.2008.09.013.
7. Meleo, L., C. Nava, and C. Pozzi. Aviation and the Cost of the European Emission Trading Scheme: The Case of Italy. *Energy Policy*, Vol. 88, 2016, pp. 138–147.
8. Hofer, C., R. Windle, and M. Dresner. Price Premiums and Low Cost Carrier Competition. *Transportation Research Part E: Logistics and Transportation Review*, Vol. 44, 2008, pp. 864–882.
9. Dresner, M., and M. Tretheway. The Effect of Market Structure on Price. *Journal of Transport Economics and Policy*, Vol. 26, 1992, pp. 171–184.
10. Zou, B., and M. Hansen. Flight Delay Impact on Airfare and Flight Frequency: A Comprehensive Assessment. *Transportation Research Part D: Transport and Environment*, Vol. 69, 2014, pp. 54–74.
11. Bailey, E., and J. Panzar. The Contestability of Airline Markets during the Transition to Deregulation. *Law and Contemporary Problems*, Vol. 44, 1981, pp. 125–145.
12. Morrison, S. A., C. Winston, E. E. Bailey, and A. E. Kahn. Enhancing the Performance of the Deregulated Air Transportation System. *Brookings Papers on Economic Activity: Microeconomics*, Vol. 1989, pp. 61–123.
13. Borenstein, S. Hubs and High Fares: Dominance and Market Power in the U.S. Airline Industry. *RAND Journal of Economics*, Vol. 20, 1989, pp. 334–365.
14. Cho, W., R. Windle, and C. Hofer. Route Competition in Multi-Airport Cities: An Analysis of US Air Fares. *Transportation Journal*, Vol. 51, 2012, pp. 265–288.
15. Drenser, M., J. Lin, and R. Windle. The Impact of Low-Cost Carriers on Airport and Route Competition. *Journal of Transport Economics and Policy*, Vol. 30, 1996, pp. 309–328.
16. Morrison, S. Actual, Adjacent, and Potential Competition: Estimating the Full Effects of Southwest Airlines. *Journal of Transport Economics and Policy*, Vol. 32, 2001, pp. 239–256.
17. Brueckner, J., D. Lee, and E. Singer. Airline Competition and Domestic US Airfares: A Comprehensive Reappraisal. *Economics of Transportation*, Vol. 2, 2013, pp. 1–17.
18. Forbes, S. The Effect of Air Traffic Delays on Airline Prices. *International Journal of Industrial Organization*, Vol. 26, 2008, pp. 1218–1232.
19. Chi, J., and K. Koo. Carriers' Pricing Behaviors in the United States Airline Industry. *Transportation Research Part E*, Vol. 45, 2009, pp. 710–724.
20. Yan, F., E. Winijkul, G. Streets, Z. Lu, C. Bond, and Y. Zhang. Global Emission Projections for the Transportation Sector Using Dynamic Technology Modeling. *Atmospheric Chemistry and Physics*, Vol. 14, 2014, pp. 5709–5733.
21. Dray, L., P. Krammer, K. Doyme, B. Wang, K. Al Zayat, A. O'Sullivan, and A. Schäfer. AIM2015: Validation and Initial Results from an Open-source Aviation Systems Model. Presented at 21st Air Transport Research Society Conference, Antwerp, Belgium, 5–8 July 2017.
22. *SABRE Market Intelligence Passenger Demand, Routing and Aircraft Schedule Databases*, 2016. http://www.sabre-global.com. Accessed November 2016.
23. FlightGlobal. *Ascend Online Fleet Database*, 2016. http://www.ascendworldwide.com/what-we-do/ascend-data/air-craft-airline-data/ascend-online-fleets.html. Accessed May 2017.
24. Sustainable Aviation. *The SA Noise Roadmap*, 2015. http://www.sustainableaviation.co.uk/wp-content/uploads/2015/09/SA-Noise-Road-Map- Report.pdf. Accessed April 2017.
25. Al Zayat, K., L. Dray, and A. Schäfer. A Comparative Analysis of Operating Cost between Future Jet-Engine Aircraft and Battery Electric Aircraft. Presented at 21st Air Transport Research Society Conference, Antwerp, Belgium, 5–8 July 2017.
26. *RDC Aviation Airport and Enroute Charges Databases*, 2007. http://www.rdcaviation.com. Accessed May 2017.
27. Breusch, T., and A. Pagan, A. A Simple Test for Heteroscedasticity and Random Coefficient Variation. *Econometrica*, Vol. 47, 1979, pp. 1287–1294.
28. Hausman, A. Specification Tests in Econometrics. *Econometrica*, Vol. 46, 1978, pp. 1251–1271.
29. Wooldridge, J. *Econometric Analysis of Cross Section and Panel Data*, 2nd ed. The MIT Press, Cambridge, MA, 2010.
30. McFadden, D. *Chapter 4. Instrumental Variables*, Berkeley Econ 240B. 1999, pp. 1–19. https://eml.berkeley.edu/~mcfadden/e240b_f01/ch4.pdf. Accessed June 2017.
31. O'Connell, J. The Rise of the Arabian Gulf Carriers: An Insight into the Business Model of Emirates Airline. *Journal of Air Transport Management*, Vol.17, 2011, pp. 339–346.
32. European Climate Exchange. *Carbon Taxes on the European Climate Exchange*, 2017. http://www.ecx.eu/. Accessed June 2017.

The Standing Committee on Aviation Economics and Forecasting (AV040) peer-reviewed this paper (18-04368).